GENERAL RELATIVITY
AND GRAVITATION

GENERAL RELATIVITY
AND GRAVITATION

Proceedings of the Seventh International Conference (GR7)
Tel-Aviv University, June 23–28, 1974

Edited by

G. SHAVIV and J. ROSEN

Department of Physics and Astronomy, Tel-Aviv University

A HALSTED PRESS BOOK

JOHN WILEY & SONS, New York · Toronto
ISRAEL UNIVERSITIES PRESS, Jerusalem

ISRAEL UNIVERSITIES PRESS
is a publishing division of
KETER PUBLISHING HOUSE JERUSALEM LTD.
P.O. Box 7145, Jerusalem, Israel

Published in the Western Hemisphere by
HALSTED PRESS, a division of
JOHN WILEY & SONS, INC., NEW YORK

Library of Congress in Publication Data

International Conference on General Relativity and
 Gravitation, 7th, Tel-Aviv University, 1974.
 General relativity and gravitation.

 "A Halsted Press book."
 1. General relativity (Physics)—Congresses.
2. Gravitation—Congresses. I. Shaviv, G. II. Rosen,
Joseph, 1933–   III. Title.
QC173.6.I57 1974   521´.1   75–33824
0–470–77939–X

Distributors for the U.K., Europe, Africa and the Middle East
JOHN WILEY & SONS, LTD., CHICHESTER

Distributors for Japan, Southeast Asia and India
TOPPAN COMPANY, LTD., TOKYO AND SINGAPORE

Distributed in the rest of the world by
KETER PUBLISHING HOUSE JERUSALEM LTD.
IUP Cat. No. 26007 4
ISBN 0 7065 1527 7

Set, printed and bound by Keterpress Enterprises, Jerusalem
PRINTED IN ISRAEL

Contents

v

Preface

It is by now a well-established tradition to start off a preface to any book on general relativity, gravitation, cosmology, and high energy astrophysics by exclaiming how rapidly the field is expanding (a big bang for relativity!). Nevertheless, this is really true (as tedious as it may be to hear again). The most exciting point is, we believe, the real possibility of checking many of the most prominent features of theory by experiment or observation. Indeed, we may be observing objects for which general relativity is no longer merely a correction.

This volume contains the invited talks and the panel discussion (on the detection of gravitational waves) of the Seventh International Conference on General Relativity and Gravitation (GR7), held in Tel-Aviv, Israel, during 23–28 June 1974. The organizers attempted to arrange for the presentation of a wide and varied spectrum of topics, which covers the many aspects of the subjects of the conference.

Since the usefulness of this material depends to a large extent on its freshness, we decided not to include the many (over 100) contributed papers in order to expedite publication. We hope that these interesting papers will be published at full length elsewhere.

It is a pleasure to thank the following for financial contributions which made the conference possible: Tel-Aviv University, Ben-Gurion University of the Negev, The Hebrew University of Jerusalem, Technion—Israel Institute of Technology, The Israel Ministry of Education and Culture, The Israel Ministry of Tourism, The Israel National Academy of Sciences and Humanities, The Israel Foundation Trustees, The International Committee on Gravitation, The International Union of Pure and Applied Physics, and Bank Leumi le-Israel.

Tel-Aviv,   G. SHAVIV and J. ROSEN
July 1975

vii

Uniqueness and Nonrenormalizability of Quantum Gravitation*

S. DESER† and P. VAN NIEUWENHUIZEN

Brandeis University
Waltham, Massachusetts 02154, U.S.A.

D. BOULWARE

University of Washington
Seattle, Washington 98195, U.S.A.

ABSTRACT

(1) The principles of Lorentz invariant quantum particle (rather than field) theory, together with qualitative empirical properties of gravitational forces, imply that the tree graphs describing gravitational interactions are necessarily generated by the Einstein action, at least for frequencies low compared to particle masses or the inverse Planck length. Consequently, the classical limit of any acceptable quantum gravitational model must be general relativity.

(2) The ultraviolet divergences of the quantized Einstein field interacting with quantized scalars, spinors, photons or vector gauge multiplets are all nonrenormalizable at the one-loop level; the required counter terms are drastically different from the original actions. We discuss the relevance of these conclusions to the unification of gravitation and quantum theory.

1. INTRODUCTION

In this report I shall be concerned with two very recent developments in quantum gravity. These developments represent explicit calculations and conclusions in a subject which has hitherto been more a "program" than an area of active results.

*Supported in part by the National Science Foundation, Nederlands Ministerie van Onderwijs en Weten-schappen, and the U.S. Atomic Energy Commission.
†Lecture by S. Deser.

The first is encouraging, the second discouraging, to our search for a consistent unification of gravitation and quantum theory. I shall summarize them briefly here, expand on each in separate sections, then attempt to assess prospects and alternatives in the concluding part. The results were obtained by the author in collaboration with D. Boulware and with P. van Nieuwenhuizen, respectively.

Special relativistic quantum theory, our most solidly established basis for understanding microscopic (rather than cosmological) phenomena, will be our guide in attempting to understand quantum gravitation. We recall that relativistic quantum theory can be divided into two different aspects: S-matrix (particle) theory, which deals with the properties of scattering amplitudes for physical particles of given spin and mass, and quantum field theory which provides a localized space-time description of interacting systems. In either category, there are two basically different classes of diagrams: trees and loops. Tree diagrams do not include vacuum fluctuations, and there is no summation over (arbitrarily high) virtual energies. This class of diagrams is sufficient for understanding the classical field limit of a quantum theory, and has no renormalization problems. The latter arise only when one encounters closed loops—whose presence is required by unitarity. We have recalled these distinctions because our two sets of results will nicely fall into the complementary categories of relativistic particle theory and low frequency structure on the one hand, and relativistic field theory in its probing of high frequency behavior through closed loops on the other.

Whatever the future fate of quantum field theory, the more modest particle description is the necessary extension of quantum mechanics to the special relativistic domain. One of its basic results is that all forces between particles are themselves mediated by exchange of quanta of appropriate spins and masses. We shall show that this principle, together with the observed qualitative structure of macroscopic gravitational forces, leads to a unique determination of the basic vertices describing gravitational interactions at wavelengths high compared to particle Compton wavelengths. This means in particular that the tree graphs, and so the classical limit of any acceptable quantum gravity model, are unique. That the interactions can be so completely determined by quantum principles is beautiful enough; that they must coincide with those of general relativity in this limit is even more gratifying. From the general properties of gravitational forces between material systems (including light), namely that they are long range, static, macroscopic and attractive, we shall show that a massless spin 2 particle ("graviton") is primarily responsible. What is more important, this particle must self-interact: gravitons emit gravitons and do so in a uniquely defined way. This follows from the fact that gravitons can only couple consistently to conserved symmetric tensor matrix elements of their sources, while a conserved tensor can only exist if the graviton content of the sources is included. (The low momentum transfer form of the "stress tensor" can actually be deduced as a low energy theorem, and this theorem *implies* the equivalence principle.) There is in fact an infinite chain of n-graviton vertices, all of which are determined uniquely by the consistency requirements of massless graviton coupling, through terms quadratic in graviton momenta.

Having determined all possible vertices, one can consider all tree graphs, constructed by joining all possible vertex configurations without forming loops (the latter would require more knowledge than we are willing to assume in this part of the discussion about the higher momentum behavior of vertices). The result is that all tree graphs can be generated by an effective action functional, which is none other than the Einstein action. But the classical limit of any theory is just that of its tree structure: it is governed by the generating action. Thus classical general relativity is really a consequence of relativistic quantum particle theory, given the observed gross character of gravitational forces. There is, then, a very satisfactory harmony between geometry and quantum mechanics in this common range of application (or at least aspiration).

Our second set of results is concerned with deeper quantum properties, involving the extrapolation of the above structure to high frequencies, namely the closed loop aspects of general relativity considered as a quantum field. We shall quantize the metric field using the same techniques (covariant quantization, gauge breaking and compensating ghost fields) which have been so successful in the recent progress of non-Abelian vector gauge theories. Our interest will focus here on the very high frequency behavior, that is on the types of divergences encountered at the one loop level in coupled Einstein-matter systems. Only if the divergences can be removed by renormalization, are these systems satisfactory at our present (perturbative) stage of understanding in field theory. Unfortunately, we shall see that all systems considered to date, which include general relativity coupled to scalars, spinors, photons, and Yang–Mills multiplets as well as some generalizations of Einstein theory, are nonrenormalizable (except for the source-free Einstein field itself). We shall discuss the significance of this discouraging conclusion, along with possible alternatives and improvements, in the final section.

Throughout our work we assume that gravitation, like all other fundamental dynamical systems, is subject to quantization. It has sometimes been suggested that this is not logically necessary either because general relativity is not a basic field or because it can somehow remain classical in contrast to its sources, or finally, because its quantum effects are not observable experimentally due to the smallness of the characteristic Planck length of quantum gravitation. These last resorts appear to us worse than the disease they would avoid, and we shall not discuss them here. Our point of view is that there is an empirical domain, large compared to particle Compton wavelengths, but small compared to cosmological distances, in which the physical world is Minkowskian, and the physical processes including gravitation are described by special relativistic quantum theory.

2. GRAVITONS: VERTICES AND TREES

The study of relativistic quantum particle theory of gravitation began about a decade ago with the work of Feynman [1] and Weinberg [2]. Weinberg especially stressed the strong constraints on the S-matrix describing interaction of massless particles, which are equivalent to gauge invariance requirements in the more familiar field-theoretical context. He showed how one could derive, for massless vector particles, both Maxwell's equations and conserved currents; that for spin greater than 2, there is no interesting low frequency theory possible at all (there being no conserved static quantities corresponding to charge or mass) while for spin 2, the low frequency form of the interactions implies amongst other things, the equivalence principle (universality of matter-graviton vertices) for all systems including the gravitons themselves. Both Feynman and Weinberg asked to what extent such arguments uniquely defined the full gravitational theory, but found the nonlinearities of the self-interactions too complicated to complete this program. The question was recently taken up by D. Boulware and myself, and the unique character of the solution will be the subject of this section; the reader is referred to the paper [3] for details.

We begin with a rapid review of why the exchanged particle responsible for gravitational forces must have (at least primarily) spin 2. From the general facts that there are static forces we may exclude $S > 2$; that they are attractive excludes $S = 1$; their macroscopic character excludes half odd integer spin, and because photons are affected in the same way as massive particles, we can exclude $S = 0$. Naturally, there are only upper limits on the possible contribution of the other spins or of short range (contact) terms but for simplicity we shall assume they are entirely absent (Brans Dicke theory could be accommodated for example). Another basic requirement of quantum theory, that only physical (positive probability) quanta exist, excludes exotic Lorentz invariant theories of gravitation which would require "ghost" behavior for the corresponding particles. Empirically, the graviton must obviously have long range (small mass); in fact, the mass dependence affects light bending discontinuously and one may conclude that the mass is strictly null [4]. We have thus specified the graviton to be a massless ($k^2 = 0$) spin 2 particle, which therefore has two helicities ($a = \pm 2$). The amplitude $T_a(k)$ for its emission by a system can be written as $T_a(k) = e_a^{\mu\nu}(k) T_{\mu\nu}(p', p)$, where $e_a^{\mu\nu}(k)$ is its transverse-traceless polarization tensor and the ten quantities $T_{\mu\nu}$, which depend on the initial and final momenta of the system, can be taken to be conserved also slightly off shell ($k^2 \neq 0$). We need not assume that they are the matrix elements $\langle p'|T_{\mu\nu}|p \rangle$ of a stress tensor operator; indeed one can *deduce* the form of $T_{\mu\nu}$ to first order in momentum transfer purely from conservation. Specifically, one finds for the amplitude $T^{\alpha\beta}_{\lambda\ \lambda}(p', p)$ for emission of a graviton by a system making a transition between physical states with momentum and polarization (p, λ) and (p', λ') the following expression [5] $(P \equiv p + p', k \equiv p - p')$:

$$T^{\alpha\beta}_{\lambda'\lambda}(p', p) = T\left[\bar{u}_{\lambda'}(p') u_{\lambda}(p) P^{\alpha}P^{\beta} + iP^{(\alpha}\delta^{\beta)}_{\eta}k_{\tau}\bar{u}_{\lambda'}(p') S^{\eta\tau}u_{\lambda}(p)\right] + O(k^2) \quad (2.1)$$

where T is a universal constant, $(u, \bar{u})$ are the initial and final state polarization spinors or tensors, and $S^{\eta\tau}$ is the Lorentz rotation operator for the system in question. The constant T is determined by correspondence with the Newtonian limit, to be (with our covariant spinor normalization) $T^2 = (2\pi)^{-5}G$, where G is the Newtonian constant. Note that the first term is just a scalar particle's stress tensor, while the second is the usual contribution of spin to the angular momentum. Thus, the equivalence principle is deduced as a low energy theorem: all systems couple in a universal way (at low enough frequencies) to gravitons through their unique symmetric conserved stress tensor and with a universal strength T. For our purposes, universality is the key, because it necessarily leads to graviton-graviton coupling. If gravitons are coupled to *any* system (and if not there is no gravitation!), they are necessarily coupled to each other. Otherwise put, if a system can emit a graviton, then every system to which it couples in any way also interacts with gravitons, and that includes the emitted graviton itself. Diagrammatically, we can deduce the existence of the basic 3-graviton vertex (Fig. 1a) from the Newtonian emission (Fig. 1b). This can be done analytically by inserting a soft (small k_{μ}) graviton into (1b); consistency forces the presence of the full set of Fig. 2, including V_3.

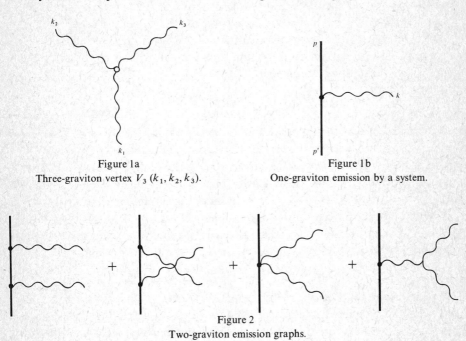

Figure 1a
Three-graviton vertex V_3 (k_1, k_2, k_3).

Figure 1b
One-graviton emission by a system.

Figure 2
Two-graviton emission graphs.

Soft graviton conditions only determine V_3 (à la (2.1)), for two gravitons on shell and the third one soft, however; and it requires a long calculation (including consideration of 4- and 6-point graviton graphs) to obtain the full constraints on V_3 for all gravitons off shell. By considering diagrams with more and more scattering gravitons, one may deduce the necessary existence of new short range vertices $V_4, V_5, \ldots, V_n, \ldots$ involving arbitrarily many gravitons. Our graviton construction has a parallel, incidentally, in the classical field theory framework in which the infinite set of self-interactions of the Einstein equations are deduced by demanding that the source of the linearized spin 2, $m = 0$ equations be the stress tensor of the linearized field and iterating this consistency demand [6]. However, no assumption about an underlying field is involved in the quantum case.

The constraints on V_3 are that it be symmetric in interchange of two gravitons (bose statistics) and conserved under appropriate contractions with external momenta. The conservation conditions require that V_3 be at least quadratic in the momenta, and the low frequency restrictions on our curiosity made us stick to quadratic momenta only (this corresponds to second derivatives only in field theory language). Then, if we express $V_3^{\mu\nu,\alpha\beta,\rho\sigma}(k_1k_2k_3)$ compactly in terms of a functional $V_3(h)$ by multiplying it with fields $h_{\mu\nu}h_{\alpha\beta}h_{\rho\sigma}$, we can show that the conservation conditions correspond to an identity of the form

$$\partial_\nu \delta V_3/\delta h_{\mu\nu} = A_\mu^{\alpha\beta}(h, \partial)(D^{-1}h)_{\alpha\beta} \tag{2.2}$$

where $(D^{-1}h)_{\alpha\beta}$ is the linearized Einstein tensor, and A is an operator linear in $h_{\mu\nu}$ and in first derivatives. Such identities generalize the Ward identity of electrodynamics, which is related to gauge invariance there. In our case, we are building up the more complicated covariant conservation identities of general relativity (2.4, 2.5). The most general solution of this identity is

$$V_3(h) = T\int(dx)(\sqrt{-g}R)^{(3)} + \int(dx)\,hh\,D^{-1}h \tag{2.3}$$

where the first term is the cubic part in an expansion of the Einstein Lagrangian $\mathscr{L} = (\sqrt{-g}R)\,T^{-2}$ in powers of $Th_{\mu\nu} \equiv g_{\mu\nu} - \eta_{\mu\nu}$. The second term stands for a sum of the type $h_{\mu\alpha}h_{\nu\alpha}(D^{-1}h)_{\mu\nu}$. These terms can be shown to represent precisely the freedom of expanding the metric (in classical language) in different variables such as $Tk^{\mu\nu} = g^{\mu\nu} - \eta^{\mu\nu}$, etc. This freedom is necessary and its presence reassures us as to the generality of our procedure. Of course, a new arbitrariness of this type will also occur at each order of V_n. To determine the general n-point vertex, one first establishes that the sum $V(h)$ of all vertices obeys the more general Ward identity

$$\partial_\nu \delta V/\delta h_{\mu\nu} = A^{\mu,\alpha\beta}(h, \partial)\left[(D^{-1}h)_{\alpha\beta} - \delta V/\delta h_{\alpha\beta}\right] \tag{2.4}$$

How unique are the $V(h)$ and A satisfying (2.4)? One may first show that the same scattering amplitude results no matter what set of vertices satisfying (2.4) (if there is any!) is used. But we know one $V(h)$ satisfying them, namely, the nonlinear part of the Einstein Lagrangian. For the Bianchi identities,

$$\mathscr{G}^{\mu\nu}_{;0} \equiv 0 = \mathscr{G}^{\mu\nu}_{,\nu} + \Gamma^{\mu}_{\alpha\rho}\mathscr{G}^{\alpha\rho} \tag{2.5}$$

together with the fact the $\mathscr{G}^{\mu\nu}_{L} \equiv (D^{-1}h)^{\mu\nu}$ is identically conserved and that $\mathscr{G}^{\mu\nu} - \mathscr{G}^{\mu\nu}_{L}$ is just the vertex $-\delta V/\delta h_{\mu\nu}$, read

$$\partial_{\nu}\delta V/\delta h_{\mu\nu} = \Gamma^{\mu}_{\alpha\beta}\left[(D^{-1}h)^{\alpha\beta} - \delta V/\delta h_{\alpha\beta}\right] \tag{2.6}$$

(This form, though not the functional form of V and $(D^{-1}h)$, is independent of what choice of $h_{\mu\nu}$ is made.) Thus, the Einstein action generates the essentially unique vertices and one can also show that it yields tree scattering amplitudes with the required properties. Finally, returning to the original matter–gravitation interactions, one can establish that the usual minimal gravitational coupling prescription for matter is also uniquely required up to higher momentum terms.

To summarize, we have shown the following: the quantum tree graphs describing all graviton scattering amplitudes are generated by the Einstein action, at least to quadratic order in momenta. Higher momentum dependence would in any case be negligible in the classical limit, which is thus necessarily governed by general relativity. This result followed entirely from empirically dictated dominance, in gravitational exchange, of massless spin two quanta together with the principles of S-matrix particle theory. General relativity is really a consequence of quantum theory! Note that our considerations began within the empirically well-established framework of Minkowski space for phenomena on scales greater than Compton wavelengths and smaller than cosmological distances. Nevertheless, they led us unavoidably to the usual Riemann geometrical interpretation of classical Einstein theory covariantly coupled to matter, with the equivalence principle as a necessary "lowest order" consequence of the graviton as mediating particle.

One could continue our particle description into the closed loop domain, which includes such classical necessities as 2 particle "ladder" exchange diagrams responsible for Keplerian orbits (together with quantum corrections $\sim (\hbar/mc/r)^2$ or $(G\hbar/c^3/r^2)$ to the dominant $1/r$ parts). However, our main interest in closed loops is that discussed in the next section, namely ultraviolet behavior of the general relativistic model taken seriously as a quantum field theory at all frequencies.

3. CLOSED LOOPS: NONRENORMALIZABILITY OF QUANTUM GRAVITATION

It has long been suspected that quantized general relativity, when ripe for calculations, would exhibit disastrous, nonrenormalizable ultraviolet behavior (at least in a perturbation expansion, which is the only method currently known). This expectation was based on the dimensional character of the Einstein constant, which is related to the rising momentum dependence of graviton vertices (for example $T_{\mu\nu}$ or V_3 both go as P^{+2}). The more loops and vertices the higher is the divergence predicted by simple power counting. Every radiative correction to a given process would give rise to new divergent counterterms, in contrast to the relatively mild logarithmic divergence in electrodynamics or Yang–Mills theory, where a finite number of renormalizations suffice. The horror of nonrenormalizability lies in the fact that a whole infinity of processes must be fixed experimentally and cannot be predicted by the theory, rather than just a small number of parameters like mass and charge. Nonrenormalizable theories are not necessarily inconsistent but they are not very predictive and one has no a priori idea of their properties.

A discussion of renormalizability would seem to require an advanced stage of development of the quantum field theory in question, yet quantization of general relativity has until now been at best a program with no well-defined calculation rules, let alone specific results. What is the nature of the obstacles to developing these rules, and how have they been overcome? The one quantization prescription we can be sure of for a dynamical system is the canonical one, in which there is a clearcut choice of basic p's and q's. While a suitable canonical formulation of classical general relativity is available, there have been two types of obstacles to progress in quantum gravity. The first is the apparent ambiguity of factor ordering in the Einstein action, both in its original form and in the later solution of its constraints. The second is simply that canonical quantization is neccessarily noncovariant in form with respect to space-time, so that the Feynman rules and any calculations would be terribly cumbersome. In electrodynamics, the possibility of canonical quantization provides the reassurance needed to perform one's calculations in covariant gauges. Now in an Abelian gauge theory such as electrodynamics, the equivalence between canonical and covariant gauges and quantization is well-known and even proved [7] within the context of perturbation theory. As soon as one gets to a non-Abelian situation such as the Yang–Mills field, equivalence is not so immediate. In practice, people went ahead, quantized covariantly and expected equivalence to be established in due course. This has in fact been accomplished [8], so that covariant rules have been validated also in the Yang–Mills case. There is one technical addendum to the covariant rules: If one just writes down the vertices and propagators in some covariant gauge as they appear in the Lagrangian, then he will obtain absurd closed loop results because wrong helicity states are also included in these gauges. These unphysical degrees of freedom (absent in canonical quantization) must be compensated for by introduction of the famous "ghost" Lagrangian. Ghosts were in fact discovered by Feynman [9] while performing the earliest loop calculations in quantum gravitation, and were later fully treated by Faddeev and Popov [10]. In electrodynamics, covariant ghosts are also present in principle but do not actually contribute due to the linearity of the theory. (Photon ghosts do contribute when coupling to gravitation is included.) In any case, canonical-covariant equivalence in gravitation is not yet established, but we will assume that here too it will work, and proceed with covariant quantization. Even with covariant rules, calculations are still a formidable task, both because of all the indices and because of the complicated form of the nonlinearities (even V_3, let alone V_4, has many terms!) of the Einstein action.

About ten years ago, Feynman [9] and then DeWitt [11] did the first calculations of gravitational closed loop effects. They computed only the contributions of closed scalar loops to gravitation–matter processes. But the full classification of divergent terms even in source-free gravitation was not accomplished until late in 1973, when the impetus of their advances in vector gauge theories led 't Hooft and Veltman [12] to a systematic attack on the problem, using covariant quantization (with appropriate ghost additions), the background field method and dimensional regularization. The first two ideas were already used by Feynman and DeWitt, the background field expansion being an especially apt tool in the gravitational context. It will be recalled that graviton vertices, such as V_3 or graviton emission by matter, grow as P^2. This means that if we insert an extra external graviton into a closed loop, we do not decrease the degree of divergence since the extra P^{-2} propagator denominator it produces is compensated by the P^2 vector numerator. Thus there are infinitely many divergent diagrams even at one-loop order. External matter line insertions also tend to proliferate, because the basic closed loop divergence is quartic (two propagators, two vertices and a d^4p integration) and insertion of say a pair of external boson lines decreases the divergence by only two powers. For fermions, one can have up to eight external lines emanating from a divergent diagram! The existence of "rings" with arbitrarily many external line attachments to a basic single closed loop is circumvented by the background field method, which recognizes that the external lines are basically classical: To obtain the scattering amplitudes, where the external lines are on shell, one lets the external fields obey the classical equations (Einstein-matter field equations). Only two powers of the field in the Lagrangian are truly quantum operators which create and then annihilate the virtual loop pair at each vertex of the bubble. Thus, we simply expand each field as the sum of a classical "background" field and a quantum operator, and keep only terms quadratic in the latter, but with arbitrary nonlinear coefficients in the background fields. The latter are then evaluated on shell, at the solutions of the field equations. For example, we write the Einstein action $I^E \equiv \kappa^{-2} \int dx\, R(\bar{g})$ in terms of $\bar{g}_{\mu\nu} = g_{\mu\nu} + \kappa h_{\mu\nu}$ and keep only the part (independent of κ)

$$I^E_{(2)} \quad \frac{1}{2} \int dx\, \delta^2 R(g)/\delta g_{\mu\nu} \delta g_{\alpha\beta} h_{\mu\nu} h_{\alpha\beta} \tag{3.1}$$

while, e.g., the Maxwell action I^M is expanded as

$$I^M_{(2)} = \frac{1}{2} \int [\delta^2 I^M(g, F)/\delta F^2) ff + (\delta^2 I^M/\delta g^2) hh + 2(\delta^2 I^M/\delta g \delta F) f h]\, dx \tag{3.2}$$

The coefficients are of course pretty messy functions, but 't Hooft and Veltman succeeded in finding an algorithm which expressed the divergent contributions in terms of these coefficients for actions which have the generic form

$$I = [A\phi \,\Box\, \phi + B^\mu \phi \partial_\mu \phi + C\phi^2] \tag{3.3}$$

including of course the relevant ghost contributions if ϕ is a gauge field. The use of dimensional regularization is a simplifying, though not essential, replacement of the usual momentum cutoff Λ by a dimensionless (effectively logarithmic) cutoff parameter ε^{-1}.

To see what happens in the case of pure gravitation at the one-loop level, we do not even need to perform the algorithm calculation explicitly. It suffices to note that the counter-Lagrangian, $\Delta\mathscr{L}$, defined as the negative of the divergent contributions, must be an invariant density function of the background metric (since no gauge choice is made for the latter, only for the quantum field $h_{\mu\nu}$), and have the usual dimension L^{-4} which makes $\Delta I \equiv \int d^4x \Delta\mathscr{L}$ properly dimensionless ($\hbar = 1$ here). But since $I_{(2)}$ is κ-independent, we are simply asking for all possible invariants of dimension 4, i.e., involving 4 derivatives, which can be constructed out of curvatures: $\Delta\mathscr{L}$ must be the linear combination

$$\Delta\mathscr{L} = (\sqrt{-g}/\varepsilon)\,[\alpha R^2_{\mu\nu\alpha\beta} + \beta R^2_{\mu\nu} + \gamma R^2] \tag{3.4}$$

where α, β, γ are (gauge invariant) numerical coefficients, determined by the algorithms, which were calculated by 't Hooft and Veltman. (Nonpolynomial expressions like $R^4_{\mu\nu}/R^2$ cannot occur in perturbation series, while explicit covariant derivatives could only contribute a total divergence here, and so are irrelevant.) But we are only interested in the infinities of S-matrix elements, for which the external $g_{\mu\nu}$ obey the on shell conditions $G_{\mu\nu}(g) = 0$. This means both $R^2_{\mu\nu}$ and R^2 vanish, leaving only the square of the full Riemann or Weyl tensor (its coefficient, α, incidentally is non-vanishing). But here a "miracle" occurs: there is an identity (generalizing the Gauss–Bonnet formula in two dimensions) which states that $(-g)^{-1/2}\varepsilon^{\mu_i\cdots\mu_n}\varepsilon^{\mu_i\cdots\nu_n}R^{(1)}_{\mu_i}\ldots R\ldots R^{(n/2)}\ldots\nu_n$ is a total divergence in an even number n of dimensions (for $n = 2$, $\varepsilon^{\mu\nu}\varepsilon^{\alpha\beta}R_{\mu\nu\alpha\beta} = \text{Div}$ is the Gauss–Bonnet formula, while for $n = 4$, we have the product of two curvatures). Using the standard expression of two ε's as a product of Kronecker δ's, it follows that, in four dimensions,

$$\int d^4x \sqrt{-g}(R^2_{\mu\nu\alpha\beta} - 4R^2_{\mu\nu} + R^2) = 0 \tag{3.5}$$

so that the α term also vanishes on shell and the scattering amplitudes of source-free gravitation are *one-loop finite* (and need no renormalization at all). At the two loop level, which no one has yet succeeded in calculating, this pleasant state is probably lost because terms such as $\kappa^2 \text{tr}(R^3_{\mu\nu\alpha\beta})$ can occur, and would not vanish on shell.

The interesting physics, of course, lies in the coupling of general relativity to matter, which involves both graviton loops, matter loops and mixed contributions. 't Hooft and Veltman therefore proceeded next to scalar field-graviton coupling, which could be cast in the general form (3.3). The net result, after setting the background fields on shell $(G_{\mu\nu} = -\frac{1}{2} T_{\mu\nu}(\phi), \Box\phi = 0)$ was that $\Delta\mathscr{L}$ did *not* vanish, but could be written as

$$\Delta\mathscr{L}_{E+S} \sim (1/\varepsilon)\sqrt{-g}R_{\mu\nu}^2 \neq 0 \tag{3.6}$$

(Throughout we deal with massless fields for simplicity. The massive case does not improve the conclusions.) This counterterm is not only not amenable to absorption by a coupling constant or field renormalization $(g \to Zg)$ but it involves fourth derivatives, with all the attendant ghost unpleasantness they bring into our second derivative world.

At this stage, an optimist would say that there are after all no basic scalar fields in nature anyway, so perhaps this is an artifact, which will disappear when one considers more realistic systems such as photons, fermions or vector gauge multiplets like the Yang–Mills field itself. In a series of papers [13], P. van Nieuwenhuizen and I proceeded to calculate the divergences of gravitationally coupled photons and fermions, and in collaboration with H.-S. Tsao, we evaluated graviton coupling to an arbitrary semisimple gauge group multiplet as well [14]. The result in each case was negative: all those models were nonrenormalizable, the only consolation being that many a priori allowed "bad" counterterms actually vanished, leaving effectively only one term (which is bad enough!) in the electrodynamics and Yang–Mills cases. In Einstein–Maxwell theory, for example, one can easily see on dimensional and invariance grounds that the general form of $\Delta\mathscr{L}$ which depends on curvature, field strength, and explicit covariant derivatives D_α is

$$\Delta\mathscr{L}_{E+M} = (1/\varepsilon)\sqrt{-g}\,[a_1 R_{\mu\nu}^2 + a_2 R^2 + a_3 R_{\mu\nu\alpha\beta}^2 + a_4 R_{\mu\nu} T^{\mu\nu} + a_5 R F_{\mu\nu}^2 +$$
$$+ a_6 R^{\mu\nu\alpha\beta} F_{\mu\nu} F_{\alpha\beta} + a_7 T_{\mu\nu}^2 + a_8 (F_{\mu\nu}^2)^2 +$$
$$+ a_9 (D_\mu F^{\mu\nu})^2 + a_{10}(D_\alpha F_{\rho\gamma})^2] \tag{3.7}$$

where $T_{\mu\nu}$ is the Maxwell stress tensor. Explicit calculation, however, yields $a_5 = a_6 = a_8 = a_{10} = 0$. Using now the field equation, $G_{\mu\nu} = -\frac{1}{2} T_{\mu\nu}$ and $D_\mu F^{\mu\nu} = 0$, we may combine all terms to the form

$$\Delta\mathscr{L}_{E+M} = (\alpha/\varepsilon)\sqrt{-g}\,R_{\mu\nu}^2 \tag{3.8}$$

where the numerical constant α has the explicit value $137/60$. In obtaining the above results, one has to include a photon ghost contribution, which is nonvanishing, in contrast to flat space. Even more surprisingly, despite the still greater possibilities for $\Delta\mathscr{L}$ in the non-Abelian multiplet case, the net result is

$$\Delta\mathscr{L}_{E+YM} = (1/\varepsilon)\sqrt{-g}\,[\tfrac{1}{60}(137 + r - 1) R_{\mu\nu}^2 - \tfrac{11}{12} C f^2 \mathbf{F}_{\mu\nu} \cdot \mathbf{F}_{\mu\nu}] \tag{3.9}$$

where r is the dimension of the gauge group, $rC \equiv C_{abc}C_{abc}$ is the square of the structure constants, and f is the self-coupling constant of the field. For $r = 1, f = 0$, we regain the Maxwell result, while the usual Yang–Mills field has $r = 3$, $C = 2$. The F^2 term is of course also present in flat space, being the usual renormalizable counter-term.

The background field method and its apparently sweeping results for $\Delta \mathscr{L}$ may perhaps be put in perspective in terms of more familiar external weak field diagrams as follows. Consider the basic vacuum polarization diagram, Fig. 3, with two external graviton lines (tadpoles with one external line and vacuum loops with no external lines vanish in the dimensional regularization scheme). This yields a divergent term for $\Delta \mathscr{L} \sim (1/\varepsilon) \int h_{\mu\nu}\theta^{\mu\nu,\alpha\beta}h_{\alpha\beta}$, where h denotes an external line and θ is quartic in derivatives. Linearized gauge invariance implies that this can be rewritten as a linear combination, with fixed coefficients, of the two invariants $(R_{\mu\nu}^L)^2$ and $(R^L)^2$ constructed out of the *linearized* curvatures. The background field method exploits the gauge invariance in the external fields[1] to predict that if one looked at the arbitrary ring diagram of Fig. 4, whose divergent part may be written as $\Delta \mathscr{L} \sim (1/\varepsilon) \int h^{(1)} \ldots h^{(n)}{}_{\mu_1 \cdots \mu_{2n}}$, then it would correspond to the nth order expansion in h of the *same* combination of the full *nonlinear* curvatures $(R_{\mu\nu})^2$ and $(R)^2$. That there can be no "new" R^3, R^4, etc., contributions from Fig. 4 (which could not be seen in Fig. 3) follows by dimension counting. Thus, there really is no "magic" to the background method which goes beyond the simple perturbative calculation (ghost loops are always needed in each, of course). When matter is present, the above correspondence still holds, but this time one would separately calculate all possible

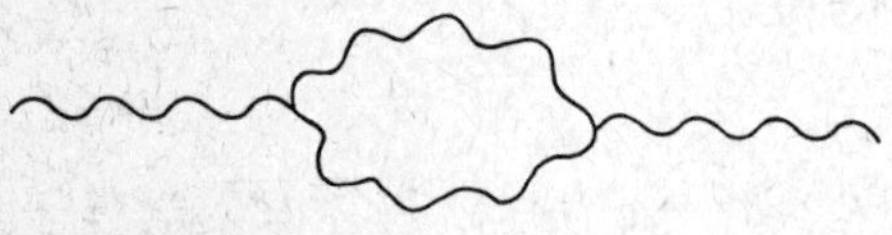

Figure 3
Graviton self-energy diagram.

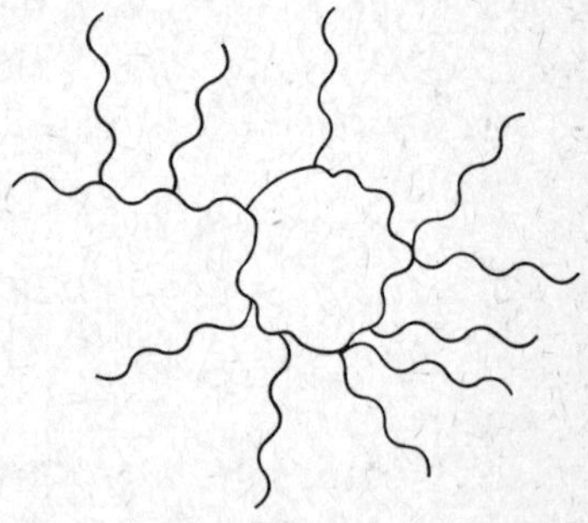

Figure 4
Ring diagram with arbitrary number of
external graviton lines.

[1] External gauge invariance of the one-loop quantum Lagrangians (such as (3.1)) and therefore of the counter-Lagrangians, can be shown as follows: The Lagrangian is manifestly invariant under combined transformations of the background and quantum fields. But the transformed quantum fields lead to the same Feynman rules as the original ones, since they just bear "primed labels." Schematically, we have $g + h \to g' + h' \to g' + h$, so that the dependence on g' is the same as it was on g—this is the required invariance.

loops with 2, 4, etc., external matter lines and either one or no external gravitons. For example, for photon-graviton coupling we could have either type of diagram of Fig. 5. The self-energy diagram corresponds to terms of the form $(\partial F)^2$, while the pictured vertex correction is the first expansion in h of the invariants $\sqrt{-g}RFF$ or $\sqrt{-g}(DF)^2$, but not of, e.g., $\sqrt{-g}F^2$, which is dimensionally wrong. All higher external graviton corrections symbolized in Fig. 5c correspond to the *same* combination of *full* invariants $\sim \sqrt{-g}(DF)^2$ and $\sqrt{-g}R\,FF$ as is determined at the linearized level by the diagrams with no and one external graviton.

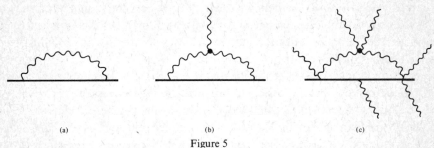

(a)   (b)   (c)

Figure 5

(a) Photon self-energy; (b) Example of photon-graviton vertex correction; (c) General ring graph with 2 external photon lines.

One regularity which arises in the various coupled systems is the non-negative sign of the coefficients of $R_{\mu\nu}^2$ and R^2 (before use of field equations). This can be understood on simple unitarity grounds by noting that $\int (R_{\mu\nu}^2 - \frac{1}{3}R^2)$ and $\int R^2$ represent respectively spin 2 and 0 states (the former combination is in fact just the square of the Weyl tensor), so that the internal loop just represents a sum over intermediate states of pure spin. One can easily check this also for any partial contribution to such terms, e.g., from a pure scalar or pure fermion or pure photon loop, and I should mention that calculations of such individual diagrams have been made by a number of workers [15], starting with the original pure scalar loop calculation by Feynman [9] and DeWitt [11]. However, it is only when *all* diagrams (including mixed loops) are added up that one can decide on nonrenormalizability, since the on shell field equations mix up a priori independent terms. The positivity of $R_{\mu\nu}^2$ alone is not sufficient to guarantee nonrenormalizability, since there are also cross terms such as $R_{\mu\nu}T^{\mu\nu}$ and also $T_{\mu\nu}^2$ (though this point deserves further consideration). Indeed, part of the interest in performing the Yang–Mills calculation was to see whether the two new parameters involved (multiplicity r and dimensionless coupling constant f) could be adjusted in a useful way, as a possible model of compensation amongst several different matter fields.

It now required a superoptimist to say that perhaps the only really fundamental matter fields are the spin $\frac{1}{2}$ fermions, and one should therefore look for renormalizability of the Dirac–Einstein system. It is a rather difficult place to look, owing to the well-known pecularities of spinors, which can only be coupled to vierbeins rather than directly to the metric. This means that there are now sixteen rather than ten gravitational variables, that there is a new gauge group, the freedom of local vierbein rotations, and unlike the scalar and vector cases, covariant derivatives are unavoidable in the matter Lagrangian. The last fact has the very unfortunate consequence that the algorithm is no longer applicable, and that direct graph calculations must be performed from scratch. The presence of sixteen variables, all of which must of course be candidates for quantization, raises an important separate question. Does quantization of general relativity in its vierbein and metric forms yield the same theory when both are permissible (i.e., in the absence of spinors)? We found that this is indeed the case, and that covariant quantization neatly disposes of the six superfluous vierbein components' apparent contributions.

The calculation of all divergent contributions to $\Delta\mathscr{L}$ is a formidable task, and there is a nightmarish number of possible terms, owing to the richness of the Dirac algebra, to the already mentioned fact that graphs with as many as eight external fermions are still divergent, and to the fact that the coupled Einstein–Dirac equations mix many invariants. Fortunately, there is one subset of divergent graphs which is well suited for a nonrenormalizability proof. These diagrams yield terms which do not mix with any others on shell, being unaffected by the field equations, yet whose presence suffices to establish nonrenormalizability. Further, they are relatively easy to calculate. We refer here to all divergent graphs with eight external fermions. Such terms are logarithmically divergent by power counting, and so by dimensions their $\Delta\mathscr{L}$ contributions are of the form $\Pi_{i=1}^{4}(\bar\psi\theta_i\psi)$, where the θ_i are numerical (Dirac) matrices, but do not contain derivatives. The field equations, on the other hand, always contain derivatives of the spinor field, so there is no mixing. (The reason such nonderivative terms are possible here, in contrast to the massless spin 0 or 1 cases is that covariant derivatives of ψ in the action break the flat space invariance under addition of a constant spinor to ψ.) Let us consider the possible graphs and their behavior. By continuity of matter lines, the external fermions will attach to the loop either as seagull (Fig. 6a) or Compton (Fig. 6b) insertions and there will be four such vertices, denoted by a black dot, in the complete diagram of Fig. 7. Wavy lines

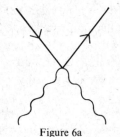

Figure 6a

A seagull fermion-graviton vertex.

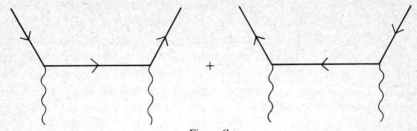

Figure 6b
A Compton vertex insertion.

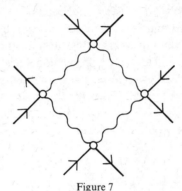

Figure 7
General eight-fermion loop diagram.

denote vierbein propagators, the fermions are solid directed lines, and we omit any external vierbein insertions, whose effect will just be to make the final $\Delta\mathscr{L}$ coordinate invariant, e.g., by adding factors $\sim \sqrt{-g}$, but never to produce dimensionally incorrect terms like $R\psi^8$. In the actual calculation, one finds the form for the effective vertex as the sum of Figs. 6a and 6b, and of course, crossing symmetry between vierbein pairs is also included. The logarithmic behavior of the corresponding integral is clear: each vierbein propagator goes as P^{-2}, while each vertex goes as P (for the seagull this is obvious, since it is part of the elementary fermion stress tensor–graviton coupling which is linear in P; the Compton insertion consists of two such P vertices with an intermediate fermion propagator $\sim P^{-1}$). Explicit calculation of the coefficient of the logarithm reduces to an algebraic trace over the four successive vertex-propagator units of Fig. 7. The net result has the very simple form [13]

$$\Delta\mathscr{L} \sim \beta(1/\varepsilon)\, \kappa^8 (A_a A_b \eta^{ab})^2, \qquad A_a \equiv \bar{\psi}\gamma_a\gamma_5\psi \qquad (3.10)$$

The counterterm is proportional to the fourth power of the axial current A_a, contracted with the local Minkowski metric η_{ab}, and β is a non-vanishing numerical coefficient. The result (3.10) is sufficient to establish nonrenormalizability of the Dirac–Einstein system, since these ψ^8 terms in $\Delta\mathscr{L}$ cannot mix even on shell with those having lower powers of ψ (and hence involving derivatives), the field equations $G_{\mu\nu} \sim \frac{1}{2}(\bar{\psi}D_\mu\gamma_\nu\psi)$, $\gamma D\psi = 0$ always involving a differentiated ψ. These other terms of $\Delta\mathcal{L}$ (which we have not calculated) may well be separately nonrenormalizable.

One final remark about the spinor case: We have treated the free Dirac field as defined in the usual second order formalism. As was noted by Weyl [16], however, a first order formalism would define a minimal coupling differing from ours by an extra (finite) contact term of the form $\kappa^2 A_a^2$. It is conceivable (the calculation has not been done) that such a term would lead to cancellation of our ψ^8 terms, although this would not necessarily imply renormalizability. In view of the importance of our conclusions, this open question should be borne in mind.

To summarize this section, the Einstein field coupled to scalars, spinors or vectors (whether singlets or gauge multiplets) is one-loop nonrenormalizable. If one had dealt with massive rather than massless fields, additional counterterms would have appeared, some of which would have been proportional to $(1/\varepsilon) m^2 R$ or $(1/\varepsilon) m^4 \sqrt{-g}$, and could have been absorbed by coupling constant or cosmological constant renormalization. Had we used the usual dimensional momentum cutoff, we would have found such terms also in the massless systems, alongside the basic nonrenormalizable ones. In either case, there would be no change in our nonrenormalizability conclusions.

4. CONCLUSIONS

Our survey of quantum gravitation has brought us to a dilemma. On the one hand, the principles of relativistic quantum particle theory unambiguously dictate the low frequency characteristics of gravitons. The vertices and tree graphs are generated by the classical Einstein action, and whatever the ultimate theory governing high frequencies, it must reduce to Einstein's in this limit. Thus, quantum gravitation must necessarily limit to general relativity. On the other hand, when we take this same general relativity seriously, as a quantum field theory, we find its high energy behavior unacceptable already at the one loop level. What alternatives are available, and what prospects can we foresee at this point?

It is quite possible that perturbative renormalizability is too restrictive a criterion in the case of dimensional coupling and when summed correctly, quantum general relativity is not only finite but, as has been suggested [17], even cuts off the divergences of flat space physics. One argument in favor of this view is that a classical point particle's gravitational self-energy in general relativity is more and more divergent in perturbation expansion, but sums to a finite closed form result [18]. Unfortunately, we have no reliable nonperturbative techniques available to decide. What about "improved" theories of gravitation? Are there any reasonable extensions of general relativity, which lead to better high frequency behavior while maintaining the desired low frequency Einstein form (and presumably its gauge invariance as well)? Several possibilities come to mind.

(a) If the root of the renormalization problem is the dimensionality of the coupling constant, then the Brans–Dicke theory [19] seems a good candidate, since one of its aims was in fact to replace κ^{-1} by a scalar field. Unfortunately, renormalizability is not achieved just by removing κ: one can transform theories of the Brans–Dicke type, by change of variable, to coupled Einstein-scalar systems, which are nonrenormalizable. It is perhaps conceivable that a more subtle treatment, using ideas of spontaneous symmetry breaking, would improve matters.

(b) One purely geometric theory in which only dimensionless constants enter is the class of Weyl models [20], with actions quadratic in the curvature, like our counterterms themselves. One would presumably obtain renormalizability here, since (by dimensions) $\Delta\mathscr{L}$ would have to reproduce the original action, but the price seems too high. First, it appears difficult to couple massive matter to a Weyl model in a reasonable way; one obvious problem is that the metric-source relation looks like $\nabla^2\nabla^2\phi \sim \rho$, with solutions $\phi \sim Mr$ rather than M/r. More drastic is the fact that the field equations are quartic in derivatives (this is what leads to renormalizability: propagators $\sim P^{-4}$), with all the attendant ghost problems of high derivative quantum theories. Finally, Weyl theory does not reduce in any straightforward way to general relativity at low frequencies. There is a possibility (first discussed by DeWitt [11] and revived in [13]) that a model of the form $\kappa^{-2}R + \alpha R_{\mu\nu}^2 + \beta R^2$, if treated in a partly nonperturbative way in κ, might be renormalizable and have proper low frequency behavior. However, it also has ghost problems, associated with the fourth order structure $(p^2 + \kappa^2 p^4)^{-1}$ of the propagators.

(c) Perhaps one should look for "improvements" in matter–graviton coupling by adding non-minimal terms (proportional to curvature) but then higher-derivative problems reappear.

(d) A more remote generalization is something like 5-dimensional theory [21]. Formally, the theory is one-loop finite, because there is no invariant $\Delta\mathscr{L}$ with dimension 5 that can be constructed from the metric or scalar or photon fields. But how does one eliminate the 5th dimension on the tree level while using its advantages in the loop?

(e) Another possibility is that an appropriately chosen set of matter fields with the right mutual interactions will lead to enough compensating divergences, despite the positive contributions to $R_{\mu\nu}^2$ and the ψ^8 free fermion terms. It is difficult to assess its value at present, but "supersymmetry" ideas might be of interest here.

(f) One very new line of development we have not mentioned is the possible connection between dual models and gauge theories. In the preprint received after this work was completed, Scherk and Schwarz [22] obtain Brans–Dicke models as zero-slope limits of certain dual models.

As of now, no model gives definite promise of providing compatibility between gravitation and quantum theory within the existing perturbative framework, while retaining the correct classical features of Einstein theory. On a practical level one could perhaps learn to live with a nonrenormalizable theory, since its difficulties are negligible at present experimental frequencies, but we will not really have understood gravitation until we can tame its infinities.

REFERENCES

1. R. P. FEYNMAN, *1962 Caltech Lectures on Gravitation* (unpublished).
2. S. WEINBERG, *Phys. Rev. Lett.* **9**, 357 (1964); *Phys. Rev.* **135**, B1049 (1964), **138**, B988 (1965); and in *Lectures on Particles and Field Theory* (S. DESER and K. FORD, eds.), Prentice-Hall, 1965.
3. D. BOULWARE and S. DESER, "Classical General Relativity Derived from Quantum Gravity," *Ann. Phys.* **89**, 193 (1975).
4. H. VAN DAM and M. VELTMAN, *Nucl. Phys.* **B22**, 397 (1970); D. BOULWARE and S. DESER, *Phys. Rev.* **D6**, 3368 (1972); P. VAN NIEUWENHUIZEN, *Phys. Rev.* **D7**, 2300 (1973).
5. C. A. ORZALEZI, J. SUCHER and C. H. WOO, *Phys. Rev. Lett.* **21**, 1550 (1968); and L. S. BROWN, private communication.
6. See, for example, S. DESER, *J.G.R.G.* **1**, 9 (1970).
7. B. Zumino, *J. Math. Phys.* **1**, 1 (1960).
8. See E. ABERS and B. W. LEE, *Phys. Reports* **9C**, 1 (1973), and S. COLEMAN, "Secret Symmetry," Harvard preprint (1974) for a discussion of this problem.
9. R. P. FEYNMAN, *Acta Physica Polonica* **24**, 697 (1963).
10. L. D. FADEEV and V. N. POPOV, *Phys. Lett.* **25B**, 29 (1967). A review has since appeared in *Soviet Phys. Uspekhi* **16**, 777 (1974).
11. B. S. DEWITT, in *Relativity, Groups and Topology*, Gordon & Breach, London, 1964; and *Phys. Rev.* **162**, 1195; 1239 (1967).
12. G. 'T HOOFT and M. VELTMAN, *Ann. Inst. H. Poincaré* **20**, 69 (1974); and G. 'T HOOFT, *Nucl. Phys.* **B62**, 444 (1973).
13. S. DESER and P. VAN NIEUWENHUIZEN, *Phys. Rev. Lett.* **32**, 245 (1974); *Phys. Rev.* **D10**, 401, 411 (1974).
14. S. DESER, H.-S. TSAO and P. VAN NIEUWENHUIZEN, *Phys. Rev.* **D10**, 3337 (1974).
15. D. M. CAPPER, M. S. DUFF and L. HALPERN, ICTP preprint 73/130; D. M. CAPPER and M. J. DUFF, ICTP 73/12; D. M. CAPPER, ICTP 73/11. D. M. CAPPER, G. LEIBBRANDT and M. R. MEDRANO, ICTP 73/26.
16. H. WEYL, *Phys. Rev.* **77**, 699 (1950); T. W. B. KIBBLE, *J. Math. Phys.* **4**, 1433 (1963).
17. C. J. ISHAM, A. SALAM and J.STRATHDEE, *Phys. Rev.* **D3**, 1805 (1971); **D5**, 2548 (1972).
18. R. ARNOWITT, S. DESER and C. W. MISNER, *Ann. Phys.* **33**, 88 (1965); Phys. Rev. **120**, 313 (1960).
19. C. BRANS and R. H. DICKE, *Phys. Rev.* **124**, 925 (1961).
20. H. WEYL, *Space–Time–Matter*, Dover, New York, 1950.
21. O. KLEIN, *Z. Phys.* **36**, 835 (1926).
22. J. SCHERK and J. N. SCHWARZ, Caltech Preprint CALT 68-444 (1974).

Canonical Quantization*

JOSHUA N. GOLDBERG

Department of Physics, Syracuse University
Syracuse, New York 13210, U.S.A.

ABSTRACT

The status of canonical quantization is reviewed. Emphasis is placed on the need for intrinsic coordinates as a means of defining either the appropriate configuration space for general relativistic observables or, equivalently, the superspace of Wheeler. Special attention is given to the scalars of the four-dimensional Riemann tensor, the transverse-traceless decomposition associated with asymptotically flat space-time, and scalars derived from the three-dimensional Ricci tensor.

1. INTRODUCTION

Twenty-five years ago a paper by Peter Bergmann entitled "Non-Linear Field Theories" appeared in *Physical Review* [1]. About the same time Paul Dirac was preparing a series of lectures on Hamiltonian theories [2]. This work inaugurated the modern attempt to join the principles of quantum theory and of general relativity [3, 4]. Quantum theory is a general method for describing the intrinsically uncontrollable and unknowable interactions among physical fields and particles. General relativity is a theory of space-time which says that geometry is not given a priori, but rather it is determined by the distribution of matter. Since the behavior of matter is restricted by quantum theory, the geometry of space-time should likewise exhibit quantum restrictions; since matter moves in space-time, its quantum interactions should similarly be limited by geometry. The unveiling and description of this mutual relationship is what interested both Bergmann and Dirac.

It is remarkable that so much time has passed and while we have learned a great deal about general relativity, geometry, space-time structure, topology, and many others matters, the original problem seems as intractable as it did originally. Perhaps

* Research supported by the National Science Foundation under Grant # GP34641X.

the conventional approach to quantum geometrodynamics is, as Pauli said in 1958 [5], a noble effort, but not wild enough. Nonetheless my role is to sketch the present status of canonical quantization and I shall leave the discussion of wilder schemes to others [6, 7, 8].

In the past two years there have been two rather remarkable review papers on the quantum theory of gravity. One is by C. Isham [9] and the other by A. Ashtekar and R. Geroch [10]. They are different, yet both list the motivation and discuss the difficulties of various approaches to the quantization program. Together they form an excellent overall summary. My own review will be narrower in scope.

I shall address the important problem of actually constructing observables using intrinsic coordinates. While considerable attention has been given to existence theorems for observables [11–13], there has been less success in actually exhibiting a non-redundant set [14–16]. Such a non-redundant set would identify a unique 4-geometry. John Wheeler has emphasized the role of "three-geometries" in the quantized theory as the appropriate configuration space for general relativity [17]. He refers to the space of unique geometries as "superspace". In terms of an intrinsic time, a unique 4-geometry is a trajectory in superspace. Therefore, the problems of finding observables and of defining the properties of superspace are intimately related.

Among the important topics I shall omit is the work by J. York [18] on a covariant decomposition of a symmetric tensor and its application to the constraint problem. There is closely related work by Y. Choquet [19], S. Deser [20] and particularly by A. Komar [21]. I shall also omit the very interesting work on model quantization particularly worked on by a group centered around C. Misner [22]. While models give us some insight into our long range goals, a reasonable discussion would be too detailed for the general discussion I have planned.

2. GENERAL DISCUSSION

Before discussing general relativity in detail, let me quickly and loosely describe a classical system of particles and the transition to quantum mechanics. Then I shall describe general relativity and the differences which lead to the difficulties in constructing a quantum theory for the gravitational field.

Start with a physical system with n degrees of freedom whose dynamical variables q_k ($k = 1, \ldots, n$) define a configuration space $\mathscr{C}$. There is a Lagrangian $L(q, \dot{q})$ whose Euler–Lagrange equations describe the time-development of the system. Through a Legendre transformation one goes over to the phase space Γ with points given by $\langle q_k, p_k \rangle, p_k \equiv \delta L / \delta \dot{q}_k$. One can then write the Lagrangian in canonical form

$$L = \sum p_k \dot{q}_k - H(q, p)$$

The Hamiltonian $H(q, p)$ generates the time-development of the system:

$$\dot{q}_k = [q_k, H] \equiv \frac{\delta H}{\delta p_k}$$

$$\dot{p}_k = [p_k, H] \equiv -\delta H/\delta q_k$$

Given two functions on the phase space $F(q, p)$ and $G(q, p)$, the Poisson bracket is defined to be

$$[F, G] = \sum \left(\frac{\delta F}{\delta q_k} \frac{\delta G}{\delta p_k} - \frac{\delta F}{\delta p_k} \frac{\delta G}{\delta q_k} \right)$$

In particular, the fundamental brackets are

$$[q_k, q_l] = \delta_{kl}$$

Through each point in phase space passes one and only one dynamical trajectory of the system. Constants of the motion map trajectories onto trajectories by means of canonical transformations.

For such a classical system, canonical quantization consists of the following prescription:

1. The symplectic form defining the Poisson brackets defines a commutator algebra such that

$$[q_k, p_l] = i\hbar \, \delta_{kl}$$

2. The physical state of the system is described by a function $\psi(q)$ on the configuration space $\mathscr{C}$.

3. The totality of allowable functions ψ defines a Hilbert space. That is, a scalar product and, therefore, a norm is defined over the configuration space.

4. The inner product is defined so that the dynamical variables become Hermitian operators on the Hilbert space.

5. The Hamiltonian becomes the time translation operator for the state vector ψ:

$$H\psi = i\hbar \, \partial\psi/\partial t$$

In general, with the necessary mathematical care in treating systems with an infinite number of degrees of freedom, these rules are applicable for the canonical quantization of a field theory. However, there are three characteristics of general relativity which come together to create extra difficulty in constructing a conventional field theory:

(1) There is no free field. Gravity is always a field in interaction because of its intrinsic non-linearity.

(2) It is a theory of space-time. Components of the metric tensor are the field variables. Therefore the gravitational field does not act within a geometrical state

which is impressed on it from outside, but rather the field itself, together with its sources, creates the geometrical framework within which it is constrained. At this time it is not clear how this dual role will actually show up in the quantized theory.

(3) The general covariance of the theory means that the Lagrangian is singular. Passage to the canonical formalism leads to constraints which implies that not all of the variables are dynamical.

Strictly from the point of view of carrying out the steps necessary for constructing a quantum theory, the existence of the constraints and the intrinsic non-linearity are the main stumbling blocks. One has not been able to solve the constraints explicitly and, therefore, we do not know an irreducible non-redundant set of dynamical variables—or observables.

Komar [23] has pointed out that in Lorentz-covariant theories we selectively carry over to the commutator algebra those classical Poisson brackets which are related to the space-time symmetry. In general relativity, the space-time symmetry is precisely what leads to the constraints and presumably that symmetry will be eliminated when the constraints are satisfied. It is expected that the algebra of observables will then be independent of representation and therefore can be carried over unambiguously to the quantum theory. This program requires, however, having the observables in hand.

Wheeler [24] and DeWitt [25] have leaped over this problem. The constraints of general relativity generate the coordinate transformations. Three of the constraints generate mappings of the surface $x^0 = c$ onto itself. The fourth constraint maps the surface onto a neighboring surface and may be referred to as the Hamiltonian constraint. Together all the constraints make up the Hamiltonian of general relativity. Wheeler and De Witt observe that if one says the magic words "three-geometry" and defines the configuration space to be the space of three-geometries, then a state vector $\psi(^{(3)}g)$ defined over that configuration space will automatically satisfy the three spatial constraints. The fourth constraint is then to be satisfied as a Schroedinger equation which Wheeler writes symbolically as

$$-\frac{\nabla^2 \psi}{(\delta^{(3)}g)} + R\psi = 0$$

The success of this program depends on being able to define and to describe explicitly the configuration space—the superspace mentioned earlier. Much work has already been done on this problem [26], but much more is needed. Then one needs to define Wheeler's symbolic equation so one can study the nature of its solutions.

As I indicated earlier, a determination of the observables will contribute to a better understanding of superspace. Therefore, the remainder of this paper will focus on the observables.

3. THE CANONICAL FORMALISM

The existence, in a field theory, of a group of transformations which are described by r arbitrary functions implies that the time development of the field is not uniquely determined by giving data on an initial surface. There exist transformations which leave the field unchanged up to and including a space-like surface $t = 0$, but change that field arbitrarily thereafter. To accomodate the transformation group, the field equations cannot determine the propagation of r field components—they may be chosen arbitrarily. In addition, the r field equations corresponding to these components impose restrictions on the initial data for the remaining components. They are the constraints in the canonical form of the theory.

Dirac [27, 28] has pointed out that the variables which may be chosen arbitrarily depend on the transformation group off the initial surface, while the remaining variables only involve the transformation group on the initial surface. Bergmann has called the latter variables "D-invariant" [29]. Only D-invariant quantities have an essential role in the dynamics of the field.

In general relativity the fundamental field variables are the components of the four-dimensional metric tensor ($\mu, v = 0, 1, 2, 3$)

$$ds^2 = {}^4g_{\mu v}\,dx^\mu dx^v$$

Under an infinitesimal coordinate transformation $x^\mu \to x^\mu + \xi^\mu$,

$$\delta^4 g_{\mu v} = -\,\xi^\rho_{,\mu}\,{}^4g_{\rho v} - \xi^\rho_{,v}\,{}^4g_{\rho\mu} - {}^4g_{\mu v,\rho}\xi^\rho$$

It is evident that $\delta^4 g_{mn}$ involve ξ^μ only on the surface $x^0 = \text{const}$, whereas $\delta^4 g_{\mu 0}$ involved time derivatives and hence require knowledge of the transformation off the surface. Therefore, the ${}^4g_{\mu 0}$ may be chosen arbitrarily; the ${}^4g_{mn}$ determine the dynamics of the field; and the field equations $G^{0\mu} = 0$ give rise to the constraints.

To emphasize the above separation, the notation below which was introduced by Arnowitt, Deser, and Misner [30], is commonly used. The signature is chosen to be -2 and Latin indices range from 1 to 3, whereas Greek indices range from 0 to 3, unless otherwise noted:

$$ {}^4g_{\mu v} = \begin{pmatrix} +N^2 + N_s N^s & N_n \\ N_m & g_{mn} \end{pmatrix} $$

$$ {}^4g^{\mu v} = \begin{pmatrix} +N^{-2} & N^{-2}N^n \\ N^{-2}N^m & g^{mn} + N^{-2}N^m N^n \end{pmatrix} $$

$$ g_{mr}g^{rn} = \delta^n_m $$

On the surface $x^0 = c$, Σ_c, g_{mn} is the negative definite induced metric, $\left\{ {}^4 \begin{array}{c} \kappa \\ \mu v \end{array} \right\}$

and $\left\{ \begin{matrix} k \\ mn \end{matrix} \right\}$ will denote the Christoffel-symbol connections for which ${}^4g_{\mu\nu}$ and g_{mn} are covariant constant, respectively; the appropriate covariant differentiation will be denoted by a semi-colon (;) or a solidus (/), respectively.

Further, we have the simple relationship

$$^4g = N^2 g$$

$$^4g = \det {}^4g_{\mu\nu} \quad \text{and} \quad g = \det g_{mn}$$

The Riemann tensor is defined by the Ricci identity:

$$2V_{\mu;[\rho\sigma]} = {}^4R^\mu_{\nu\rho\sigma} V^\nu$$

the Ricci tensor by

$$^4R_{\mu\nu} = {}^4R^\kappa_{\mu\nu\kappa}$$

and the scalar curvature by

$$^4R = {}^4g^{\mu\nu}\, {}^4R_{\mu\nu}$$

These conventions agree with those of Bergmann and are opposite to those of DeWitt [25] and Arnowitt, Deser, and Misner [30].

The same definitions and signs will apply for the Riemann tensor defined on the 3-space Σ_c in terms of g_{mn} and $\left\{ \begin{matrix} k \\ mn \end{matrix} \right\}$. Of course, in that case the superscript 4 is omitted and the indices will be Latin.

As noted above, the spatial metric g_{mn} is D-invariant whereas N and N_s (or ${}^4g_{0\mu}$) are not. Hence the latter variables are not dynamical and may be chosen arbitrarily. In the language of Wheeler [24, 31], N is the lapse, the proper time of a unit x^0-displacement measured along the normal to the surface Σ_c; N^s is the shift so that the corresponding proper displacement tangent to Σ_c is $(-N_s N^s)^{1/2}$.

Arnowitt, Deser, and Misner [30] have shown that the Lagrangian has the canonical form

$$\mathcal{L} = p^{mn} g_{mn,0} - N\mathcal{H}_L - N^s \mathcal{H}_s \tag{1}$$

$$p_{mn} = \sqrt{-g}\,(g^{mn}g^{rs} - g^{mn}g^{ns})\,K_{rs} \tag{2}$$

$$\mathcal{H}_L \equiv 2\sqrt{-{}^4g}\,G^0_\rho l^\rho \equiv (1/\sqrt{-g})\,(p^{mn}p_{mn} - \tfrac{1}{2}p^2) + \sqrt{-g}\,R \tag{3}$$

$$\mathcal{H}_s \equiv 2\sqrt{-{}^4g}\,G^0_s \equiv -2g_{mn}\pi^{mn}/n \tag{4}$$

where l^ρ is the unit normal to Σ_c,

$$l_\rho = N\delta^0_\rho$$

and K_{mn} is the second fundamental form or extrinsic curvature of Σ_c,

$$K_{mn} = - l_{m;n}$$

$$= \tfrac{1}{2} N(N_{m/n} + N_{n/m} - g_{mn,0})$$

The field equations are obtained from the action principle

$$\delta S = \delta \int \mathcal{L} \, d^4 x \qquad (5)$$

by variation of g_{mn}, p^{mn}, N, and N^s:

$$g_{mn,0} = 2N(\sqrt{-g})^{-1} (p_{mn} - \tfrac{1}{2} g_{mn} p) + N_{m/n} + N_{n/m} \qquad (6)$$

$$p^{mn}_{,0} = - N \sqrt{-g} (R^{mn} - \tfrac{1}{2} g^{mn} R) + \tfrac{1}{2} (-g)^{-1/2} g^{mn} (p^{rs} p_{rs} - \tfrac{1}{2} p^2) - $$
$$- 2N(-g)^{-1/2} (p^{mr} p^n_r - \tfrac{1}{2} p p^{mn}) + \sqrt{-g} \, (N^{mn}_{/} - g^{mn} N^r_{/r}) + $$
$$+ (p^{mn} N^r)_{/r} - N^m_{/r} p^{rn} - N^n_{/r} p^{rm} \qquad (7)$$

$$\mathcal{H}_s = 0 \qquad (8)$$

$$\mathcal{H}_L = 0 \qquad (9)$$

Eqs. (8) and (9) contain no time derivatives and are the four constraints per space point which exist among the variables g_{mn} and p^{mn}.

The configuration space $\mathcal{C}$ for the dynamical system described by the Lagrangian density (1) is the space of all negative definite metric tensors $g_{mn}(\mathbf{x})$ on Σ_c, Riem (Σ_c) [26]. The phase space Γ is just the cotangent bundle of $\mathcal{C}$. A point of Γ is the set $\langle g_{mn}(\mathbf{x}), p^{rs}(\mathbf{x}) \rangle$ of covariant symmetric tensors and contravariant symmetric tensor densities of weight 1 on Σ_c. However, the constraints $\mathcal{H}_s = 0$ and $\mathcal{H}_L = 0$ define a hypersurface $\bar{\Gamma}$ in Γ to which the trajectories of the dynamical system are confined. Indeed one can show that initial data on $\bar{\Gamma}$ is propagated by Eqs. (14) and (15) to remain on $\bar{\Gamma}$.

It follows from Noether's identity [29] that

$$C = \int \left\{ \xi^L \mathcal{H}_L + \xi^s \mathcal{H}_s \right\} d^s x \qquad (10)$$

generates the transformation of D-invariant quantities for mappings such that

$$\delta x^\mu = \xi^L l^\mu + \delta^\mu_s \xi^s$$

where $\xi(\mathbf{x})$ and $\xi^s(\mathbf{x})$ are arbitrary functions. Thus the $\mathcal{H}_L$ and $\mathcal{H}_s$ form a Lie algebra of the above transformation [32]. In particular, the Hamiltonian is obtained for $\xi^L = N$ and $\xi^s = N^s$:

$$H = \int \left\{ N \mathcal{H}_L + N^s \mathcal{H}_s \right\} d^3 x \qquad (11)$$

Thus we see that the constraints map $\bar{\Gamma}$ onto itself and, indeed, trajectories onto trajectories. But, because we cannot distinguish between a coordinate mapping (10) and the Hamiltonian mapping (11), there does not exist a unique trajectory through each point of $\bar{\Gamma}$. Rather $\bar{\Gamma}$ divides up into equivalence classes, each one of which is representative of a whole space-time. The factor space of $\bar{\Gamma}$ by the equivalence classes is what Bergmann calls the reduced phase space [11, 12].

Points of the reduced phase space are constants of the motion—they are invariant under all mappings of the form (10). Symbolically the points of the reduced phase space are ^4g, a whole 4-geometry. Unfortunately, however, this argument is not constructive. It does not suggest an algebra for the variables (observables) in ^4g. Therefore, while the transition to quantum theory can be made merely by writing $\Psi(^{(4)}g)$ for the state vector, there is no way at this point to give it meaning.

What the argument does tell us, however, is that the dimensionality of the reduced phase space is four per space point and that is the number of independent observables to be expected.

4. INTRINSIC COORDINATES

In order to be able to describe the reduced phase space in a meaningful way, one must have an explicit invariant method for identifying points in space-time. Variables which depend only on the space-time points, and not on their coordinatization, will have vanishing Poisson brackets with the constraints. If not identically zero themselves, such quantities will be observables [29].

General covariance tells us that the labeling of points in space-time by coordinates has no physical significance. There is no extrinsic way to locate an event in space-time. The only way to mark a position is in terms of the matter and geometry of the world. In an empty space-time, only the geometry is available. It is this property of general covariance that led Arnowitt, Deser, and Misner [30] to describe general relativity as "already parametrized".

The language implies that some of the information carried by the metric tensor is just that needed to give an invariant prescription for the location of space-time points. This prescription is what we shall call "intrinsic coordinates". The search for a suitable set of intrinsic coordinates has taken different forms. Penrose, for example, would like to use intersecting light rays or twistors to locate points. Wheeler's symbol $^{(3)}g$ which denotes a unique three-geometry presupposes in an unspecified way that intrinsic coordinates can be introduced on a space-like hypersurface. Likewise, his Schroedinger equation, in some sense, again unspecified, describes the intrinsic time evolution of the state vector. In what follows I shall describe three specific and related approaches to the problem.

Bergmann and Komar [16] have suggested using the four scalars derived from the Riemann tensor as intrinsic coordinates. The components of the metric tensor and their canonical momenta in terms of these intrinsically defined coordinates will then be observables, but a redundant set. It is a redundant set for two reasons. First of all, there are six components of the metric tensor and six components of the canonical momenta, thus 12 rather than only four per space point. Secondly, they are not defined on a hypersurface, but in all space-time. However, the latter property is no problem for we may define the hypersurface to be $x^0 = 0$, where x^0 is the intrinsic time. The variables at other x^0-values are equally good observables and, in fact, this redundancy shows that the "frozen" formalism is not frozen at all [16]. In quantum mechanics we want to take expectation values at $x^0 = t_0$ and again at $x^0 = t_1$. In this formalism, assuming it could be carried out explicitly, one does not need a time-development operator because the complete solution is laid out at once. The reduction from 12 to 4 components per space point, however, is a serious problem which I shall return to again shortly.

Arnowitt, Deser, and Misner [30] take a different approach to intrinsic coor-dinates. They begin by assuming space-time to be asymptotically flat. Therefore, they wish to introduce coordinates which are asymptotically Minkowskian and they assume this can be done in a manner which is unique up to a Poincaré trans-formation. Actually, the group is considerably bigger [33], but that is not a major concern. They need asymptotic flatness in order to decompose the metric tensor into transverse-traceless, longitudinal, and trace parts. Several different decomposi-tions have been suggested, both non-covariant and covariant. What is of interest here is that the decomposition is used to aid the explicit solution of the constraints. One wants to solve the constraints for four functions in terms of the remaining eight. Symbolically, let me write

$$\mathcal{H}_s = 0 \Rightarrow \psi_s = P_s(g_A, \pi^A, \phi^\mu)$$

$$\mathcal{H}_L = 0 \Rightarrow \psi_0 = P_0(g_A, \pi^A, \phi^\mu)$$

where g_A and $\pi^A (A = 1, 2)$ represent the sought for set of observables and the ϕ^μ, which may not be fully identified at this point ($\mu = 0, 1, 2, 3$), will be fixed eventually by coordinate conditions. ADM substitute this solution into the Lagrangian of Eq. (1), and then consider the change in the action assuming there exists an invariant transformation. In the usual way [39], $\delta S = 0$ for an invariant transformation gives constants of the motion which are the generators of the invariant transforma-tions. The generator takes the form [30]

$$G = \int d^3x \{\pi^A \delta g_A - \mathcal{T}_\mu \delta \phi^\mu\} \tag{12}$$

By identifying ϕ^μ with the coordinates x^μ which are asymptotically Minkowskian one then finds that the $\mathcal{T}_\mu(g_A, \pi_A)$ are the densities for the generators of the asymptotic symmetry group. As noted, this group is actually larger than the Poincaré group. It has the same structure as the BMS group at null infinity [33]. That is, it contains non-rigid translations.

At this point one has in principle a set of intrinsic coordinates based on an asymptotic symmetry group. The configuration space is defined by $g_A(x)$ and the commutation relations should be defined by the asymptotic symmetry group. Given that one is willing to settle for asymptotically flat space, this structure has a certain appeal. The biggest problems, which remain, are the actual solution of the constraints and the rigorous demonstration that the coordinate conditions can indeed be satisfied modulo an asymptotic symmetry group.

I now want to present an attempt which straddles the Bergmann-Komar and ADM approaches just described. This work was done in part with Fred Klotz [35] and was inspired originally by John Klauder's work on positivity conditions in general relativity [36] and Asher Peres' paper on canonical quantization [37] which introduces three scalars as intrinsic coordinates on a three manifold. Peres' scalars satisfy the harmonic equation in an asymptotically Euclidean space. He uses Fourier analysis, however, which leads to infinite convolutions in a non-linear theory. Klotz and I propose to use the three scalars of the Ricci tensor R_{mn} which always exist in the generic case. Define intrinsic coordinates [39]

$$y^1 = R, \qquad y^2 = R_n^m R_m^n, \qquad y^3 = R_n^m R_r^n R_m^r \tag{13}$$

The metric

$$\bar{g}^{mn} = \frac{\partial y^m}{\partial x^r} \frac{\partial y^n}{\partial x^s} g^{rs} \tag{14}$$

are invariants with respect to transformations on $x^0 = c$, but in general will depend on x^0. Again we have the problem with redundance. In Peres' work the harmonic coordinates plus Fourier analysis eliminates the redundance. Here we eliminate the redundance by fiat. Choose as new configuration space variables the y^m and the three diagonal components of the metric $(\bar{g}^{11}, \bar{g}^{22}, \bar{g}^{33}) \equiv (\bar{g}^m)$ from Eqs. (13) and (14). Consider the canonical transformation

$$F = \int d^3x \left[y^m(g_{mn}) \pi_m + \bar{g}^m(g_{mn}) \bar{p}_m \right]$$

which leads to the equations

$$p^{rs} = \frac{\delta y^m}{\delta g_{rs}} \pi_m + \frac{\delta \bar{g}^m}{\delta g_{rs}} \bar{p}_m$$

The spatial constraints take the form [38, 39]

$$\mathscr{H}_s = - y_{,s}^m \pi_m = 0$$

because the y^m are scalars and the $\bar{g}^m$ are invariants. In the generic case this implies

$$\pi_m = 0 \tag{15}$$

The remaining constraint takes the form

$$\mathscr{H}_L = (1/\sqrt{-g})\, \Gamma^{mn} \bar{p}_m \bar{p}_n + \sqrt{-g}\, y^1 = 0 \tag{16}$$

$$\Gamma^{mn} = (g_{jk}\, g_{ks} - g_{jk}\, g_{rs}) \frac{\delta \bar{g}^m}{\delta \bar{g}_{jk}} \frac{\delta \bar{g}^n}{\delta \bar{g}_{rs}}$$

The calculation for Γ^{mn} is extremely long and not very pretty. Although it can be carried out, the calculation has not been completed even at the origin of Riemannian coordinates. Therefore, the structure of Γ^{mn} is not known. The idea would be to solve this constraint, Eq. (16), presumably for $\sqrt{-g}$. Following the ADM prescription outlined above, one would then substitute Eq. (15) and the solution for (16) into the Lagrangian of Eq. (1). Variation of the action would now give a generator corresponding to (12)

$$\bar{G} = \int d^3x \, \{ \bar{g}^A \bar{p}_A - \mathscr{P}\, \delta T \} \tag{17}$$

T would define an intrinsic time to go along with the already defined intrinsic spatial coordinates.

The difficult problem one faces here is not in solving the constraints. That part is easy. However, given the three diagonal components of the spatial metric in intrinsic coordinates, we must construct the three off diagonal components. That must be done through the coordinate conditions

$$x^m = y^m \tag{18}$$

where the y^m are the three scalars given in Eq. (13). Thus, once again we are led to solve partial differential equations and to the dependence on boundary conditions to which they are subject. There is one major gain, however. For if we can establish the conditions under which the coordinate conditions, Eq. (18), have unique solutions, we will have gone a long way toward defining the configuration space for the observables.

5. CONCLUSION

The procedure outlined above for the solution of the constraints by the introduction of intrinsic coordinates is strictly classical. Nonetheless, we see that the scalars defining the intrinsic coordinates appear to depend on the dynamical field variables through the Ricci tensor. While the non-covariant decomposition of ADM [30] separate the dynamical from non-dynamical parts of the field, the covariant decompositions do not. In a quantized theory, then, it is not evident whether the intrinsic coordinates can be introduced as c-numbers or whether they become necessarily q-numbers. With the latter possibility one has the interesting conjecture that even in this conventional approach to a quantized theory points of the space-time manifold lose their identity. Only their expectation values would be observable.

Unfortunately, we have not yet obtained a complete description of observables in the classical theory and much less is known about a truly quantized theory.

REFERENCES

1. PETER G. BERGMANN, *Phys. Rev.* **75**, 680 (1949).
2. P. A. M. DIRAC, *Canad. J. Math.* **2**, 129 (1950).
3. L. ROSENFELD, *Ann. Physik* **5**, 113 (1930).
4. L. ROSENFELD, *Ann. Inst. Henri Poincaré*, **2**, 25 (1932).
5. W. PAULI in brief remarks before a working group in Neuchatel, Switzerland, July 1958.
6. R. PENROSE, *J. Math. Phys.* **8**, 345 (1967).
7. R. PENROSE, "Angular Momentum: An Approach to Combinatorial Space-Time," in *Quantum Theory and Beyond* (TED BASTIN, ed.), Cambridge University Press, Cambridge, 1971.
8. D. FINKLESTEIN, *Phys. Rev.* **184**, 1261 (1969).
9. C. ISHAM, "Quantum Gravity," ICTP/72/8, a report based on a lecture given at the Conference on Quantum Gravity, Boston University, 1972.
10. ABHAY ASHTEKAR and ROBERT GEROCH, "Quantum Theory of Gravitation," to be published.
11. PETER G. BERGMANN, *Rev. Mod. Phys.* **33**, 510 (1961).
12. PETER G. BERGMANN, "Status of Canonical Quantization," in *Relativity and Gravitation* (C. KUPER and A. PERES, eds.), Gordon and Breach, New York, 1971.
13. A. KOMAR, *Phys. Rev.* **D4**, 923 (1971).
14. A. KOMAR, *Phys. Rev.* **111**, 1182 (1958).
15. PETER G. BERGMANN and A. KOMAR, *Phys. Rev. Lett.* **4**, 432 (1960).
16. PETER G. BERGMANN and A. KOMAR, "Quantization of the Gravitational Field," in *Recent Developments in General Relativity*, Pergamon Press, London, 1962.
17. J. A. WHEELER, "Particles and Geometry," in *Relativity* (M. CARMELI, S. FICKLER and L. WITTEN, eds.), Plenum Press, New York, 1970.
18. J. YORK, *J. Math. Phys.* **14**, 456 (1973).
19. Y. CHOQUET-BRUHAT, *GRG* **5**, 49 (1974).
20. S. DESER, *Ann. Inst. Henri Poincaré* **7**, 149 (1967).
21. A. KOMAR, "Canonical Decomposition of the General Relativistic Initial Value Problem," preprint (to be published).
22. C. MISNER, *Phys. Rev.* **D8**, 3271 (1973).
23. A. KOMAR, "The Quantization Program for General Relativity," in *Relativity* (M. CARMELI, S. FICKLER and L. WITTEN, eds.), Plenum Press, New York, 1970.
24. J. A. WHEELER, "Superspace and the Nature of Quantum Geometrodynamics," in

Batelle Rencontres (C. DeWitt and J. A. Wheeler, eds.), W. A. Benjamin, Inc., New York, 1968.

25. B. S. DeWitt, *Phys. Rev.* **160**, 1113 (1967).

26. J. Marsden, D. Ebin and A. Fischer, "Diffeomorphism Groups, Hydrodynamics, and Relativity," in *Proceedings of the Thirteenth Biennial Seminar of the Canadian Mathematical Congress* (J. R. Vanstone, ed.), Canadian Mathematical Congress, Montreal, 1972.

27. P. A. M. Dirac, *Proc. Roy. Soc.* **A246**, 326 (1958).

28. P. A. M. Dirac, *Proc. Roy. Soc.* **A246**, 333 (1958).

29. Peter G. Bergmann, *Encyclopedia of Physics*, Vol. 4 (S. Flugge, ed.), Springer-Verlag, Berlin, 1962.

30. R. Arnowitt, S. Deser and C. Misner, "The Dynamics of General Relativity," in *Gravitation* (L. Witten, ed.), John Wiley, New York, 1962.

31. C. Misner, K. Thorne and J. A. Wheeler, *Gravitation*, W. H. Freeman and Company, San Francisco, 1973.

32. Peter G. Bergmann and A. Komar, *Int. J. of Theor. Phys.* **5**, 15 (1972).

33. R. Geroch, *J. Math. Phys.* **13**, 956 (1972).

34. Peter G. Bergmann, *Phys. Rev.* **124**, 274 (1961).

35. F. Klotz, "A Study of General Relativity in Canonical Form," dissertation, Syracuse University, August 1970.

36. E. W. Aslaksen and J. R. Klauder, *J. Math. Phys.* **9**, 206 (1968).

37. J. N. Goldberg and F. Klotz, *Int. J. Theor. Phys.* **7**, 31 (1973).

38. A. Peres, *Phys. Rev.* **171**, 1335 (1968).

39. F. Klotz, *Int. J. Theor. Phys.* **6**, 251 (1972).

Modern Theoretical and Observational Cosmology

Thomas B. Criss, Richard A. Matzner, Michael P. Ryan, Jr. and
L. C. Shepley

Center for Relativity Theory
The University of Texas at Austin
Austin, Texas 78712, U.S.A.

ABSTRACT

A review is given of recent observational and theoretical results in cosmology in light of general relativistic models of the universe.

Classes of Models	Observations	Theoretical Topics
Isotropic	Direct Optical	Hamiltonian Cosmology
Anisotropic	Mass	Singularities
Homogeneous	QSO's	Galaxy Formation
Inhomogeneous	X-Ray, γ-Ray	Thermal History
	Microwave	Black Holes
		Particles
	Element Abundances	Elements

Figure 1
Topics.

Note: Many of the figures are taken from articles published by others. For permission to use these figures and for their comments, we wish to thank the respective journals, Lick Observatory, and our colleagues, especially H. Arp, G. R. Burbidge, G. deVaucouleurs, R. H. Dicke, S. A. Fulling, R. Hagedorn, M. Kruskal, J. G. Miller, C. W. Misner, R. B. Partridge, P. J. E. Peebles, M. Rees, P. G. Roll, D. N. Schramm, D. W. Sciama, P. A. Strittmatter, B. Tinsley, and D. T. Wilkinson. Figures 2, 5, 8, 9, 24, 25, 26, 27, 28, 36, 37, 38, 39, and 40 were published in the *Astrophysical Journal* by the University of Chicago Press, © American Astronomical Society.

1. INTRODUCTION

This paper intends to give a rough overview of the field of modern cosmology, as it has evolved in the last decade or so, accentuating the connections that exist between theory and observations in this field. We will in general take the "conven-

tional" path through the jungle of competing theories and interpretations of observations. (At the same time, we will post road signs large enough—and intelligible
enough—to guide the explorer on the alternate paths.) Thus we will give only the
briefest mention of steady state cosmologies which seem to be cast out on the face
of it by the fact that QSO's (Schmidt, 1972) and elliptical galaxies (Sandage, 1973)
evolve. Our main emphasis will be on solutions obtained from Einstein's theory
of gravitation, without cosmological term (except in one section, 2.4 below) in conjunction with physically reasonable types of matter.

Sec. 2 will restrict itself to fitting observations to an isotropic homogeneous
model. The metric in such a cosmology is

$$ds^2 = -dt^2 + R^2(t)\, d\sigma^2 \tag{1.1}$$

where $d\sigma^2$ is the unit 3-space of the appropriate symmetry. Subsequent parts will
reduce the symmetry of the model and look for predictions of new features in the
observations.

2. ISOTROPIC COSMOLOGY

2.1. Classical observational cosmology. The Hubble recession

Observational and theoretical cosmology are as old as the human ability to regard
and remember the plan of the sky and to predict the sunrise. The heavens changed,
but their character did not. This concept of basic unchangingness (stationarity) of
the universe was held until the studies of Hubble (1929), long after the first cosmological models of Einstein. In order to obtain static models to correlate with
his world picture, Einstein (1917) had to incorporate the "cosmological" Λ term,
which (depending on its sign) can retard collapse. It was immediately clear that this
cosmological term has no effect in planetary dynamics. Other models, such as
deSitter's (1917a, b) had some dynamics, and initially these results were taken to
be weaknesses of the teleology of the theory since Einstein had hoped it would lead
to only a single cosmology. The non-existence of a $\Lambda = 0$ static solution could have
been realized as predicting the Hubble expansion, and the observation of the expansion would then have taken its place as another classical test of Relativity.

At any rate, Hubble (1929) discovered the redshift of distant galaxies and determined the proportionality constant which bears his name. We forbear presenting
a graph which plots the accepted value of his constant $H_0 = (\dot{R}/R)_{\mathrm{now}}$ against the
year the value was published; suffice it to say that since the original publication of
Hubble, $H_0 \sim 500\ \mathrm{km} \cdot \mathrm{sec}^{-1} \cdot \mathrm{Mpc}^{-1}$, the value has monotonically dropped to that
obtained most recently by Sandage (1972a), $H_0 \sim 55\ \mathrm{km} \cdot \mathrm{sec}^{-1} \cdot \mathrm{Mpc}^{-1}$.

One of the important results is that the Hubble parameter tends to be more or
less isotropic, with a scatter of perhaps 15% over the regions of sky considered by
Hubble (Trendowski, 1971). A complete search over the whole optical sky is im-

possible since the Milky Way obscures distant galaxies in a large fraction of the sky.

However, this apparent isotropy, coupled with the evidence that the nebulae are not local (i.e., Milky Way) phenomena, strongly influenced the early designs for model universes. An appeal to the non-uniqueness of our position in spacetime (and a desire for mathematical simplicity) led to the universal consideration of cosmologies that were spatially homogeneous and isotropic.

Take, as your model of the universe, such a run-of-the-mill homogeneous, isotropic model with $\Lambda = 0$, and idealize the matter to a fluid with a known equation of state (usually $p = 0$). Then the Einstein equations require only two parameters: the density ρ, and Hubble's constant $H_0 = (\dot{R}/R)_{now}$ to specify the model. More complication enters if $\Lambda \neq 0$.

We have mentioned the current best determination of the Hubble constant, $H_0 \sim 55 \text{ km} \cdot \text{sec}^{-1} \cdot \text{Mpc}^{-1}$. To determine H_0, essentially what is found is a redshift vs apparent brightness curve for certain galaxies. The lowest order results (for all cosmological models) gives a luminosity distance

$$d_2 = H_0^{-1}[z + \tfrac{1}{2}(1 - q_0)z^2 + ...]$$

The deceleration parameter $q_0 = (-\ddot{R}R/\dot{R}^2)_{now}$ thus is a coefficient in a higher order term than is H_0.

The difficulty in determining q_0 and H_0 is that local factors affect the recession velocity of nearby galaxies, and several steps of inference are required to obtain the distance yardstick for the very distant galaxies. It has been strongly argued (deVaucouleurs, 1970) that clustering of galaxies exists on the largest scales that can be observed optically. Hence, except for determination using the most distant galaxies, the Hubble and deceleration parameters found may not be those appropriate to the universe, but only those appropriate to our particular cluster. In particular, the universe may be quite inhomogeneous, or might be cast into a hierarchical structure which has a different average density on different scales. Then models like the isotropic homogeneous ones become very poor indeed. Sandage, Tammann and Hardy (1972) have shown that the observed clusters do not affect H_0. The situation for determining q_0 is quite difficult, because we are searching for the derivative with time (i.e., with distance) of the Hubble parameter. Sandage (1972a, b, 1973) and Sandage and Hardy (1973) have given a very exhaustive determination of all the difficulties and corrections necessary to derive the value he has recently published $q_0 = 0.96 \pm 0.4$ (p.e.). Evolutionary effects of galaxies as he points out could lower this to $q_0 \simeq 0$. Moreover, it should be pointed out that Gott, Gunn, Schramm and Tinsley (1974) suggest that corrections for these evolutionary effects do in fact shift the result to a small value of $q_0 \simeq 0.1$ (the separation between open and closed cosmologies in which the pressure is negligible is $q_0 = 1/2$, always assuming $\Lambda = 0$). Gott et al. (1974) have other reasons for believing $q_0 \simeq 0.1$, since this seems to give the correct primordial deuterium abundance, and we return to these points shortly.

2.2. Direct mass measurements and missing mass. The topology of the universe

In the discussion above we passed very briefly over the Hubble determination for the density of luminous matter, from galaxy counts, which led to estimates of the order of $\sim 10^{-31}$ gm $\cdot$ cm^{-3}. With the currently observed values of Hubble's constant, this density is too low (by about a factor of 100) to lead to recollapse of a $\Lambda = 0$ homogeneous-isotropic cosmology.

It is relatively clear that luminous matter does not nearly account for all the matter in the universe. Most direct evidence for this comes from observations of the dynamics of clusters of galaxies, where observed masses are usually too low by factors of 10 to 50 to account for the dynamics of the cluster. (Higher factors, up to 100, are associated with some clusters which have suspicious identification.) The mass determined by such methods is still too low for closure of the universe. However, a number of alternative methods of hiding the missing mass have been proposed over the years. These have been discussed to some extent by Gott et al. (1974), who conclude that none of them are likely to lead to a sufficiently high density to close the universe. The reader should refer for details to their excellent review paper devoted to just this question.

Figure 2 (after Gott et al., 1974) shows conservative estimates of bounds on the observables determining the universe ($\Lambda = 0$). The Hubble parameter is bracketed by 120 km $\cdot$ sec$^{-1} \cdot$ Mpc$^{-1} > H_0 > 30$ km $\cdot$ sec$^{-1} \cdot$ Mpc^{-1}. The age of the universe, t_0, is calculated assuming that the pressure is currently negligible. The bounds 8×10^9 yrs $< t_0 < 18 \times 10^9$ yrs arise from considerations of globular cluster evolution. The deceleration parameter q_0 is set at $q_0 < 2$, a value two standard deviations above Sandagés $q_0 = 0.96$. Finally the mass of the universe puts a lower limit on the deceleration; the parameter Ω is twice q_0 in a pressureless $\Lambda = 0$ cosmology. The Ω^* plotted as vertical lines are estimates based on estimates of mass *in galaxies*. It should be noted that deVaucouleurs (1970) has argued that the observed mass density in galaxies could be just an upper limit, since there is no clear evidence that the observations reach deep enough to determine the ultimate number density of galaxies (Fig. 3).

This question of the total mass in the universe relates also to the topology of the spatial (constant-time) sections of the cosmology. The isotropic homogeneous models which recollapse have finite-volume space sections with (in the simplest case) the topology of a 3-sphere, S^3. Those that expand with an excess of velocity in the infinitely dilute state in the far future have in the simplest case the topology of an (infinite) 3-hyperboloid, H^3. The limiting case between the two has the topology of Euclidean space. Hence knowledge of global properties follows, not surprisingly, from local observations and assumptions about global symmetry.

The idea that we may be able to determine the global topology of the universe by measurement has been considered by Dautcourt (1971), Ellis (1971), and Zel'dovich

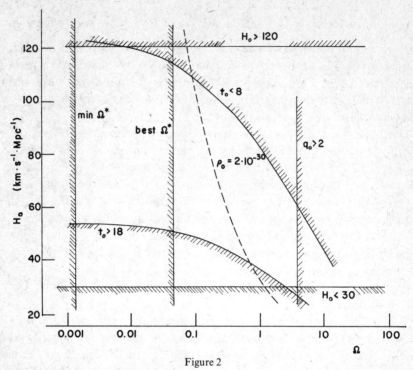

Figure 2

Observations of the parameters determining the universe (after Gott, Gunn, Schramm and Tinsley, 1974). Ω is the ratio of observed matter to that for closure of the universe. H_0 is Hubble's parameter, q_0 the deceleration parameter, t_0 the age of the model. Ω^* is the observed mass in galaxies. Any homogeneous-isotropic model is constrained to lie in the curved-trapozoid in the center of the figure.

and Novikov (1967). Zel'dovich and Novikov use parity and *CP* violations to argue that space sections of our universe must be orientable (Geroch, 1967a, has a similar argument). Ellis has a long discussion based on the homogeneous-isotropic models. If one assumes that these models represent our universe, then there are a number of global topologies that are compatible with them in addition to the simplest ones mentioned above. Since the Friedmann–Robertson–Walker (FRW) models have 3-spaces of constant curvature for $t = $ constant sections, we need only list all such orientable 3-spaces that allow an FRW metric (such three-spaces are studied in detail in Wolf, 1967). For $k = +1$ the possible homogeneous 3-spaces are 1) S^3, 2) P^3 (real projective 3-space), 3) $S^3 Z_n$ ($n > 2$), 4) S^3/D_m^* ($m > 2$), 5) S^3/T^*, 6) S^3/O^*, 7) S^3/I^*, where Z_n is the cyclic group of order n, D_m^* is the binary dihedral group, and T^*, O^*, and I^* are the binary tetrahedral, octahedral, and icosahedral groups respectively. For $k = 0$, the homogeneous spaces are 1) R^3, 2) $R^1 x$ (cylinder), 3) $R^1 \times T^2$, 4) T^3. T^2 is the 2-torus and T^3 the 3-torus. For $k = 1$ the only homogeneous 3-space is H^3, the usual 3-space associated with $k = -1$ FRW models. Ellis also

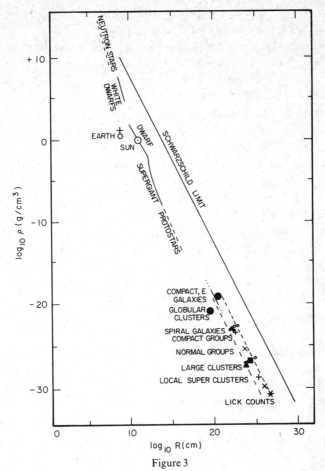

Figure 3

Measured density contained in a sphere of indicated radius. The usual determination of density of galactic matter corresponds to the deepest observations (after de Vaucouleurs, 1970).

lists inhomogeneous and non-orientable examples. The only spaces listed above that are isotropic are R^3, S^3, P^3, and H^3.

As Ellis points out, we cannot exclude inhomogeneous and anisotropic spaces, because locally they could be the same as the usual FRW models. Even if we take only the homogeneous spaces, requiring isotropy is probably too stringent. Consider the $k = 0$ "cylinder model",

$$ds^2 = -dt^2 + R^2(t)\left[dx^2 + dy^2 + a^2\, d\varphi^2\right], \quad a = \text{const}, \quad 0 \le \varphi < 2\pi$$

It is not *strictly isotropic* because if we look around the cylinder (if a is small enough) we see multiple images, while if we look along the axis we do not. However, it is *locally isotropic*. The redshift would be isotropic, and the various images of a galaxy would be seen at different times in its evolution and might be difficult to identify. The gross features of this model would be the same as the usual $k = 0$ isotropic model, but detailed measurement could distinguish the two. At least the nearby galaxies are well enough identifiable to put a lower limit on a. It is obvious that all the $k = 0$ models listed above are almost isotropic, and the $k = +1$ models should be, but some kind of proof of this would be needed.

There are three further areas in which topological questions in general relativity have touched on cosmology: 1) In the foam geometry of Weyl (1949) and Wheeler (1962). 2) As a vital element in the proof of general singularity theorems. 3) In the important theorem that there is no way for the topology of the universe to change in a non-singular way.

The first two of these subjects we shall not discuss in detail. The first postulates that on a microscopic scale spacetime is no longer a smooth manifold but is a sponge-like construct that is "boiling" with changes in geometry and topology. This idea might help prevent a singularity by changing the behavior of the universe when it had contracted to the scale of the fluctuating foam, and if it has any validity at all, it would be most important during the quantum era of the universe near the singularity. This will be discussed with regard to avoidance of singularity. The second application is more important, but it will be dealt with in the section on singularity theorems.

The third subject has been considered by Geroch (1967b) and Kundt (1967), who showed that a 3-space evolving in a four-dimensional manifold that has a light-cone structure cannot change its topology without some sort of singularity. The basic theorem is due to Geroch (1967b):

> Let M be a compact geometry whose boundary is the disjoint union of two compact, spacelike 3-manifolds, S and S'. Suppose M is isochronous (a continuous choice of the forward light-cone exists) and has no closed time-like curve. Then S and S' are diffeomorphic and M is topologically $S \times [0, 1]$.

Thus a $k = 1$ universe cannot become a $k = -1$ universe without some drastic change occurring. Kundt shows that a universe cannot break up into separate pieces. In quantum general relativity we might expect that the requirement of isochronicity might be relaxed and the theorem invalidated. The basic idea, then, is *no change of topology in non-quantum cosmology*.

To close this section, we mention that Dautcourt (1971) has discussed the possibility that we may be able to find out whether the universe is $k = 0$ or $k = \pm 1$ from the effect of the 3-space metric on the local d'Alembertian.

2.3. Quasar–galaxy associations. Anomalous redshifts

Of all the optically accessible objects of observation, the quasi-stellar objects (QSO's or quasars) have the highest observed redshifts. From the beginning there has been an attempt to use the galaxy derived Hubble relation to place the quasars at cosmological distances. The large amount of energy required from a small region of space (small because of the short-term variation observed in the radio and optical flux from them (Hoyle and Burbidge, 1966)) poses theoretical difficulties. As yet there have been no conclusive results indicating the cosmological nature—or lack of it—of quasars. There have, however, been some interesting results on the association between quasars and galaxies. For instance Burbidge et al. (1972) have found a correlation between certain galaxies and QSO's, with the angular separation decreasing in the appropriate $1/d$ way for the galaxies with higher redshift. This is indicative of a cosmological origin; the QSO's are all at about the same linear separation from their associated galaxies, and the more distant ones thus have smaller angular separation from their galaxies. There, however, is no correlation with the quasar redshift (Fig. 4, after Burbidge, 1973). Further, if larger samples are taken, the quasar galaxy associations cease to be statistically significant (Burbidge and O'Dell, 1973; Ozernoi, 1972; Bahcall et al., 1972).

One of the associations listed by Burbidge et al. (1971) is the galaxy IC 1746 ($z = 0.026$) and the QSO PHL 1226 ($z = 0.404$). They are separated by $0'.8$. Figure 5

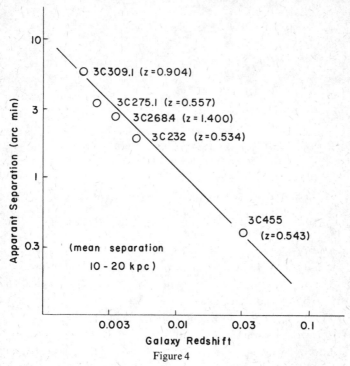

Figure 4

Plot showing angular separation between selected 3C quasars and nearby galaxies. The angular size decreases with galaxy redshift as if the quasar-galaxy distance were constant. There is no correlation with the quasar redshifts in parentheses (after Burbidge, 1973).

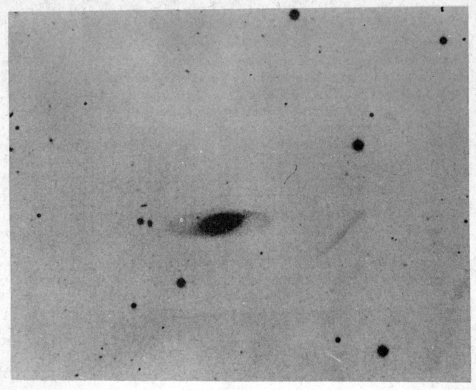

Figure 5

Example of an association between galaxy IC 1746 ($z = 0.026$) and radio-quiet QSO PHL 1226 ($z = 0.404$). The quasar is the second object from the wisp of the galaxy; a non-stellar object is between. On Palamar sky plates the wisp seems to reach to the QSO and the intermediate object does not appear (photograph by H. Arp, published in Burbidge et al., 1971).

shows that the nebulosity of the galaxy extends partway to the QSO. A compact non-stellar object is placed between the QSO and the galaxy (Fig. 5; photograph by H. Arp, published in Burbidge et al., 1971).

Kristian (1973) has studied a series of quasars, and within his sample those which have low redshifts show faint images of what could be background galaxies, while the larger redshift ones do not. This again suggests a cosmological interpretation. Stockton (1973) has found that the quasar 4C37.43 ($z = 0.37$) lies within 11′ of a faint galaxy. Stockton measured the galactic redshift to be 0.3736.

Such a QSO–galaxy coincidence would be taken as strong evidence that the QSO is at the distance of the galaxy and both are at the distance indicated by the redshift. However, the QSO distance problem has caused a restudy of redshifts even for galaxies, and a number of "anomalous" redshift situations have been found. The first example discussed by Arp (1969) is that of M51 (Fig. 6). The knot on the end

Figure 6

The "whirlpool" galaxy M51. The (apparently connected) knot has 200 km/sec redshift difference from the main galaxy. Estimated lifetime of the association if this velocity difference is correct is 2×10^8 yrs, approximately one rotation period of M51 (Lick Observatory photograph).

of the arm has a velocity relative to the large galaxy of 200 km/sec. Although this is not large as galactic velocities go, the difference is enough to disrupt the system in 2×10^8 years, approximately one rotation period of the main galaxy. Arp has published several other interesting associations of this type. In one of them (Arp, 1971), there is an apparently associated small galaxy in the system NGC 7603 which has a different redshift amounting to about 8100 km/sec (Fig. 7). A second system, NGC 772 (Arp 1970), has three small apparently associated galaxies. One has the same redshift as the large galaxy (2450 km/sec), while the other two have redshifts of 20,200 and 19,700 km/sec. Another interesting system (Fig. 8) is Stephan's quintet which was also investigated by Arp (1973) (see also Burbidge and Burbidge, 1961). Here the system appears to be very strongly interacting. Four of the five members have redshifts of 6417 km/sec while the fifth has $z = 795$ km/sec. Sargent (1968) has investigated the system VV 172 (Fig. 9, from Arp, 1966). There the redshift of

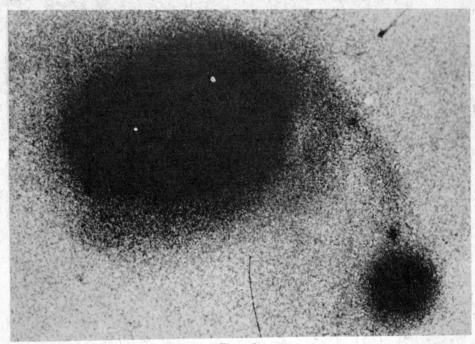

Figure 7

NGC 7603. Redshift of large object is 8800 km/sec, that of smaller comparison is 16,900 km/sec (Arp, 1971).

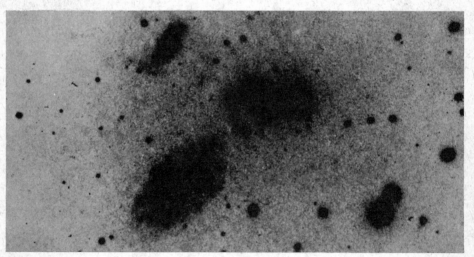

Figure 8

Stephan's quintet. Four of the galaxies lie on an approximately straight line (the central one is double). They are NGC 7319 (redshift 6700 km/sec), NGC 7318B (redshift 5700 km/sec), NGC 7318A (redshift 6700 km/sec) and NGC 7317 (6700 km/sec). The fifth member, NGC 7320, has redshift 800 km/sec (Arp, 1973).

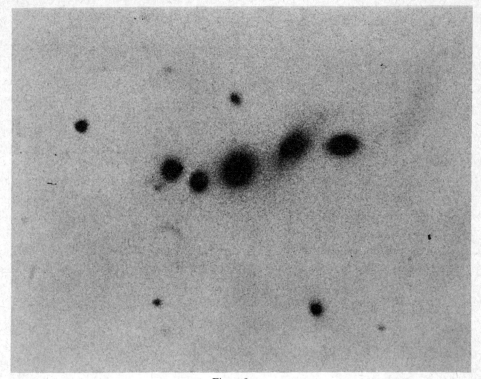

Figure 9

The quintet VV 172. The small object second from the north end has the discordant redshift. Starting from the north, the redshifts (Sargent, 1968) are 16,070 km/sec, 36,880 km/sec, 15,820 km/sec, 15,690 km/sec and 15,480 km/sec (photograph from Arp, 1966).

all but one of the member galaxies is ~ 16,000 km/sec, while the fifth has redshift 36,880 km/sec.

All of this indicates that for QSO's at least, the redshift must be suspect as a cosmological distance determiner. Additionally, it appears that some intrinsic redshift mechanism is at work even in galaxies, although (as verified by the consistency of different measures, Sandage, 1972a,b, 1973, Sandage and Hardy, 1973) for normal galaxies it does not distort the Hubble relationship.

The cosmological assumption for quasars seems to be held by most workers in the field, but as pointed out by Burbidge (1973) in an excellent review article, this assumption is chiefly buttressed by the lack of any definitive evidence excluding it.

In the next section the quasar redshift is assumed cosmological. We shall, after that section, not mention QSO's again, and we shall assume that the galactic redshift-distance relation is a correct cosmological yardstick.

2.4. Quasars and the cosmological constant Λ

We have so far assumed that the so-called cosmological constant Λ has been set equal to zero. As is well known, choice of nonzero Λ (of the appropriate sign) leads to a repulsive interaction, while the other sign leads to an added gravitational attraction. The expansion equation for the radius $R(t)$ of the universe is modified by this added constant term whereas terms involving ordinary matter fall off with increasing R at least as R^{-3}. The Λ-terms thus are never important near the singularity. We then must expect that if Λ is negligible *now*, it can safely be ignored for retrodiction, and will enter in a very simple way for prediction in the universe.

One feature of quasar observations is that there seems to be a break in the number with redshifts just over $z = 2$. Very few are seen with larger redshifts (Burbidge and Burbidge, 1969). It appears that selection effects can have a strong influence in this question (Sandage, 1972b), although none has been definitely implicated as causing the cutoff. Further, larger redshifts have been found, so less weight is currently being given to these observations. They did, nonetheless, lead to several suggestions that the cosmological constant was not, in fact, negligible for a redshift for $z = 2$. The Lemaitre (1927) universe model, which has a Λ-term, can exhibit a long "resting" period during which the size of the universe does not change very much. Since redshift in a homogeneous isotropic model is a function only of radius:

$$1 + z = R_f/R_i \tag{2.1}$$

where R_f is the final (observation time) radius and R_i is the initial (emission time) radius, a resting period with $R \sim \frac{1}{3}R_{\text{now}}$ would lead to a lot of quasar redshifts of order 2, if quasars were formed continuously in time, and if their redshift is cosmological.

Now Λ is a constant while both the matter and curvature terms decrease as the universe evolves. Hence, subsequent to a resting period, Λ very quickly becomes dominant compared to the other terms determining the expansion H_0:

$$3H_0^2 = \rho + \Lambda - 3k/R^2 \tag{2.2a}$$

(where $k = +1$ is necessary for a resting period).

A second field equation may be cast in the form (Bondi, 1960)

$$\tfrac{1}{2}\rho - 3q_0 H_0^2 = \Lambda \tag{2.2b}$$

which shows that a positive Λ is necessary for a resting period to have occurred. (If $\Lambda = \frac{1}{2}\rho$, q_0 vanishes from (2.2b). If then $3\Lambda = 3k/R^2$, from (2.2a) the universe is exactly static. The "resting" period in a continuously expanding Lemaitre universe requires that these conditions be very closely (but not exactly) met. In particular, $\Lambda \simeq R_{\text{rest}}^{-2}$ to a very good approximation, for a significant resting period to exist. An example of a Lemaitre model is in Fig. 10.)

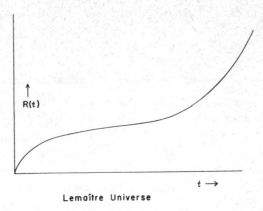

Lemaître Universe

Figure 10

Schematic diagram of radius vs. time behavior of a Lemaitre model which exhibits a "resting" phase. Such models have been called on to explain certain quasar observations.

At present, Sandage's determination of q_0 gives $q_0 \simeq 1$, and it seems very safe to assume $q_0 > -1$ for the estimate which follows. (Sandage and Hardy, 1973 consider a q_0 so negative to be completely excluded by the data.) Further, we have seen that observations make it extremely unlikely that $\rho/3H_0^2$ exceeds unity. Hence

$$\frac{\Lambda}{3H_{0\text{now}}^2} \gtrsim 1.5$$

From (2.2a) we then have

$$\Lambda < 1.5\left(\frac{3k}{R_{\text{now}}^2}\right) = \frac{1.5}{9}\left(\frac{3k}{R_{\text{rest}}^2}\right) = \frac{1}{2}\left(\frac{k}{R_{\text{rest}}^2}\right)$$

compared to the value k/R_{rest}^2 required for the resting phase.

On the face of it, $\frac{1}{2}$ is certainly not close enough to the value unity for a resting period to have taken place. Additionally we have taken estimates in a way which makes Λ larger (i.e., in a way which is biased toward producing a resting period). Nonetheless it can be claimed that slight errors in our estimates could at least make the condition plausible. (The matter density can also be found to fit roughly into the correct range.) This evidence, shaky as it is, can be interpreted as saying that Λ could have been important around a redshift of 2, although to claim it was important earlier would contradict the estimates made above.

Another feature of the slowed-down expansion of the Lemaitre model is that photon circumnavigation of the universe is possible, and it has been suggested that quasars in roughly opposite directions would show correlation because some of them are the *same* quasar seen from two ways around the cosmos (Solheim, 1968a,b). It appears that this requires a long resting period and the value of Λ is still constrained

by the arguments mentioned above. For the remainder of the paper we will return
to the convention

$$\Lambda = 0$$

2.5. X-ray and γ-ray observations and the matter–antimatter ratio of the universe

Prior to this point we have been principally concerned with optical observations
in what might be called the classical manner. These observations perforce do not
reach back too far into the past of the universe. On the other hand, X-ray and γ-ray
observations may extend our view. This can be important theoretically in many
ways. For instance, observations indicate that there is an excess positive baryon
number in our regions of the universe. One possibility which has been suggested
by Omnes (1969a,b) is that this is only a local fluctuation in a universe with net
baryon number zero. A fairly elaborate analysis is then performed to demonstrate
the possibility that instabilities lead to large scale separation of the two types of
matter. At lower temperatures ($\leq 10^{10}$ K) the driving force behind such separation is pressure arising from particle-antiparticle annihilations and under some
conditions this can lead to the desired separation. An apparently inescapable
problem, however, arises from the large number of very energetic X-ray and γ-rays
this would produce. Jones and Jones (1970), and Steigman (1973) have studied this
problem in view of current X-ray and γ-ray observations. They conclude that the
annihilation would lead to unacceptably large fluxes of high energy photons (which
are not observed). Also, Steigman notes that *no* antimatter cosmic rays have been
found. The conclusion is that matter predominates very strongly over antimatter
for distances out to a large fraction of the observable universe, so a matter-only model
is the appropriate one. Figure 11, from Longair and Sunyaev (1969), shows all of
the observations of electromagnetic radiation, including X-rays and γ-rays.

2.6. 3 K microwave radiation

In 1965, Penzias and Wilson detected excess electromagnetic noise, in a microwave
communications antenna, which corresponded to a sky temperature at a wavelength
of 7.35 cm, of approximately 3 K (Penzias and Wilson, 1965). Since then there have
been many determinations of the microwave background, by direct measurement at
wavelengths between ∼50 cm and ∼3 mm. Wavelengths longer than 50 cm are
blanketed by radiation from the galaxy; the atmosphere imposes an upper limit of
transmission, although observations above the atmosphere can go to shorter wavelengths. For these shorter wavelengths, there are indirect radio observations which
determine the temperature of the microwave background by its effect on the rotation
levels of interstellar radicals. Peebles (1971) estimates that thermalized starlight
would swamp the microwave background for very short wavelengths, i.e., for
$\gtrsim 0.04$ cm.

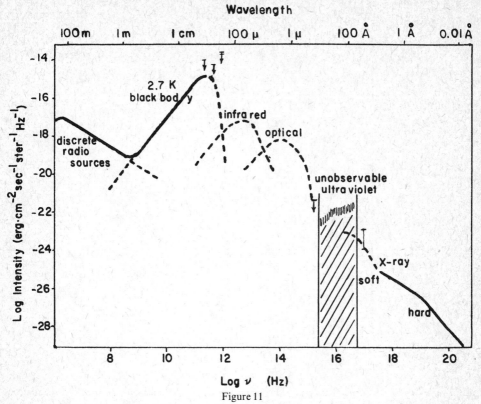

Figure 11

The measured (solid lines) and predicted (dashed lines) background intensity of the night sky as a function of frequency (from Longair and Sunyaev, 1969).

Table 2.1 lists some of the observations of the microwave radiation versus wavelength. These results are also plotted in Fig. 12.

The spectral peak of a 2.7 K blackbody lies at about 0.15 cm, depending on precisely what function is being measured. Although the results of Muehlner and Weiss (1970) suggested a high temperature compared to 2.7 K, the subsequent results by Muehlner and Weiss (1973a,b) were consistent with 2.7 K and identified several sources of error, from the atmosphere and from the instrumentation. However, initial rocket flights (Shivanandan, Houck and Harwit, 1968; Pipher, Houck, Jones and Harwit, 1971) which presumably were free from atmospheric contamination and which had bandpasses reaching even shorter wavelengths found excessive fluxes.

At present, however, all groups agree the microwave flux is consistent with 2.7 K. The 1972 rocket experiment of Houck et al. now shows a flux corresponding to at most ~ 4 times that expected for 2.7 K (for a temperature $\gtrsim 3.7$ K). These results are plotted in Figure 12. Shortward of 0.04 cm have been other measures (Pipher et al.,

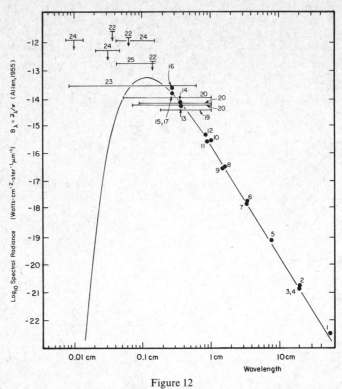

Figure 12

Measurements of the isotropic background in the microwave region. The solid curve is the spectrum of a 2.7 K blackbody. The solid dots are individual measurements at specific frequencies while the horizontal bars indicate average intensity over the frequency range indicated. The numbers refer to the list of measurements in Table 2.1.

1971) which place upper bounds on the radiation, consistent with a 2.7 K background. However it is in this region that radiation from interstellar grains should begin to dominate the cosmic radiation. Thus the shortward side of the blackbody curve will probably remain unobserved.

The isotropy of the radio background has also been studied extensively. Some determinations of the isotropy are listed in Table 2.2.

Figure 13 shows one plot of the isotropy (Wilkinson and Partridge, 1967; see also Partridge, 1969).

The surprising isotropy of the microwave radiation suggests it is a quite ancient fossil. Equilibrium with the matter filling the universe (presumably hydrogen) could be achieved at temperatures only above 3000 K. The 3 K radiation has undergone a redshift of 1000 since that epoch. Some efforts have been made to suggest reionizing mechanisms which would allow the radiation to undergo a few scatterings at smaller redshifts (such as possibly the creation of super massive stars as a preliminary to

TABLE 2.1

Observations of the Microwave Background Spectrum

	Wavelength, cm	Thermodynamic temperature, K	Comments	Reference
1	50–75	3.7 $\pm$ 1.2	Large galactic correction	Howell and Shakeshaft (1967)
2	21.2	3.2 $\pm$ 1.0		Penzias and Wilson (1967)
3	20.9	2.8 $\pm$ 0.6		Howell and Shakeshaft (1966)
4	20.7	2.5 $\pm$ 0.3		Pelyushenko and Stankevich (1969)
5	7.35	3.1 $\pm$ 1.0		Penzias and Wilson (1965)
6	3.2	3.0 $\pm$ 0.5		Roll and Wilkinson (1966)
7	3.2	2.69 $\pm$ 0.16		Stokes, Partridge and Wilkinson (1967)
8	1.58	$2.78 {+ 0.12 \atop - 0.17}$		Stokes, Partridge and Wilkinson (1967)
9	1.50	2.0 $\pm$ 0.8		Welch, Keachie, Thornton and Wrixon (1967)
10	0.924	3.16 $\pm$ 0.26		Ewing, Burke and Staelin (1967)
11	0.856	$2.56 {+ 0.17 \atop - 0.22}$		Stokes, Partridge and Wilkinson (1967)
12	0.82	2.9 $\pm$ 0.7		Puzanov, Salmonovich and Stankevich (1967)
13	0.33	$2.46 {+ 0.40 \atop - 0.44}$	Large atmospheric correction	Boynton, Stokes and Wilkinson (1968)
14	0.33	2.61 $\pm$ 0.25	Large atmospheric correction	Millea, McColl, Pederson and Vernon (1971)
15	0.264	3.2 $\pm$ 0.5	Interstellar CN	Field and Hitchcock (1966)
16	0.264	3.7 $\pm$ 0.7	Interstellar CN	Peimbert (1968)
17	0.264	2.83 $\pm$ 0.15	Interstellar CN	Bortolot, Clauser and Thaddeus (1969)
18	1 cm − 0.5 cm	7.4 $\pm$ 0.2 (5.5)	Ballon radiometer. Temperature in parentheses is an estimated corrected temperature	Muehlner and Weiss (1970)
	1 cm–0.08 cm	8.0 $\pm$ 0.5 (7)		
	1 cm–0.1 cm	4.7 $\pm$ 0.3 (3.6)		
19	1 cm–0.19 cm	$2.7 \pm {0.4 \atop 0.2}$	Balloon radiometer. Large atmospheric correction shorter than 0.09 cm	Muehlner and Weiss (1973a)
	1 cm–0.128 cm	2.8 $\pm$ 0.2		
	1 cm–0.126 cm	2.8 $\pm$ 0.2		
	1 cm–0.09 cm	< 2.7		
	1 cm–0.055 cm	$\leq$ 3.4		

TABLE 2.1 (cont.)

	Wavelength, cm	Thermodynamic temperature, K	Comments	Reference
20	1 cm–0.09 cm	$2.55 \begin{array}{c} +0.45 \\ -0.45 \end{array}$	Balloon radiometer	Muehlner and Weiss (1973b)
	1 cm–0.075 cm	$2.45 \begin{array}{c} +0.45 \\ -1.05 \end{array}$		
	1 cm–0.055 cm	$2.75 \begin{array}{c} +0.8 \\ -2.75 \end{array}$		
21	0.04 cm–.13 cm	$8.3 \begin{array}{c} +2.2 \\ -1.3 \end{array}$	Rocket radiometer	Shivanandan, Houck and Harwit (1968)
22	0.13 cm	< 4.74	Interstellar CN	Bortolot, Clauser and Thaddeus (1969)
	0.06 cm	< 5.43 K	Interstellar CH	
	0.036 cm	< 8.11 K	Interstellar CH^+	
23	0.08 cm–0.6 cm	$3.1 \begin{array}{c} +0.5 \\ -2.0 \end{array}$	Rocket radiometer	Blair, Beery, Edeskuty, Hiebert, Shipley and Williamson (1971)
24	0.04–0.15 cm	Large flux, ~ 32 times expected from 2.7 K black-body; see text and graphs	Rocket radiometer	Pipher, Houck, Jones and Harwit (1971)
	0.02 $\simeq$ 0.45 cm			
	0.07–0.013 cm			
25	0.04–0.13 cm	< 3.7 K, ~ 4 times flux expected from 2.7 K blackbody background	Rocket radiometer	Houck, Soifer, Harwit and Pipher (1972)

TABLE 2.2

Isotropy of the Radiation Background

Reference	Wavelength	Angular resolution	$\Delta T/T$
Parijskij and Pyatunia (1970)	4 cm	$1'.4 \times 20'$	$< 2.6 \times 10^{-4}$
Parijskij (1973b)	4 cm	$12'.5$	$< 1.3 \times 10^{-4}$
Conklin (1969)	3.75 cm	24-hour	~ 0.0006
Carpenter, Gulkis and Sato (1973)	3.56 cm	$2'.3$	$< 7.15 \times 10^{-4}$
Partridge and Wilkinson (1967)	3.2 cm	24-hour	< 0.0008
Partridge and Wilkinson (1967)	3.2 cm	$15°$	< 0.005
Conklin and Bracewell (1967)	2.8 cm	$10'$	< 0.002
Parijskij (1973a)	2.8 cm	$3'$ to $1°$	$< 0.8 \times 10^{-4}$
Boughn, Fram and Partridge (1971)	0.86 cm	24-hour	< 0.0045
Penzias, Schraml and Wilson (1969)	0.35 cm	$2'$	< 0.01
Boynton and Partridge (1973)	0.35 cm	$80''$	$< 2 \times 10^{-3}$
Epstein (1967)	0.34 cm	$12'$	< 0.025

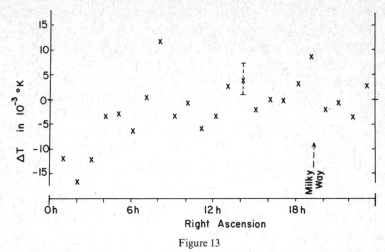

Figure 13

The isotropy of the microwave background. The crosses indicated the difference of the temperature of the sky from 2.7 K versus hour angle around a small circle of the celestial sphere (after Wilkinson and Partridge, 1967; see also Partridge, 1969).

galaxy formation, but this idea is difficult to maintain because of the precise isotropy and the difficulty of adjusting mechanisms to cause the reheating at the correct epoch.

Nonetheless, it is true that the energy density in a 3 K blackbody flux is approximately the same as the integrated starlight (so that the microwaves could be fitted into a steady-state model as degraded starlight). Layzer and Hively (1973), for example, point out that the radiation could be essentially starlight, thermalized by dust grains. The inability to observe quasars with large redshifts may then be caused by the very large dust grain extinctions for distant sources. And the energy density of the background microwave radiation is also approximately the same as the energy density of the galactic magnetic field.

The small scale limits on the anisotropy of the background tend to support a cosmological origin. Boynton and Partridge (1973) point out that the smoothness of their result can be explained in terms of discrete sources only if the sources are about as numerous as galaxies, and they have measured M32 and found it does not seem to be one such source.

An interesting result is that of Carpenter et al. (1973) who point out that their result gives a small enough limit to eliminate anisotropy due to galaxies forming in the recombining plasma. They compare with Peebles and Yu (1970) and conclude that if the radiation has been free since $z \sim 1000$, the only homogeneous isotropic model consistent with galaxy formation at recombination and with their result is a flat scalar-tensor model (Brans and Dicke, 1961). Parijskij (1973a,b) points out that the level of results currently being reached requires also that the ~ 1 Mpc gravitational waves postulated by Rees (1971a) be absent. Although the lack of any galaxy formation evidence in the anisotropy is troublesome, the simplest picture seems to have the radiation as described above, a relic of the era $z \sim 1000$.

The X-radiation is isotropic also. Schwartz (1970) puts the variation in flux as less than 1% on a 24 hour (dipole) basis and less than 4% on scales of the order of 28° (in the 0.3–1.6 Å band).

3. ANISOTROPIC HOMOGENEOUS COSMOLOGY

The remarkable discovery of the microwave radiation and the early measurements of its isotropy led very quickly to studies of the behavior of collisionless radiation in cosmology. Particular interest centered in anisotropic cosmologies because the background radiation gives infinitely better statistics for the determination of isotropy than does the galactic redshift relation.

We have mentioned above the current determination of isotropy of the 3 K radiation and of the X-ray background. Theoretical cosmology expanded dramatically as a result of these observations. Some of the most important papers in this field have been Belinskii, Lifshitz, and Khalatnikov (1971); Chernin (1972a, 1972b); Chitre (1972a, 1972b); Collins and Stewart (1971); Dicke (1968); Doroshkevish, Zel'dovich and Novikov (1967a, 1967b); Ellis and MacCallum (1969); Ellis (1971); Gowdy (1971); Harrison (1967); Hawking (1966a, 1966b, 1966c, 1969); Hawking and Ellis (1968); Hughston (1969); Hughston and Shepley (1970); Jacobs (1967, 1969); Liang (1971, 1972); MacCallum and Ellis (1970); MacCallum, Stewart, and Schmidt (1970); Matzner (1969a); Matzner, Shepley, and Warren (1970); Misner (1967a, 1967b, 1968a, 1969a); Novikov (1968); Rees and Sciama (1968); Ryan (1969, 1971a, 1971b); Sachs and Wolfe (1967); Shepley (1969); Stewart (1969); Thorne (1967).

Misner's (1968a) approach was to introduce the anisotropic generalization of the "flat" homogeneous isotropic models. These are called type I according to a listing by Bianchi (1897) of the properties of the symmetries of 3-dimensional spaces. Investigations have often concentrated on this model, and also on type IX, which is the anisotropic generalization of the closed homogeneous isotropic models, but the other types have not been ignored. If we represent a homogeneous isotropic model as an expanding and perhaps recollapsing balloon as in Fig. 14, anisotropic models are represented by Fig. 15 (non-rotating) or by Fig. 16 (rotating). Collins and Hawking (1973b) give an excellent survey of cosmological types (they argue that galaxy formation can occur in only a few subtypes, so that, e.g., we observe the universe to be nearly isotropic because only such a model could give use to intelligent life in galaxies).

In the notation of Misner (1968a) the metrics of such homogeneous but anisotropic

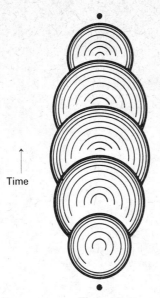

Time

Figure 14

The time evolution of a closed homogeneous, isotropic model. Each of the spheres is a radius of a three-sphere space section at a particular time.

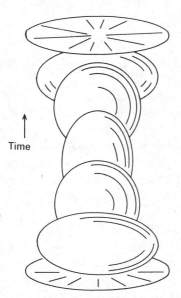

Time

Figure 15

The evolution of the T–NUT–M universe. The surfaces represent an elementary volume of fluid in the model.

Figure 16

The evolution of a rotating type IX model. The surfaces represent an elementary volume of fluid as in the T–NUT–M model.

models can be written

$$ds^2 = -dt^2 + R^2(t)\, e_{ij}^{2\beta(t)} \sigma^{(i)}\sigma^{(j)}, \qquad i = 1, 2, 3 \tag{3.1}$$

Here the spatial metric has been expressed in a non-holonomic basis

$$\sigma^{(i)} = f_j^{(i)}(x^k)\, dx^j \tag{3.2}$$

The x^i are some coordinates in the space section of the spacetime. Two frequently used alternate notations are $R = R_0 e^\alpha = R_0 e^{-\Omega}$; $\alpha = \alpha(t), \Omega = \Omega(t)$. The advantage of this notation is that it incorporates the non-Euclidean homogeneity of the model into the basis, and the metric hence is only a function of time, $g_{ij}(t)$. In Eq. (3.1) the spatial metric has been factored by defining the traceless matrix β_{ij} (so that the matrix exponential $e_{ij}^{2\beta}$ has unit determinant), thus letting the single function R define the overall expansion while β_{ij} describes the anisotropy. It should be apparent that $\beta_{ij} = 0$ leads to a model like the homogeneous isotropic model of Eq. (1.1), and in fact the expansion equation analogous to Eq. (2.2a) becomes

$$3(\dot{R}/R)^2 = \rho + \tfrac{1}{2}\sigma_{ij}\sigma_{ij} - \tfrac{1}{2}{}^*R \tag{3.3}$$

where ${}^*R$ is the 3-space scalar curvature ($=6/R^2$ for the 3-sphere). Here σ_{ij} is a matrix which depends on the rate of change of β_{ij}, so this equation demonstrates that a "shearing" cosmology, one in which the distances are evolving differently in different directions, has an average Hubble parameter which is increased by the presence of shear but decreased by the presence of a positive scalar curvature.

3.1. Observations in an anisotropic cosmology

3.1.1. Anisotropy of the Hubble law

Before continuing a discussion of the dynamics of β_{ij} in the various possible 3-spaces, let us indicate how the recent history of β_{ij} (since $z \sim 1000$) affects observations now, in a universe where $\sigma^i = dx^i$ and $*R = 0$ (the type I models). This treatment follows Misner (1968a).

Suppose the metric $d\beta_{ij}/dt$ is diagonal. Then $d\beta_{ij}/dt$ gives the differential rate of expansion along the principal axes of the metric; the Hubble parameter viewed in the x^1 direction is then calculated to be

$$H_1 = \dot{R}/R + d\beta_{11}/dt \tag{3.4}$$

with similar expressions for the Hubble parameter along the other axes. Since β_{ij} is traceless, we have

$$3\langle H \rangle = H_1 + H_2 + H_3 = 3\dot{R}/R \tag{3.5}$$

so $\dot{R}/R$ has the meaning of the average Hubble parameter, as R has the meaning "the average size of the universe".

On the other hand, the r.m.s. variation in H is

$$3\langle \Delta H \rangle^2 = (H_2 - H_1)^2 + (H_3 - H_2)^2 + (H_1 - H_3)^2 =$$

$$= 3\left[\left(\frac{d\beta_{xx}}{dt} \right)^2 + \left(\frac{d\beta_{yy}}{dt} \right)^2 + \left(\frac{d\beta_{zz}}{dt} \right)^2 \right] \tag{3.6}$$

so the current rate of shear of the universe determines the minimum spread in this value of H. It should be noted that this result assumes vanishing velocity of the observer with respect to the surfaces of homogeneity; otherwise there would be a "dipole" Doppler shift superimposed on the quadrupole result. This quadrupole result, with the dipole term added, is found in lowest order in all anisotropic cosmologies. As we noted above,

$$\frac{(\Delta H)^2_{\text{rms}}}{H^2} < 10^{-2} \tag{3.7}$$

(Trendowski, 1971), which means that shear motions now are hardly affecting the average expansion via (3.3) above.

3.1.2. Anisotropy of the microwave background. Rings of fire

The aspect of observational cosmology in which the isotropy can be most readily tested, presumably back to $z = 1000$, is in the microwave radiation. We assume a type I background. In such a background it can be shown that three constants of the motion p_i exist for null geodesics, which are related to the physical (3-) momentum components P_j by

$$P_k = R^{-1} e_{kj}^{-\beta} p_j \tag{3.8}$$

Hence the photon energy at any particular epoch is

$$E^2 + P_k P_k = R^{-2} e_{kj}^{-2\beta} p_k p_j \tag{3.9}$$

Now the blackbody equilibrium photon distribution is a function only of E/T. (The same applies to a neutrino distribution with vanishing Fermi energy.) If the microwave radiation was last scattered and thus isotropic at a redshift $z = 1000$, its distribution at that time was a Planck distribution

$$f(E/T) = f(R_0^{-1} [e_{kj}^{-2\beta_0} p_k p_j]^{1/2} / T_0) \tag{3.10}$$

where the subscript 0 refers to the $z = 1000$ instant. The function f is a Planck distribution:

$$f(x) = (e^x - 1)^{-1} \tag{3.11}$$

and (3.10) has been written expressly in terms of the constants of the motion and of the parameters of the $z = 1000$ surface.

Since the radiation subsequently travels without scattering, the distribution remains the same function of the constants p_i and of the parameters T_0, R_0 and β_{ij0} until it is ultimately detected.

The detecting apparatus responds to P_i or to E. Thus at the detection instant the distribution is

$$f(R_0^{-1} [e_{kj}^{-2\beta_0} p_k p_j]^{1/2} / T_0) = f(R_0^{-1} [e_{kj}^{-2\beta_0} R e_{km}^{\beta} p_m R e_{jn}^{\beta} p_n]^{1/2} / T_0) \tag{3.12}$$

If $P_i = n_i E$ for photons and n_i is a direction cosine, $T(\Omega)$ is defined as

$$T(\Omega) = \frac{T_0 R_0}{R (e_{ij}^{2(\beta - \beta_0)} n_i n_j)^{1/2}} \tag{3.13}$$

Hence for any direction we have a blackbody distribution but with an angle dependent temperature given by (3.13). It is apparent that along each axis this temperature is simply given by the total expansion (due to R and e^{β}) along that axis. Eq. (3.13) shows how to interpolate between the axes.

Assuming $(\beta - \beta_0)_{ij}$ is small,

$$T(\Omega) = \frac{T_0 R_0}{R} [1 - (\beta - \beta_0)_{ij} n_i n_j] \tag{3.14}$$

which shows that the lowest order term in this model is given by a quadrupole (12-hour) distribution on the sky, and $\Delta T/T$ is of the order of the change in β_{ij} since the radiation decoupled. This is schematically shown in Fig. 17.

This type I model includes the flat Robertson–Walker isotropic model, and tends to it for large t ($d\beta/dt \to 0$ and β can be rescaled to zero although some residual anisotropy in the radiation temperature may remain). In more complicated cos-

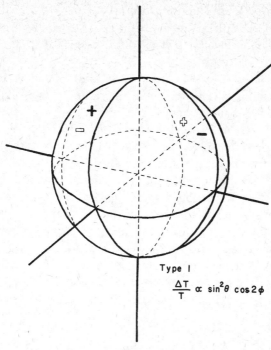

Figure 17
The anisotropy in the temperature of the microwave background in a type I model plotted on the celestial sphere. A plus indicates a hotter than the average, and a minus colder.

mological models, more complicated (than quadrupole) temperature anisotropy patterns may occur, and thus may yield some information about the symmetry structure (hence the topology) of the spatial sections of the universe.

To begin with, there is the possibility of relative motion between the observer (us) and the source of the microwave radiation. One possibility has the emitting matter (and hence the radiation in contact with it) at rest in the homogeneous space slices, and the observer moving through the radiation. The second has the matter generating the gravitational fields itself moving with respect to the surfaces of homogeneity. In either case there will be a dipole (24-hour) term in the temperature anisotropy. Conklin (1969) and Conklin and Bracewell (1967) have searched for this effect. As with all anisotropy sweeps this one was over one particular circle in the sky. Hence this scan could miss the dominant region in a 24-hour pattern. Nonetheless Conklin (1969) finds a velocity just at the limits of his detectability which corresponds to ~ 300 km/sec.

An interesting result obtains for the temperature anisotropy in type IX. Here the spatial sections are distorted spheres, and the absolute value of β_{ij}, rather than just the difference $(\beta - \beta_0)_{ij}$ enters into the anisotropy, because β_{ij} modifies the equation of motion for the photons. The result is a distorted quadrupole pattern, shown in Fig. 18, which looks as if the boundary dividing the areas of higher and lower (than average) temperature have been twisted (Matzner, 1970, 1971a,b). (This has nothing to do with rotation, however.)

In type V cosmologies, which are a generalization of the open Friedmann–Robertson–Walker model, an even more interesting phenomenon occurs. To an extent it depends on the importance of matter in determining the expansion of the universe. Matzner (1969a, 1970), Hawking (1969), Collins and Hawking (1973a) and MacCallum and Ellis (1970) have studied the temperature anisotropy in such models. Take first the models in which the matter flows normal to the space sections of the cosmology. Then the effect is that the temperature anisotropy appears only around a pole of the celestial sphere (a pole picked out by the group symmetry of the model). The anisotropy is a function of *changes* in the metric matrix β, and depends on how far in the past β was last significantly changing. The more distant the past, the more the anisotropy is concentrated near the pole. Matzner (1969a, 1970) assumed a moderate deceleration parameter. In such a case the rate of change of β drops abruptly at the point where the curvature term begins to dominate the expansion, and thus the angle θ into which the anisotropy is crowded, the "Ring of

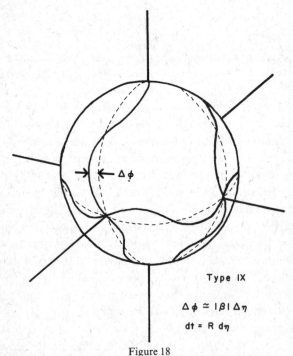

Type IX

$$\Delta\phi \simeq |\beta|\,\Delta\eta$$

$$dt = R\,d\eta$$

Figure 18

The anisotropy in the temperature of the microwave background in a type IX model plotted on the celestial sphere. The anisotropy pattern is distorted from a simple quadrupole form.

Fire" (Wheeler, 1973) is given by the time since this curvature domination began, and this can be put in terms of the current value of the deceleration parameter, q_0. The table below gives some indication of the opening angle calculated assuming a decoupling at $z \sim 1000$ (Matzner, 1970; Collins and Hawking, 1973a; see Figs. 19, 20).

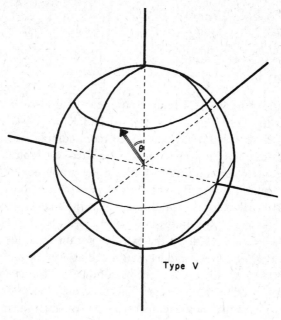

Type V

Figure 19

The anisotropy in the temperature of the microwave background in a type V model plotted on the celestial sphere. The temperature is uniform except around the "ring of fire" at the angle θ. Around the ring the temperature varies above and below the average.

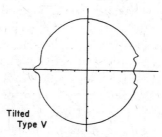

Tilted
Type V

Figure 20

The anisotropy pattern of a type V model in which the matter velocity has non-zero space components. Temperature is indicated by radius from the center. Near one pole is a "ring of fire" as in Fig. 19, while at the other is a temperature "spike" arising from a Doppler shift (after Ellis and King, 1974).

TABLE 3.1

Deceleration parameter q_0	Diameter of "Ring of Fire"
0.39	40°
0.15	9.4°
0.1	7°
0.02	2.1°
0.01	49′
0.005	30′
0.001	3′

If there is relative motion, a 24-hour dipole term will also be present, and will also be shifted to the pole. Fig. 20 (from Ellis and King, 1974) shows this effect.

Collins and Hawking (1973a) have discussed more complicated models in which the anisotropy pattern can be even more complicated, e.g., spiral.

This discussion has given nothing to indicate the amplitude of the temperature anisotropy in the region where it is important (see the section on dissipation of anisotropy, below). Further, for the small values of q_0 the angles into which the anisotropy is squeezed in type V are quite small, and a search for such features would require a very narrow beam width if the extreme parameter values held. For some other models (such as the type IX model mentioned above) the anisotropy measurements already made essentially exclude the use of this tool for determining the symmetry (hence the topology) of the universe. This is because the antenna scans already made show that gross features of the temperature anisotropy, which covers the whole sky, are so weak that subtle features such as those mentioned above would be indetectable. MacCallum and Ellis (1970) have outlined these and similar results for the other symmetry types, and for some, such as the type V, mentioned above, they still hold out the possibility of an anisotropy determination of the spatial topology.

We have so far ignored any discussion of the rotation of the universe. The most straightforward method of determining this is by considering the proper motion (or transverse Doppler shifts) of galaxies. By this method Kristian and Sachs (1966) estimate the rotation as $<7 \times 10^{-11}$ rad/yr, roughly equal to a transverse velocity of c at a Hubble distance. On the other hand, in some models, such as the type IX closed models, rotation can be estimated by noting the connection between rotation and the dipole term in the microwave anisotropy, which is on the order of 300 km/sec (Conklin, 1969), and reduces the rotation by a factor of 10^3 or so ($\sim v/c$) over the Kristian–Sachs estimate (Hawking, 1969). However, an even better limit, 2×10^{-21} rad/yr, can be obtained if the universe is type IX, because the specific form of one of the Einstein equations relates the rotation to the (very small) quadrupole anisotropy (Collins and Hawking, 1973a; see also Matzner, Shepley and Warren, 1970).

3.2. Dynamics of anisotropic models. Hamiltonian cosmology

Because the basis used to express the metric of homogeneous cosmologies can be adjusted to absorb the symmetry of the space, the metric coefficients are functions of time only. Hence Einstein's equations are (coupled) ordinary differential equations. This fact allows them to be written in the form of single-particle motion inside a potential in multi-dimensional space, and allows easy visualization of the qualitative behavior of the solution.

A new calculational tool due to Misner (1969b, 1970) is the use of Hamiltonian methods in general relativity. It is applicable to some cosmological models. The direct applicability depends on the vanishing of certain boundary terms in the derivation of the Hamiltonian variational principle. Those models in which these terms vanish and the Hamiltonian model works, unmodified, are called Class A. MacCallum and Taub (1972) have pointed out that the Hamiltonian techniques do not work straightforwardly in Class B. Ryan (1974) has developed a quasi-Hamiltonian approach which allows the potential concept to be retained in Class B. We will briefly display the Hamiltonian analysis for Class A models only.

One begins by writing the Einstein action in the form

$$I = (16\pi)^{-1} \int \left[\pi^{ij} \, \partial_t g_{ij} - C^0 N - C^i N_i \right] d^4 x, \qquad i, j = 1, 2, 3 \qquad (3.15)$$

where g_{ij} is the metric on a $t = $ constant surface, N and N_i determine g_{00} and g_{0i} respectively, and the π^{ij} may be thought of as nothing but momenta conjugate to g_{ij}. One must vary this action with respect to π^{ij} and g_{ij} to obtain a set of first order equations for each of these quantities, and with respect to N and N_i to get

$$C^0 = (\det g_{ij})^{1/2} \left\{ R^* + (\det g_{ij})^{-1} \left[\tfrac{1}{2} (\pi^k_k)^2 - \pi^{ij} \pi_{ij} \right] \right\} = 0 \qquad (3.16)$$

$$C^i = -2\pi^{ij}_{|j} = 0 \qquad (3.17)$$

where R^* is the scalar curvature on $t = $ constant surfaces and $|$ means covariant derivative on $t = $ constant surfaces. If the model contains matter there must be added a matter Lagrangian to (3.15). This results in nonzero right-hand sides in (3.16), (3.17) (see below).

We want to apply this to Bianchi-type homogeneous models,

$$ds^2 = -dt^2 + R_0^2 e^{-2\Omega(t)} e^{2\beta(t)}_{ij} \sigma^i \sigma^j \qquad (3.18)$$

where β_{ij} is a 3×3 matrix and the forms σ^i obey $d\sigma^i = C^i_{jk} \sigma^j \wedge \sigma^k$ (models where $C^i_{ji} = 0$ are Class A, while if $C^i_{ji} \neq 0$ they are Class B).

To continue we make the coordinate transformation $t \to \Omega(t)$ and parametrize β_{ij} by

$$\beta = e^{-\psi\kappa_3}e^{-\theta\kappa_1}e^{-\phi\kappa_3}\beta_d e^{\phi\kappa_3}e^{\theta\kappa_1}e^{\psi\kappa_3}$$

$$\beta_d = \text{diag}\,(\beta_+ + \sqrt{3}\beta_-, \beta_+ - \sqrt{3}\beta_-, -2\beta_+) \tag{3.19}$$

$$\kappa_3 = \begin{bmatrix} 0, & 1, & 0 \\ -1, & 0, & 0 \\ 0, & 0, & 0 \end{bmatrix} \qquad \kappa_1 = \begin{bmatrix} 0, & 0, & 0 \\ 0, & 0, & 1 \\ 0, & -1, & 0 \end{bmatrix}$$

Because the Bianchi-type models are homogeneous, we can integrate over the 3-spaces (artificially closing open models). Doing this, making the proper parametrization of π^{ij}, and assuming $C^0 = 0$ has been solved, (3.15) becomes

$$I = \int p_+\,d\beta_+ + p_-\,d\beta_- + p_\phi\,d\phi + p_\Psi\,d\psi + p_\theta\,d\theta - 2\pi\,(\pi_k^k)\,d\Omega \tag{3.20}$$

If we define $H = 2\pi(\pi_k^k)$, $C_0 = 0$ gives us H as

$$H^2 = 6p_{ij}p^{ij} - 24\pi^2\,(\det g_{ij})\,R \tag{3.21}$$

where

$$6p_{ij} = e^{-\psi\kappa_3}e^{-\theta\kappa_1}e^{-\phi\kappa_3}\left\{ \alpha_3\,\frac{3p_\psi}{\sinh(2\sqrt{3}\beta_-)} + \right.$$

$$+ \alpha_4\,\frac{3(p_\phi\sin\psi - p_\psi\cos\theta\sin\psi + p_\theta\cos\psi\sin\theta)}{\sin\theta\sinh(3\beta_+ - \sqrt{3}\beta_-)} +$$

$$\left. + \alpha_5\,\frac{3(p_\theta\sin\psi\sin\theta - p_\phi\cos\psi + p_\psi\cos\psi\cos\theta)}{\sin\theta\sinh(3\beta_+ - \sqrt{3}\beta_-)} \right\} e^{\phi\kappa_3}e^{\theta\kappa_1}e^{\psi\kappa_3} \tag{3.22}$$

$$\alpha_1 = \text{diag}\,(1, 1, -2), \qquad \alpha_2 = \text{diag}\,(\sqrt{3}, -\sqrt{3}, 0),$$

$$\alpha_3 = \begin{bmatrix} 0, & 1, & 0 \\ 1, & 0, & 0 \\ 0, & 0, & 0 \end{bmatrix}, \qquad \alpha_4 = \begin{bmatrix} 0, & 0, & 1 \\ 0, & 0, & 0 \\ 1, & 0, & 0 \end{bmatrix}, \qquad \alpha_5 = \begin{bmatrix} 0, & 0, & 0 \\ 0, & 0, & 1 \\ 0, & 1, & 0 \end{bmatrix}$$

In general one leaves $C^i = 0$ as a constraint on (3.22).

In some cases, such as type IX universes with β non-diagonal, it is impossible to satisfy $C^i = 0$. Then we must allow matter and add a matter Lagrangian density to the integrand of (3.15). Ryan (1972a,b) gives a matter Lagrangian for fluids in Bianchi-type universes with $p = (\gamma - 1)\rho$ that has $\mathscr{L}_\mathscr{M} = N\mathscr{L}_\mathscr{M}^0 + N_i\mathscr{L}_\mathscr{M}^i$ and is valid for Class A models. The addition of such a matter Lagrangian changes (3.16), (3.17) to

$$C^0 + \mathscr{L}_\mathscr{M}^0 = 0 \tag{3.23}$$

$$C^i + \mathscr{L}_\mathscr{M}^i = 0 \tag{3.24}$$

modifying the Hamiltonian (3.21) and the constraint on the momenta, $\pi_{1j}^{ij} = 0$.

The Hamiltonian (3.21) is the same as that for a particle moving in a five dimensional space. The second term in (3.21) plays the role of a potential, and in general is Ω-dependent (time-dependent). The fact that the potential is time-dependent is the only departure from an elementary problem in Hamiltonian dynamics. If we let β be diagonal, $\phi = \psi = \theta = p_\psi = p_\theta = p_\phi = 0$, then H^2 becomes (in vacuum)

$$H^2 = p_+^2 + p_-^2 + 36\pi^2 R^4 e^{-4\Omega}(V(\beta_+, \beta_-) - 1) \tag{3.25}$$

Figures 21, 22 give a representative equipotential of the potential V for each of the Bianchi types.

The Hamiltonian description is useful in quantum cosmology (discussed below) and in qualitative cosmology. In qualitative cosmology one notices that the potentials $V(\beta_+, \beta_-)$ in (3.25) are exponentially steep, and in most cases can be replaced by infinitely hard walls. Because the potentials are time-dependent, these walls move in Ω-time. One obtains an approximate solution by allowing the point that describes the universe to bounce around in a moving potential well. Figure 23 shows several bounces of the universe point in a type IX potential.

Misner (1972) has pointed up the role of superspace in Bianchi-type universes by rewriting (3.20) as

$$I = \int p_+ \, d\beta_+ + p_- \, d\beta_- + p_\phi \, d\phi + p_\psi \, d\psi + p_\theta \, d\theta - p_\Omega \, d\Omega - \mathscr{H} \, d\lambda \tag{3.26}$$

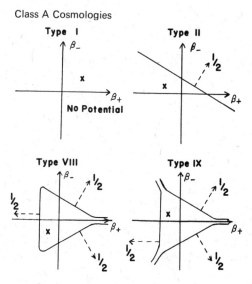

Figure 21

The anisotropy potentials for Ellis–MacCallum Class A Bianchi-type models. A representative equipotential is displayed. The other equipotentials are replicas of the given one (from Ryan and Shepley, 1975).

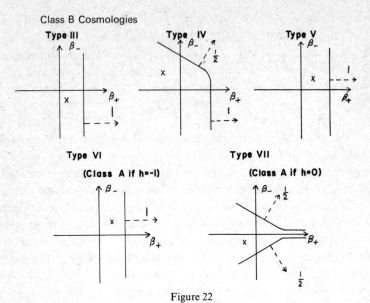

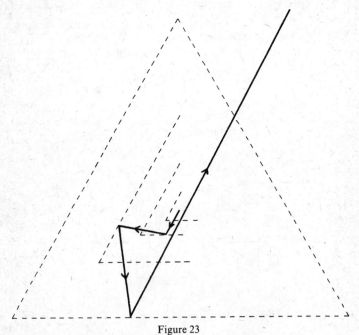

Figure 22

The anisotropy potentials for Ellis–MacCallum Class B Bianchi-type models. A representative equipotential is displayed. The other equipotentials are replicas of the given one (from Ryan and Shepley, 1975).

Figure 23

Schematic diagram showing the evolution of the anisotropy parameters (coordinates in the plane) in an expanding type IX potential (from Ryan and Shepley, 1975).

thus defining the *supertime* λ. The path of the universe in $\beta_+,\beta_-\phi,\psi,\theta$-space becomes a "world line" in $\beta_+,\beta_-\phi,\psi,\theta,\Omega$-space. Misner (1972) has found this approach useful in several applications.

3.3. Damping of anisotropy

We saw in Sec. 3.1 that anisotropic universes can give rise to observable anisotropies, as in Hubble law and blackbody temperature. However, the universe as observed is remarkably isotropic. This led several authors (Misner, 1968a; Thorne, 1967) to consider how the anisotropy may have evolved away by the current epoch.

Misner's article (1968a) was designed to demonstrate the inevitability of small values of $\Delta T/T$ in any universe which contains normal matter (his mechanism is damping of anisotropy by means of a "viscosity" due to almost collisional neutrinos (as the universe passed through a temperature of $\sim 10^{10}$ K)). Some of the techniques making use of potentials can be carried over to this case, even though the system is dissipative. Misner's calculations did show that the anisotropy was dissipated.

Three principal objections have been raised against these early calculations.

(i) The viscosity approximation would be inapplicable to more realistic treatments since the ratio of collision time t_c to expansion time t_{exp} would be in the range 1–100 in important epochs (Doroshkevich, Zel'dovich, and Novikov (hereafter DZN), 1967b, 1968, 1969; Stewart, 1969).

(ii) Arbitrarily large amounts of dissipation can apparently never occur between any two fixed epochs in the expansion of the universe (fixed volume ratio) because of a theorem of Stewart (1969) that *the rate of work (heating) done by the tangential stresses is never greater than one-half of the rate of adiabatic cooling (i.e., cooling due to the increase in the total volume).*

(iii) Only solutions which were collision dominated (fluid) near the singularity were studied (DZN, 1967b, 1968, 1969; Stewart, 1969; Matzner, 1969b).

Objections (i) and (ii) can be overcome in the sense that (i) intense dissipation is possible under conditions where $t_c/t_{exp} \gg 1$ even though the viscous approximation would lead to substantial quantitative errors, and (ii) dissipation as measured by the rate of entropy generation can proceed arbitrarily rapidly even with no change in the energy density T^{00}. This does *not*, however, guarantee small anisotropy at the current epoch as Misner (1968a) predicted.

The content of objection (iii) is as follows. If the universe is sufficiently anisotropic (and of low enough density now) then for reasonable cross section behavior the matter in the universe could have been always non-collisional, and dissipative processes would have to be excluded. The original calculations assumed complete collision dominance near the singularity so that this objection is fatal to Misner's argument (1968a) (see Matzner and Misner, 1972 and Matzner, 1972). Figure 24 shows a model which starts collision dominated.

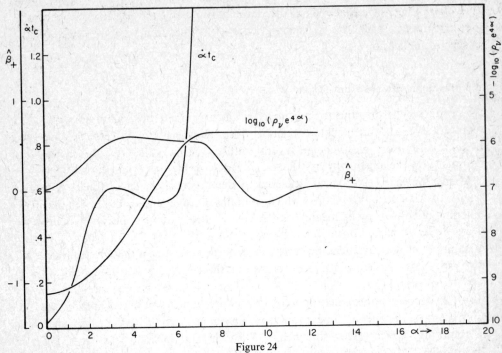

Figure 24

A calculation showing the damping of anisotropy $\hat{\beta}_+$ and the increase in fluid density ρ_ν over its adiabatic value $\rho_\nu \propto R^{-4} \propto e^{-4\alpha}$. The physical process is irreversible due to moderate collision time particles ("neutrinos"). Here $\dot{\alpha} t_c$ measures the ratio of collision time t_c to the e-folding time for the model, $\dot{\alpha}^{-1}$. In this case the matter begins in a collisional state and the collisions keep the physical anisotropy $\hat{\beta}_+$ at a small value, and also heat the matter present sufficiently rapidly that the ratio $\dot{\alpha} t_c$ does not increase dramatically, until essentially all the metric anisotropy (not plotted) has been dissipated. Then the particles cool, the mean free time becomes long and undamped oscillations are seen in $\hat{\beta}_+$ (after Matzner and Misner, 1972).

Using a collision time approximation to the Boltzman equation, more accurate studies of neutrino dissipation have been carried out (Carswell, 1969; Matzner and Misner 1972; Matzner 1972) for type I cosmologies (taken diagonal for simplicity). Because the neutrinos can be partly collisionless, they can be blueshifted to have a much higher average energy than the collision-dominated electrons. Thus DZN (1967b) point out that the cross section for the process $\nu\bar{\nu} \to e^+ e^-$ is much higher than the cross section for the inverse process. It may be conjectured that because of this disparity the neutrinos will be removed completely, soon after their mean free time becomes appreciable. However, this is found not to be the case (Matzner, 1972). The neutrino number stabilizes, at a level much lower than the equilibrium value, but at such a number that the two processes—neutrino destruction and production—balance one another. Neutrinos are able to continue to dissipate anisotropy over many epochs of expansion.

The physical mechanism by which the dissipation takes place is fairly simple to describe. When anisotropy is large, so that T^{00} is negligible in the Einstein equation

$$T_{00} = 3\dot{\alpha}^2 - \tfrac{1}{2}\dot{\beta}_{ij}\dot{\beta}_{ij}, \qquad \text{where } \dot{\alpha} = \dot{R}/R$$

then there is always one axis along which expansion does not occur. Between collisions, particles moving along this direction suffer blueshifts, not redshifts, even though there is a net volume expansion which would lower the temperature of any collision-dominated fluid. Thus the average energy of any long-mean-free-path particle which has a momentum component along the stationary or contracting axis will be larger than the average energy of thermalized particles, and collisions between particle populations with these significantly different average energies will be highly dissipative. If $t_c/t_{exp} \simeq t_c(\dot{R}/R)$ is small (viscous regime), then collisions are frequent; but only small nonthermal particle energies can be acquired between collisions, so the entropy generated,

$$\Delta S = \Delta Q \left(\frac{1}{T} - \frac{1}{T_b} \right) = \frac{(\Delta T)^2}{T T_b}$$

is small. Here $\Delta Q = \Delta T$ is the average energy transfer in a collision, and T_b is the average energy of the blueshifted particle. This situation also obtains in solutions which are only slightly anisotropic. If the expansion rates are roughly the same in all directions, then the blueshifted particles have only slightly greater energy than the collision-dominated (fluid) particles.

The situation is very different if one is willing to discuss large expansion ratios. For large t_c/t_{exp} one can then have

$$\Delta Q = \Delta T \approx T_b \gg T$$

so

$$\Delta S \approx \Delta T/T \gg 1$$

gives a large entropy generated per collision. It is therefore very important not to neglect collisions completely even if the mean free path is very long, since we therefore neglect very dissipative processes. The rare collisions have large effects. Figure 25 shows a system where long mean free paths are still effective in damping anisotropy.

The above picture of streams of "hot" blueshifted nearly collisionless particles ("neutrinos") immersed in an expansion-cooled fluid of other, collision-dominated, particles ("electrons") shows us how even rare collisions could give dissipation. Thus we understand how objection (i) is circumvented in the calculations which will follow.

This same simple picture shows that limits (ii) on the rate of increase of local energy T^{00} do not imply limits on the dissipation. One need only imagine that all hot particles suddenly decide to collide; then entropy is greatly increased and the anisotropy in the momentum disappears without any change in the local energy density

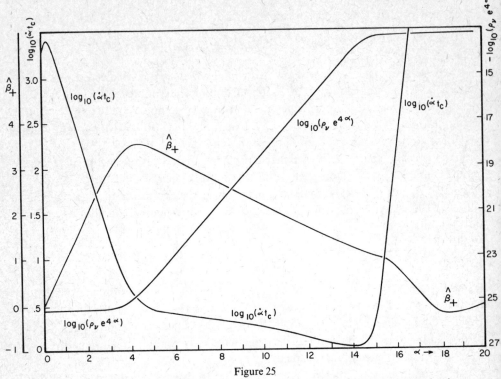

Figure 25

An example which starts off collisionless, and hence has an initial increase in the physical anisotropy. During this time the mean free path shortens, and as it nears unity heating begins, and the physical aniso- tropy begins to decrease. This continues again until the geometrical anisotropy has been almost completely dissipated, at which time the mean free path becomes very long and free mean time oscillations are again seen (Matzner and Misner, 1972).

T^{00}. These arguments and the calculation presented in Fig. 25 are also partial, but inadequate, answers to point (iii).

Stewart's form (1969) of the heating rate limit is based on a general stress tensor. It is derived from the requirement that the spatial stresses be non-negative, and reads

$$\left| \frac{d}{d\alpha} \ln \left(T_{00} e^{4\alpha} \right) \right| \leq 2$$

This equation exhibits a limit in the rate at which anisotropy can increase the matter energy density; shear energy density cannot be transformed into the matter energy density T_{00} at an arbitrarily high rate. The conclusion is at first sight distressing: it appears that if the initial anisotropy is sufficiently large, it cannot decay or dissi- pate away by any particular epoch—such as the present. However, anisotropy energy density can be stored in a "potential" form, in which it resides in the energy of distortion necessary to change an isotropic distribution of collisionless particles to the anisotropic form typical of an anisotropic universe. In this stored form it resides already in T_{00}. It thus exists as potential anisotropy energy which can go back into the kinetic (shear) motion of the system. The suddenly colliding particles mentioned above (and more realistic processes involving neutrinos) can degrade this ordered potential energy to thermal energy with the production of large amounts of entropy in arbitrarily short times. (This potential due to collisionless particles acts in exactly the same way as the potentials due to geometrical distortion in the empty models described above.)

Figure 26 shows an example of a choice of initial conditions which does not lead to significant dissipation because the system passes too quickly through the moderate mean-free-path stage. Figures 27 and 28 show the subsequent evolution of such a free-neutrino model.

The inevitability of dissipation of anisotropy by a given finite epoch has not been

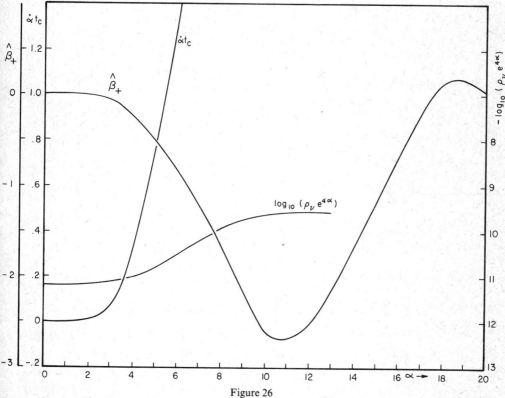

Figure 26

An example in which heating does not occur, because of a different behavior of the mean free path. The system passes from collision dominated to free behavior with essentially no heating, and large oscillations are subsequently seen (Matzner and Misner, 1972).

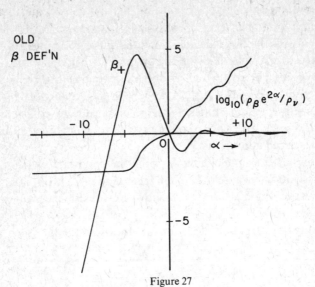

Figure 27

Further example of free-neutrino evolution. For large anisotropy, the anisotropy bounce is sharp (as at $\alpha \simeq -4$). For smaller anisotropy, the behavior is a damped oscillation (after Misner, 1968a).

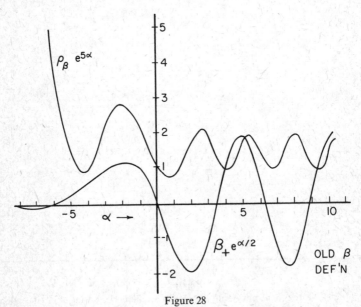

Figure 28

Continuation of previous example showing the small anisotropy behavior with oscillations proportional to $e^{-\alpha/2} \propto R^{-1/2}$ (after Misner, 1968a).

demonstrated. Arguments such as the one mentioned above about a high anisotropy universe with rarely colliding particles indicate, in the words of Misner (1969b), that

> "*the equations which govern the problem are regular, well-posed differential equations, so that the simple continuity of the solutions as functions of the initial conditions shows that no finite limit on the present anisotropy can result if arbitrary anisotropy is admitted at some finite initial epoch, whether that be 10^{14} K, or even higher. The continuity requires that the differential equations be regular on a finite interval, and serves to point up the essential contribution which a singularity brings. For equations which are singular at the initial time (or which set the initial conditions in the infinite past), an infinite range of initial conditions could evolve into a finite range of possible present conditions. Thus any argument that some features of the present universe are independent of most parameters specifying the initial conditions (Misner, 1968b) could only succeed if initial conditions are specified at a true singularity, or in the infinite past, but not at any finite and regular past era*".

Any mechanism which occurs near the singularity must occur in the strongly quantum first moments of the universe. Zel'dovich (1970, 1971) has discussed effects which might occur in this highly quantum epoch, and one of the most intriguing is that of the production of particles.

The intuitive mechanism in particle creation is the very rapid expansion during the early phases which leads to a gravitational field changing very rapidly, on times faster than the crossing time for an elementary particle (10^{-23} sec). Anisotropic models can have very violently different expansion rates, and the energy associated with this expansion (or the contraction in some directions) may exceed that available in isotropic expansion. Hence one could expect rapid particle production in anisotropic models, and in addition could have very rapid dissipation of the anisotropy during the quantum era of the universe. That particles are created at all in isotropic models depends on breaking the conformal invariance (e.g., there being a mass term), so the relevant equation is not conformally invariant in the (conformally flat) homogeneous isotropic models (Parker, 1972). Although these ideas are very suggestive, no one has yet succeeded in carrying out a rigorous calculation which demonstrates the accuracy of the idea, and dissipation of anisotropy is still very much an unanswered question. We will return to the problem of particle production per se below, and more references will be found there.

3.4. Mixing

Mixing is the ghost of an idea that failed. It originated in the concept of "chaotic cosmology" (Misner, 1967a): The universe is as homogeneous and isotropic as it is now because any universe, no matter how inhomogeneous and anisotropic, will settle down to the observed universe by means of natural processes. Misner (1969a) showed that diagonal type IX universes have possible evolutions in which the distance around the universe in one direction is very small for a reasonably long time. In such a case light and even sound (shock) waves could propagate completely around the universe. He then conjectured that disturbances propagating around the universe could effectively smooth out condensations and homogenize the universe. If the evolution of the universe then allows light to circle it in different directions, inhomogeneities will be smoothed out everywhere. This process is called *mixing*.

To show the existence of mixing consider a diagonal type IX model,

$$ds^2 = -dt^2 + R^2 e^{-2\Omega} e_{ij}^{2\beta} \sigma^i \sigma^j, \qquad \beta = \text{diag}\,(\beta_+ + \sqrt{3}\beta_-, \beta_+ - \sqrt{3}\beta_-, -2\beta_+)$$

The evolution is identical to that of a particle (the *universe point*) moving in β_+,β_--space under the influence of a potential $e^{-4\Omega}(V(\beta_+, \beta_-) - 1)$, where $V(\beta_+, \beta_-)$ is shown in Fig. 29. If the universe point enters one of the channels almost directly, the rapid narrowing of the channel causes an analogue of the magnetic bottle effect;

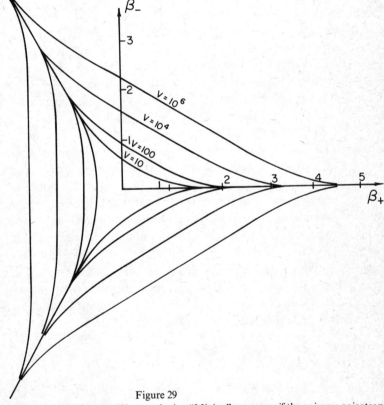

Figure 29

More detailed view of the potential for type IX cosmologies. "Mixing" can occur if the universe anisotropy takes on values corresponding to motion up one of the corners of the potential.

after a number of oscillations the universe point leaves the channel. A typical excursion into a channel is shown in Fig. 30.

Chitre (1972a) has shown that light can circle the universe within a few of the oscillations of the universe point shown in Fig. 30. He also showed (Chitre, 1972b) that sound waves can travel around the universe in roughly the same time. Matzner and Chitre (1971) have shown that rotation makes little difference in this process.

The results of Misner and of Chitre show that mixing certainly can occur. However, to use it in chaotic cosmology one must show that no matter what the initial state of the universe (no matter where on the β_+,β_--plane the universe starts and in what direction), that the universe point will enter all the channels before the present. It is at this stage that the idea breaks down. Doroshkevich, Lukash, and Novikov (1971) and Chitre (1972a) have shown that mixing in even one direction

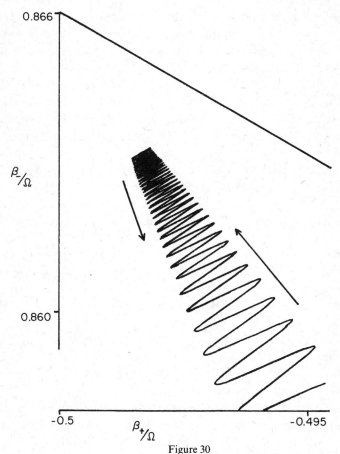

Figure 30

Example of mixing motion in one of the corners of the potential of Fig. 29 (from Moser, Matzner, and Ryan, 1973).

occurs for less than 2% of the possible initial configurations of the universe. This happens because for mixing to occur the universe point has to enter the channel with the direction of its motion in a very narrow band of angles about the center of the channel. This band shrinks rapidly near the singularity, and makes it nearly impossible for mixing to occur. The 2% limit considers the entire span of the anisotropy dominated epoch. If anisotropy is reduced suddenly by damping processes, the limit becomes even less than 2%.

4. INHOMOGENEITIES AND GALAXY FORMATION

4.1. Inhomogeneous cosmological models and their effect on observations

The standard cosmological models reflect the observed homogeneity and isotropy of the universe on the largest scale. Inhomogeneous models have been investigated by several authors for four reasons: (1) To describe and explain small-scale inhomogeneity—galaxies and cluster of galaxies; (2) To find the effect of inhomogeneity on observations and refine our knowledge of the homogeneity of the universe; (3) To provide models for early stages of the universe that may have been inhomogeneous (we include models that are purely mathematical exercises in general relativity); (4) To study galaxy formation. The fourth subject is considered in the next section.

The first and second reasons are the basis of the Einstein and Straus (1946) "Swiss cheese" model in which all the mass inside a sphere in an otherwise homogeneous space is concentrated at the center of the sphere. They demonstrated that the gravitational fields outside the sphere are the same as in a perfectly uniform model so that these holes may be placed in the space arbitrarily as long as no two regions overlap. The effect on astronomical observations in a Swiss cheese universe has been studied by Kantowski (1969) (and previously by Bertotti, 1966 for linearized gravity and also by Refsdal, 1970). They found that the distance-redshift relation could in some cases be altered enough to double the apparent value of the deceleration parameter, q_0.

Perhaps the most basic article on observations in inhomogeneous models is that of Kristian and Sach (1966). They assume only that the Riemann tensor varies slowly in time and compute series expansions in luminosity distance of such properties as redshift, angular diameter of galaxies, etc.

Rees and Sciama (1967, 1968) have shown how large scale inhomogeneities indicated by possible quasar clustering would affect the temperature of the background microwave radiation (Fig. 31). Regions on the scale of 750 Mpc with $\delta\rho/\rho \sim 1$ would cause temperature fluctuations of 0.2 to 2% across an angular scale of $20°$ and should be detectable. Such measurements were undertaken by Wilkinson and

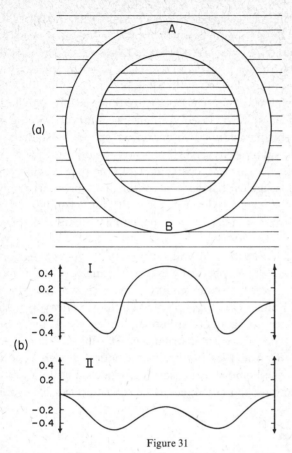

Figure 31

Illustration of temperature profiles across sherically symmetric cluster with ~ 3 times the background density. The profiles are drawn for an undecelerated background (I) and an Einstein−de Sitter background (II) (after Rees and Sciama, 1968).

Partridge (1967) who compared the radiation temperature at 3.2 cm along a circle 8°S. of the celestial equator with the radiation temperature at the north celestial pole. The largest temperature variation found, 0.016 K, was only a few times the level of noise ($\pm$ 0.003 K), and apparently has smoothed out with further observation (Partridge, 1969). It may be more than coincidence that the variation fell at the location of a possible quasar cluster suggested by Rees and Sciama (1968) (Fig. 13).

In the section on galaxy formation it is shown that perturbations grow very slowly from any statistically expected fluctuations, so it is reasonable to study the nature of structure which may be present at the initial singularity. Belinskii, Lifshitz and Khalatnikov (1971) have studied "velocity-dominated" solutions by ignoring the effects of spatial curvature near the big bang. Eardley, Liang, and Sachs (1972) have used these solutions in irrotational models to invariantly define the metric of a three-dimensional manifold identified as the cosmological singularity. Liang (1972) has extended this analysis to more general universes although it is difficult to define the singularity manifold for mixmaster-like type IX solutions. Recently, Liang (1975) has discussed shock formation in the early stages of cosmologies with cylindrical symmetry.

Irregularities which may be present from early times include gravitational waves. Gowdy (1971, 1974) has studied closed vacuum spacetimes containing gravitational radiation. These spacetimes admit two-parameter spacelike isometry groups and have compact regular spacelike hypersurfaces with topology S^3, $S^1 \otimes S^2$, or $S' \otimes S' \otimes S'$. The topology is prevented from changing with time by Einstein's equations in a manner analogous to classical barrier penetration. The three-torus universe begins with a singularity and expands forever while the three-handle and three-sphere solutions expand from a singularity to a maximum volume and then collapse to a singularity again.

The hierarchical universes of the deVaucouleurs (1970) type are, of course, examples of inhomogeneous models. A spherically symmetric analogue could have a density function that looked like a set of descending steps, each step longer and shallower than the one before. Bonnor (1972) has pointed up this similarity by finding an exact solution (a special case of the Tolman–Bondi model, see Tolman, 1934; Bondi, 1947) that has this type of density distribution. The Bondi–Tolman models have also been studied by Callan (1964). Other mathematical models are due to Edelen (1968) (conformal to the homogeneous isotropic models) and to Ryan (1972b) who considered space sections that are three-dimensional surfaces of revolution.

4.2. Galaxy formation

It is in the problem of galaxy formation where modern cosmology finds its greatest troubles. The early calculations of Jeans (1929) were hopeful: Using Newtonian theory, and assuming a static, uniform, fluid cloud as a starting point, the equation for perturbations of wavelength λ can be cast into the form

$$\ddot{\delta} = (4\pi G\rho - 4\pi^2 c_s^2 \lambda^{-2})\,\delta$$

where $\delta = \Delta\rho/\rho$, ρ = background density, $\Delta\rho$ = perturbed density, $c_s = (dp/d\rho)^{1/2}$ = speed of sound, p = pressure. Let λ_J be the Jeans' length:

$$\lambda_J = (\pi c_s^2 G^{-1}\rho^{-1})^{1/2}$$

If $\lambda > \lambda_J$ the density contrast δ grows exponentially fast. For large λ, the time scale for the exponentially growing disturbance in the Jeans theory becomes asymptotic to τ_J defined by

$$\tau_J = (G\rho)^{-1/2}$$

In an isotropic expanding cosmological model there is a second time scale, the Hubble time, defined by the inverse of the logarithmic derivative of the cosmic

"radius" function:
$$\tau_H = R/\dot{R}$$

The equation for δ is modified by to (Peebles, 1971)

$$\ddot{\delta} + 2\tau_H^{-1}\dot{\delta} = (4\pi\tau_J^{-2} - 4\pi^2 c_s^2 \lambda^{-2})\delta$$

In general relativity cosmology, however, τ_H is a function of time. Further, when ρ is approximately the average cosmic density, ρ depends on R (for example, in a dust-filled model $\rho \propto R^{-3}$), so that τ_J is also t-dependent, and τ_J has approximately the magnitude of τ_H.

As an example, in a dust-filled, isotropic type I model, $R \propto t^{2/3}$, so that $\tau_H \propto t$. Since $\rho \propto R^{-3}$, $\tau_J \propto t$ also. It turns out that the growing mode of δ behaves as $t^{2/3}$ rather than growing exponentially with t. In fact, perturbations in the isotropic models typically behave like the models themselves (Fig. 32). Each model has two basic modes of density perturbations, one acting as if the expansion had started at a slightly later time and the other acting as if the expansion had started with a slightly lower energy. Pressure modifies this statement, of course, but the power law behavior of density perturbations is a typical result of calculations (begun by Lifshitz, 1946; Lifshitz and Khalatnikov, 1963; and continued in many special circumstances by several others such as Harrison, 1967, 1973).

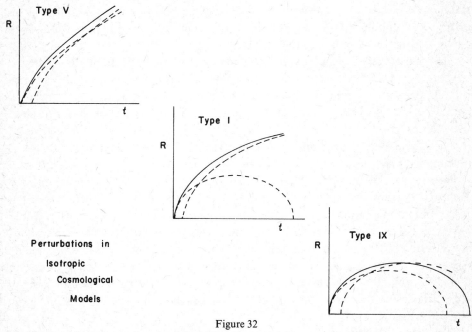

Figure 32

Perturbations in homogeneous cosmologies. Elementary density perturbations amount to changing the initial time or the initial energy. Thus a slightly more bound region in a closed (type IX) model recollapses before the universe does.

Let us denote by n the power law index for density contrast perturbations:

$$\delta = \Delta\rho/\rho \propto t^n$$

The index n depends, of course, on the background model, but can be as high as 8/3 in a dust-filled, anisotropic, type I model. It is helpful to bear in mind the differences in gravitational perturbations illustrated by Fig. 32. In each case a growing mode exists, the density contrast δ growing with time. In the type V case, however, δ grows in spite of the fact that the density itself may be falling because the perturbation is expanding in an unbounded fashion. In the type I and type IX cases, the perturbation eventually recollapses due to its own gravitational forces. Observation of galaxies and clusters could be important here, for if a gravitationally distinct system is found to be expanding, the expansion rate presumably sets a lower bound on cosmic expansion. In general, it is assumed that when $\delta = 1$, the perturbed region can act relatively independently of the cosmic background.

Galaxy formation probably took place over a time span of about one galactic year (10^8–10^9 years). This time is roughly the free-fall time for a particle to fall from one galactic radius to the center of a galaxy. When galaxy formation started, however, is unknown, and it is also unknown whether galaxies formed before or after smaller collections of stars. It is also not known whether it were gravitational instabilities which started the formation process or whether non-gravitational instabilities or conditions existing at the initial singularity caused perturbations large enough to proceed by gravitational collapse.

The cosmic age when radiation decoupled from matter was about 10^5 years. At that time the universe was probably quite isotropic (but see Collins and Hawking, 1973b). The horizon length calculated in an exactly isotropic model included perhaps 10^{18} $M_\odot$. The lowering of the effect of radiation pressure caused the computed Jeans length to drop to well below galaxy size (see below). If δ was then about 1% for a galaxy-sized distribution, δ would grow to the value 1 in about 10^8 years (assuming a power law growth rate of $t^{2/3}$).

Peebles and Dicke (1968) have pointed out, however, that at decoupling the Jeans length probably included 10^5–10^6 $M_\odot$, about the size of a globular cluster. If perturbations at all length scales had existed prior to this time, pressure effects would have kept them from growing strongly. It may be that such primordial perturbations would cause globular clusters to form first when the radiation pressure became ineffective, and that galaxies are made out of these clusters. The attractive feature of this idea is that globular clusters in real life seem to have fairly standard properties. How galaxies would form and why galaxies are much more massive has not been described, though.

The Jeans length at the decoupling epoch could instead imply that primary condensations are in the form of superstars (Doroshkevich, Zel'dovich, and Novikov, 1967c), rather than stellar clusters. These superstars would explode, heating their environment to perhaps 10^6 K (if, say, 10^{-4} of cosmic matter acted in this way, Rees, 1971b). This heating would be irregular, presumably resulting in perturbations sufficient to form galaxies by gravitational instability. Further, the high resulting temperature could result in thermal instabilities with characteristic growth time comparable with the radiative cooling time (see Field, 1975 and Rees, 1971b). Such an instability was not effective previously when matter and radiation were strongly coupled. If the matter is so heated, a region which is slightly compressed may indeed cool more efficiently than its surroundings and result in instabilities further enhanced by the irregularity of the heating.

Whether globular clusters, superstars, stars, or galaxies form first, there still remains the necessity of postulating large perturbations at the decoupling epoch. To some extent this problem can be alleviated by postulating a highly anisotropic cosmology before that time.

To see the effect of anisotropy, it is helpful to count the physical modes present in a metric perturbation tensor (Perko, Matzner and Shepley, 1972). This symmetric tensor has 10 components, but four components may be set by coordinate conditions. The remaining six components form a symmetric 3×3 tensor which obeys differential equations which determine the density and velocity perturbations and the way they develop in time. The time development equations are six in number and are second-order. There are therefore 12 initial conditions to be set for each given wavelength. Four of these numbers are simply functions of the initial orientation and coordinates of the initial hypersurface (gauge terms). Four more are intensity and polarization of gravitational waves. Two more are rotation perturbations, and the remaining two are density-pressure perturbations (including both collapsing perturbations and if pressure terms are strong, sound waves).

In an isotropic cosmology, gravitational wave perturbations and density perturbations are decoupled. When anisotropy is included, as in a general Bianchi type I model, gravitational wave energy can enhance the growth of a density perturbation. If the wave front associated with a Fourier-analyzed perturbation is in an eigendirection of the anisotropy matrix, one of the two gravitational modes is free—not coupled to density perturbations. Figure 33 is a computed example in such a case, pressure postulated to be zero for convenience, showing that the free and the coupled waves are not strongly different in their development (Perko, 1971).

The coupled wave, however, does strongly affect the density perturbation mode, as shown in Figs. 34 and 35 (Perko, 1971). In Fig. 34 the perturbation is started by postulating only an initial gravitational wave. In Fig. 35 the perturbation is started not with an initial value of δ but with a small "kick": an initial value of $\dot{\delta}$. The figures give values for the power law index n. In all cases n approaches the isotropic value $2/3$ as the universe expands. The gravitational wave provides an effective pressure, too, causing oscillatory terms in the density even if there is no pressure. The values of the index are in some cases vastly different from $2/3$ at early times. These calcula-

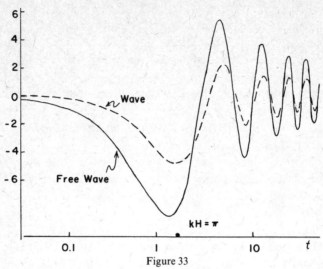

Figure 33

Comparison of the uncoupled and matter-coupled gravitational waves in a homogeneous anisotropic cosmology. The point of this graph is that the coupling to the matter does not significantly affect the period or rate of damping of the wave oscillation (from Perko, 1971).

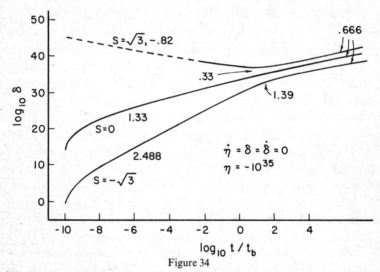

Figure 34

Example of the behavior of density perturbations in anisotropic cosmologies given only an initial wave amplitude for the coupled gravitational waves. $S < 0$ implies the universe is contracting along the direction of propagation of the (coupled) wave. These directions show the fastest perturbation growth; the wavelength initially is much greater than the horizon size for those cases. Those waves which propagate in a direction in which expansion occurs ($S > 0$) have a strong coupling to the matter as soon as the (long) horizon size is greater than the wavelength. Their average behavior (over several periods) is shown dotted. The slopes of $\delta = \Delta\rho/\rho$ are indicated; they tend to the isotropic value 2/3 for $t > t_0$, when the model approaches isotropy (after Perko, 1971).

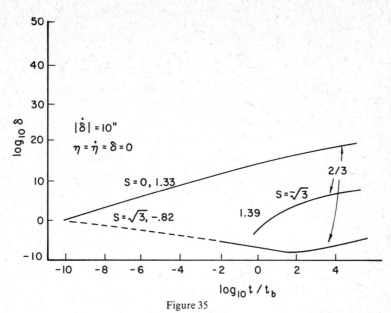

Figure 35

As Fig. 34, but here the initial conditions have all variables but the time derivative of δ initially zero (after Perko, 1971).

tions were for a hypothetical model in which the anisotropy is important only before $t \sim 100$ yrs.

Although the law of growth is still only a power law, n can be as high as 8/3. The index n depends on direction, as does the expansion rate, n being largest in directions of slow expansion, or initial contraction, and small horizon. (A small horizon corresponds to a later onset of gravitational wave induced oscillations, also.) In mixmaster models, type IX models, directions of rapid and slow expansion can alternate, and the effect could be a more rapid growth of perturbations in all directions. This effect and effects caused by cosmic rotation have yet to be calculated, but Hu and Regge (1972) have developed a formalism for handling such computations. Further problems, even in the type I model illustrated, concern the detailed damping effects of pressure, which at the early times pictured in the figures is high.

The problem of large perturbations thus could be reduced to a problem of smaller, earlier perturbations. It does not appear likely that purely random perturbations are sufficient, even so. Non-random mechanisms, such as thermal instabilities, also appear to be ineffective at these early times, when matter and radiation are strongly interacting (see Rees, 1971b). At present, therefore, the galaxies we see seem to be a direct, if confusing, picture of the initial singularity.

5. SINGULARITIES AND THE ORIGIN OF MATTER

5.1. Helium production. A thermal history of the universe, $T < 10^{12}$ K

Assume a homogeneous isotropic model in which there was a big bang, and assume
the 3 K microwave radiation is a relic of the big bang. This will give us a way of
correlating the temperatures with redshift (R/R_{now}). A temperature of 10^{12} K corresponds to roughly 100 MeV. Hence at the temperature at which this discussion
begins, (thermally produced) heavy particles such as pions and nucleion-antinucleon
pairs have annihilated, and the muons have been almost completely annihilated.
The locality we inhabit in the universe has non-zero net baryon number, and at the
temperature of 10^{12} K, these are present in the form of protons and neutrons.

The other particles present are electrons, neutrinos which we assume non-degenerate, photons and gravitons (which would have decoupled at much higher temperatures, $T \sim 10^{20}$ K (Matzner, 1968)). The gravitons will be ignored in this history
but will be taken up again below. At a temperature of 10^{12} K the age of the universe
is of the order of a tenth of a millisecond. The presence of the e^- neutrinos keeps
the protons and neutrons in equilibrium at these high temperatures; the mean free
time t_c for $v + n \to p + e^-$, for instance, is of the order of the expansion time t_{exp}
at a temperature of the order of $T \sim 10^{10}$ K. The collision rate increases rapidly (as
T^5) for higher temperatures, while the expansion rate increases only as T^2. Hence
the reaction goes rapidly for $T > 10^{10}$ K, but the proton-neutron ratio is fixed at the
value given by thermal equilibrium at $T \sim 10^{10}$ K as the reaction is cut off by the
expansion. This gives $n_n/n_p \sim 0.2$, when the universe was ~ 1 sec old. In general
$n_n/n_p = e^{-Q/kT}$ for thermal equilibrium, with $Q = (m_m - m_p)$. The ratio remains at
this value until the neutrons begin to decay (Peebles, 1971). These observations are
based on an assumption that the neutrinos involved in the reactions are non-degenerate. This is not an observationally accessible fact. By demanding that the neutrino
energy density be less than that to close the universe, one concludes that the Fermi
level is ~ 0.0075 eV, or about 45 times the energy $kT_{v0} \sim 1.9$ K. Since the Fermi
level is red-shifted like other energies, degeneracy is maintained as the universe
expands. This cosmological estimate is a more sensitive test than observation of
cosmic rays spectra for instance (Weinberg, 1972). If the current estimates for primordial helium are correct (see below), an even tighter limit on the neutrino density
is possible, putting the Fermi energy at $\lesssim 10^{-3}$ eV since degenerate neutrinos would
affect nucleosynthesis (Wagoner, Fowler and Hoyle, 1967).

The present neutrino temperature T_{v0} is not the same as the present photon temperature. The precise behavior of the neutrinos in this model depends on the still
uncertain details of the behavior of weak interactions. The difference hangs on
whether $ev_e \to ev_e$ scattering is possible in addition to reactions like $ev_\mu \to \mu v_e$. If
only the μ-mediated interaction occurs, the disappearance of μ-mesons means that

both muon and electron neutrinos decouple at a rather high temperature $T \sim 10^{11}$ K. If direct $e v_e$ interactions are allowed, then the electron neutrinos will be in equilibrium with the electron-photon gas down to $\sim 5 \times 10^9$ K. (A recent result of Gurr, Reines and Sobel, 1972 may have some bearing on this subject.) Although this model is supposedly an isotropic one, we note again that the mechanism first proposed by Misner (1968a) for damping of anisotropy was the neutrino produced irreversibility which occurs during this epoch.

Since most of the electron-positron pairs annihilate at temperatures near 5×10^9 K, electron and muon neutrino densities follow a simple redshift law $T \propto R^{-1}$ starting from 10^{12} K all the way down to the present. The electron-positron annihilation, when it occurs, dumps almost all the energy of the annihilating pairs into the photon density. By mode counting at a temperature near 10^{11} K when the muons have annihilated but electrons have not, one finds an apportionment between the models (neglecting gravitons) of (this follows Weinberg, 1972):

$$\rho_{v_e} = \rho_{\bar{v}_e} = \rho_{v_\mu} = \rho_{\bar{v}_\mu} = \tfrac{7}{16} \rho_\gamma$$

$$\rho_{e^+} = \rho_e \cong 2\rho_v$$

$$T_e = T_v = T_\gamma$$

The baryon matter is totally negligible in these estimates.

After the temperature has dropped, low enough for the electron-positron annihilation to occur, this balance will be upset. The annihilations occur when the mean free paths are very short, so the transition can be considered thermodynamically reversible. Conservation of the specific entropy gives:

$$(RT_\gamma)_{\text{after}}/(RT_\gamma)_{\text{before}} \cong 1.4$$

The neutrinos will remain isolated from this annihilation because of the smallness of the weak coupling. Subsequent to the electron-positron annihilation the neutrinos will continue to evolve independently of the photons, and in parallel with them, down to $T_v \sim 1.9$–2.0 K now.

Because deuterium is bound (though not very strongly) there is a tendency for the neutrinos and protons to combine to form ^{2}H, as the temperature drops. If all the neutrons frozen out as estimated above combined to form deuterium, we would find a mass fraction

$$\frac{2n}{n+p} = 0.33$$

in deuterium, and once this is formed, it very quickly will burn to helium-four, (with the same mass fraction) which gives a "ball park" agreement with the "normal" helium abundance ratio ~ 0.24. As Peebles (1971) points out, the thermodynamic equilibrium shifts toward deuterium production at $T \simeq (0.8$–$0.9) \times 10^9$ K, when the universe is already old enough (300 seconds) so that the neutrons are beginning to be lost by free decay. At 10^9 K the rate is fast enough to allow almost complete deuterium production, although if the present temperature were higher by a factor ten or so ($T_{\text{microwave}} \sim 30$ K) the baryon density near $T = 10^9$ K would have been too low to give much deuterium production.

After the completion of helium formation, the temperature of the universe continues to drop. Since the hydrogen and helium are ionized the photons remain collisional and in temperature equilibrium with the matter and the whole assemblage follows the $T \propto R^{-1}$ law with the neutrinos at their lower (by factor $1/1.4$) temperature. This continues until the hydrogen recombines. Peebles (1968) and Peebles and Yu (1970) have shown that the ratio of t_c/t_{exp} for photons goes from 10^{-4} at $T = 5 \times 10^3$ K to ~ 5 at 2×10^3 K. It is thus at this epoch that the photons comprising the microwave radiation are least scattered in the usual çosmological interpretation. At approxi-

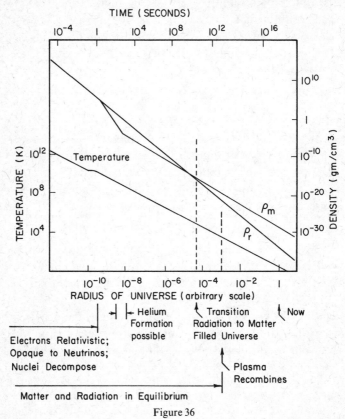

Figure 36

Thermal history of an isotropic-homogeneous cosmology. The temperature indicated is the photon temperature, the neutrino temperature is a factor $\sim (1.4)^{-1}$ smaller. Electron-positron annihilation causes the upward kink in the temperature at $\sim 10^{10}$ K. Helium production occurs near 10^9 K. The somewhat uncertain transition from radiation to (baryon) matter dominance occurs in the range 10^2–10^4 K, and the recombination of hydrogen occurs near 2×10^3 K. Subsequent to hydrogen recombination the universe is transparent, and hence the redshifted photons in the microwave background could have traveled freely to us since then (after Dicke, Peebles, Roll and Wilkinson, 1965).

mately this temperature also, the baryon matter finally becomes dominant over the radiation in determining the overall density (since the radiation energy density $\propto R^{-4}$ while baryon matter density $\propto R^{-3} \propto T^{-3}$). And because of the disappearance of the drag force due to the radiation, perturbations can grow to give rise eventually to galaxies (see Sec. 4). Finally, at redshifts no greater than 2–3 ($T \le 10$ K) the cosmological term may have become important in affecting the overall expansion as may (perhaps) the spatial curvature. Figure 36 (after Dicke et al., 1965) summarizes this conventional thermal history.

A basic aspect of this discussion has been the assumption that the "normal" helium abundance as observed currently reflects a primordial abundance. Wagoner (1973) has assembled references which give estimated pregalactic ^{4}He abundance ratios ∼0.22–0.32.

There have been some observations of halo stars (which were presumed not-too-evolved from their primordial constitution) which showed lower than normal helium abundances (Greenstein, 1966; Sargent and Searle, 1966). However, recent analysis has shown that the stars showing low helium abundances are very abnormal in other ways also, often showing metals heavier than helium with weak helium lines (Sargent

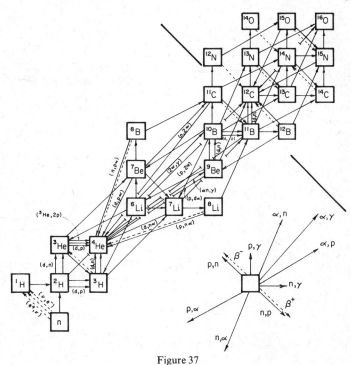

Figure 37

Diagram showing isotopes and reactions considered by Wagoner (1973) for calculation of elemental abundances.

and Searle, 1967; Baschek, Sargent and Searle, 1972). Also, Searle and Sargent (1972) have discovered two dwarf blue galaxies. These are strange systems because they show a large number of blue, young stars. Either their present rate of star formation greatly exceeds the average rate in the past, or alternatively they have always produced very massive (observably blue) stars. In any case the stellar evolution within them would have been quite different from normal stellar evolution, yet they show a normal helium abundance. Wagoner (1973) takes these results and attempts to find model theories which predict such abundances. He takes account of 26 isotopes up to ^{16}O (Fig. 37). He finds that

"*Standard big-bang models in which the present baryon density is (1–3) $\times$ 10^{-31} g/cm^3 agree best with the probable pregalactic abundances of ^2H, ^3He and ^4He, if the galactic production of ^3He is also assumed negligible*".

Figure 38 shows the evolution of the elements as a function of time and temperature. Of more interest is Fig. 39 which shows that the deuterium (pregalactic abundance (0.3–5.0) $\times$ 10^{-4}) and ^3He (pregalactic abundance $\leq 10^{-4}$) lead to the small range in current density, at a temperature of 3 K.

Gott et al. (1974) have taken these results as indicating that the universe is open.

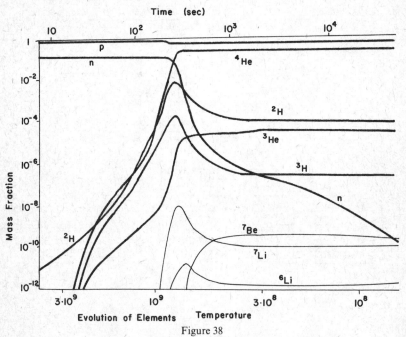

Figure 38

The production of the elements in a standard big bang cosmology for a model with present density $\sim 2 \times 10^{-31}$ g/cm^3 (after Wagoner, 1973).

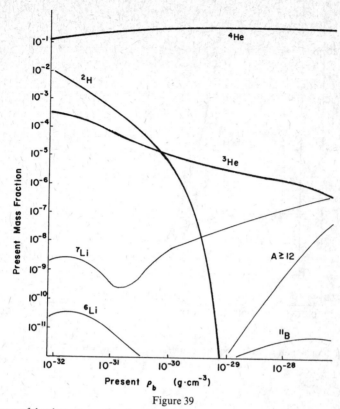

Figure 39

Final abundances of the elements as a function of present baryon density. Estimated pregalactic abundance ratios of ^2H $((0.3-5) \times 10^{-4})$ and ^3He $(\leq 10^{-4})$ require $\rho_{b/now} = (1-3) \times 10^{-31}$ g/cm³ (after Wagoner, 1973).

Figure 40 (after Gott et al., 1974) is Fig. 2 with the deuterium estimate included. The ratio ρ/ρ_c estimated above is of the order of 5×10^{-2}, where ρ_c is the mass density needed to close the universe. This number is also in close agreement with the determinations of the density in galaxies. Of course other matter could be distributed in a way that escapes observation but has enough density to close the universe. However, Gott et al. argue that it can be reasonably assured that hidden matter does not exceed the density to close the universe. This is based on the assumption that any low temperature mass will be accreted into galaxies (increasing the masses determined from virial studies) or accreting, giving excessive radiation as they fall into the galaxies. If the galaxies were *not* formed by gravitational instability, it is still possible that substratum of such matter exists. Similarly, a background fluid, e.g., neutrinos, must have a high pressure to avoid condensation into the galaxies or

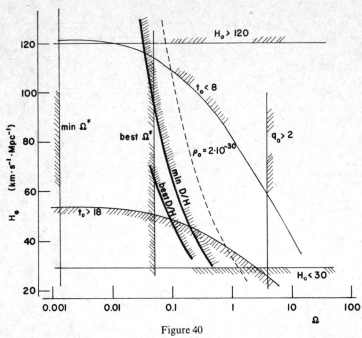

Figure 40

Figure 2 with additional lines added for correct deuterium abundance, strongly suggesting the universe is open (Gott, Gunn, Schramm and Tinsley, 1974).

clusters, so cold hydrogen gas seems to be ruled out. Hot hydrogen ($T > 10^8$ K) sufficient to close the universe faces the problem of finding energy to give this level of ionization. As we pointed out above, degenerate neutrinos cannot be excluded on any other basis than that they would strongly modify the nucleosynthesis, thereby destroying the agreement with Wagoner's results (1973). Although the arguments against a closed universe could each be circumvented by careful selection of initial conditions, Gott et al. consider that, taken with the positive criteria a) correct ^{1}H and ^{2}H production; b) a mean predicted density consistent with (slightly greater than) current estimates and c) an age of the universe consistent with globular star cluster ages, d) agreement with a value of q_0 obtained from observations by applying evolutionary corrections, the case is strongly in favor of the open types. A careful reading of their paper indicates that they rely strongly on the primordial deuterium abundance to fix q_0. Since Colgate (1973) has raised the possibility of deuterium production in supernovae, this point is somewhat weakened. Nonetheless, Gott, Gunn, Schramm and Tinsley (1974) conclude:

"*Loopholes in this reasoning may exist, but if so are primordial and invisible, or perhaps just black.*"

5.2. $T > 10^{12}$: ultimate temperature? Particle production

The presentation in the preceding section has been based on physics that occurs at low enough energies to be completely "known". Above 10^{12} K there is enough uncertainty in the behavior of elementary particles that we can postulate two extremes in the behavior of the equation of state of matter near the singularity. If the number of elementary particles and resonances increases sufficiently slowly for $E \gtrsim 100$ MeV, then as the matter is compressed near the singularity, these particles become relativistic and the equation of state approaches that of a radiation fluid, $p = \rho/3$. If the density of states increases fast enough, however, there may be some ultimate temperature T_u such that adding energy to the system does not raise the average kinetic energy, but simply boils out more new particles so the temperature is bounded by T_u. A possible description of the density of states which produces such an ultimate temperature has been given by Hagedorn (1970). He fits a formula for the density of states:

$$n(m) \simeq Am^{-B} \exp \left(\frac{m}{T_m} \right) \tag{5.1}$$

to the experimental results for parameter values:

$$B \simeq 2\text{--}4, \qquad T_m \simeq 1.7 \times 10^{12} \text{ K} \tag{5.2}$$

Figure 41 displays how the data are an improving fit to this idea. (However, see also Leung and Wang, 1973, who find a polynomial fit.) This gives an equation of state (best displayed in a form parametrized by l): for $l \to 0, p, \rho \to \infty$:

$$p \propto \frac{l^{-2}}{|\ln l|}, \qquad \rho \propto \frac{l^{-2}}{|\ln l|^{1/2}} \tag{5.3}$$

as the density tends to infinity near the singularity. The solution for an isotropic-homogeneous cosmology using this equation of state gives:

$$R(t) \propto t^{2/3} |\ln t|^{1/2} \tag{5.4}$$

For comparison, the dust ($p = 0$) models have

$$R(t) \propto t^{2/3} \tag{5.5}$$

while radiation filled models have

$$R(t) \propto t^{1/2} \tag{5.6}$$

It is not surprising that the soft infinity in the pressure (5.3) gives a behavior nearly like the $p = 0$ models. Such behavior would give rise to a "warm" big bang, with temperatures never rising above T_m ($\simeq 10^{12}$ K according to Hagedorn's fit; see also Carlitz, Frautschi and Nahm, 1973). None of the discussion in the preceding section concerning the evolution of the universe for $T < 10^{12}$ K is affected by this possibility.

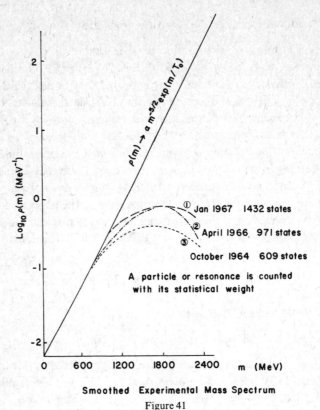

Smoothed Experimental Mass Spectrum

Figure 41

The fitting of elementary particle data to a theoretical mass spectrum, showing better fit to eq. (5.1) as more data are obtained (after Hagedorn, 1970).

The existence of residual thermal gravitons from the early ages of the universe depends critically on the existence of temperatures at least as high as 10^{20} K (10^7 GeV), however. This has been calculated (using linearized theory) as the temperature necessary to bring gravitons into equilibrium with the matter of the universe in a radiation dominated homogeneous isotropic model (Matzner, 1968). The age of the universe at this temperature is

$$t \sim 10^{-6} \, \text{cm} \sim 10^{-16} \, \text{sec} \qquad (5.7)$$

The radius of the causally connected parts of the universe (i.e., the part within a horizon) is of this order also; accordingly, the sizes and the energies are not in the region where fundamental problems of quantum gravity arise since the wavelength of 10^7 GeV gravitons, 10^{-20} cm, is much larger than the characteristic length $G^{1/2} h^{1/2} c^{-3/2} \sim 10^{-33}$ cm. It can be verified that linearized theory is sufficiently accurate for this calculation.

As pointed out by Alpher et al. (1953), the number density of the gravitons is approximately $1/n$ times the total number density, where n is the total number of modes available to carry the energy. These modes consist of two polarizations for photons, four for neutrinos, as well as those for mesons, and presumably baryons. The rest of the energy (not initially in gravitons) eventually ends up in photons (γ) or in neutrinos, so

$$\left(\frac{T_g}{T_\gamma}\right)^3_{\text{now}} \sim \frac{2}{n} \tag{5.8}$$

The number n in this equation is difficult to estimate. If one should count all possible hadron states with mass less than 10^7 GeV, then n will be very large and T_g(now) will be very small. If there exists an ultimate temperature $T_u < 10^{20}$ K, as suggested above, then gravitons were never in equilibrium with the matter in the universe, and so there are no thermal relict gravitons. On the other hand, one might conjecture that the hadrons would decompose, at sufficiently high temperatures, into quarks, and then n will be small, $n \sim 14$. Whether this idea is valid at temperatures $T \sim m_q$ is certainly questionable, because the mean free path would be much smaller than a Compton wavelength. Accepting $n \sim 14$, however, gives $T_g \sim 1.6$ K for the present temperature of blackbody gravitational radiation. As discussed above, Misner's analysis (1968a) has shown that this amount of collisionless gravitational radiation would strongly affect the subsequent dynamics of homogeneous models with large amounts of anisotropy. The gravitons would behave similarly to the collisionless neutrinos in modifying the evolution of the anisotropy.

We mentioned above the fact that rapidly changing gravitational fields such as near a cosmological singularity (big-bang) should give rise to particle production. There have been investigations of this possibility by Parker (1969, 1972), by Zel'dovich and Starobinsky (1971), by Parker and Fulling (1973), and also by Berger (1972, 1974a,b) and by Misner (1974). This is not a question of quantum gravity, which becomes important only on a scale of $\sim 10^{-33}$ cm, but one of quantum fields in a classical background geometry, and could be important for universe sizes $\sim 10^{-13}$ cm and smaller.

The mass of the particles involved is important in considerations of particle production if the equations for massless particles are conformally invariant, since no creation occurs in *conformally* flat cosmologies, such as the homogeneous isotropic models we have discussed. If this conformal invariance is broken for the scalar equations then large estimates for particle production can be obtained in homogeneous isotropic models (Parker, 1972).

Estimates of the rates involved can be found in the references mentioned but no one has yet produced a definite number for the particles produced in the universal experiment. The principal difficulty in a calculational approach to the particle production problem is the question of defining the particle. The separation into positive and negative energy states required to define particle and antiparticle number is an explicitly special-relativistic one. It depends on the existence of time translations in the Poincaré group. In a highly curved space this separation is often impossible and when not, it is ambiguous. In flat space, one has the possibility of basing the separation into positive and negative frequencies on timelike Killing vectors other than the usual one. For instance, in two-dimensional Minkowski space, one can use the generators of the Lorentz transformations which do not move the origin. The constant time surfaces are lines, while the trajectories of ∂_t are hyperbolae. This is a good covering of 1/4 of M^2, except for the null cone (and its apex) through the origin. The particle-antiparticle separation here is different from the usual one. A moment's reflection shows that this second frame is accelerated, and one might expect effects—like particle production—due to the equivalence principle. The particle-antiparticle separation apparently requires a global Minkowski space to be invariantly defined.

Because of this difficulty, the results on cosmological particle production have remained ambiguous. One observes that far enough in the future of a homogeneous cosmology, say, the geometry changes sufficiently slowly that particles can be defined and the calculated particle production *rate* tends to zero. Hence the number of particles existing in the late universe tends to a constant. If one could postulate initial conditions that no particles at all existed at some early time t_0 in the universe (and take the limit when $t_0 \to 0$), one could compute at some late time the number of particles produced, and could predict the total number of particles to be expected in a typical universe. (The fact that the net baryon number is zero is a minor irritant at this stage.) The catch comes because the particle number cannot be defined in the early evolving model. The technique used by Parker (1969), Zel'dovich and Starobinsky (1971), and Berger (1972) has been to stop the expansion at some early point to say the universe was flat and empty before t_0. At t_0 the initial condition of particle field defining a vacuum (in the flat manifold) is imposed, and the solution is then allowed to evolve. Particle production occurs, and some net number N_{t_0} of particles are produced to limit the final universe. Unfortunately N_{t_0} is far from independent of t_0 ($N_{t_0} \to \infty$ as $t_0 \to 0$), so this method must rely on the ad hoc choice of an initial time t_0.

Misner (1974) and Berger (1974a,b) have analyzed this problem in a different manner, using the solutions of Gowdy (1971, 1974). Gowdy's solutions are inhomogeneous and anisotropic, depending on two variables. For large times, they are approximately homogeneous and isotropic with (approximately) zero spatial curvature *R, but containing small amounts of gravitational radiation. This radiation consists of one polarization of radiation traveling in one direction around these closed models. (These space sections have the topology of a flat 3-torus.) For large times this radiation has short wavelength compared to the horizon size. If this cosmology is evolved back toward the singularity, the wavelength becomes longer compared to the horizon size, and near the singularity these waves behave like the anisotropy in one of the homogeneous anisotropic models. That is, near the singularity the model looks, within each causally connected region (within each horizon size) like a type I (flat 3-space) anisotropic homogeneous cosmology. These models can be solved exactly classically (Gowdy, 1971). They can also be solved exactly (up to factor ordering questions) within the truncated quantum mechanics which suppresses the zero-point motion of the modes which vanish classically due to symmetry (Berger, 1974a). This model, from viewpoint adopted by Berger and by Misner, should be an example of particle (graviton) production, since there is initially only an empty anisotropic universe, while finally there are gravitons present. Berger first treated this model using the techniques of Parker (1969) stopping the evolution at some early time t_0. As in the treatment of Parker, the final graviton number depends in an unsatisfactory way on t_0. In subsequent discussion, Berger (1974b) and Misner (1974) have argued that this model actually indicates *no* particle production. First of all one can classically prescribe the number of gravitons found as $t \to \infty$ by specifying some particular initial motion of the anisotropy. The quantum analysis smears out the classical motion very little, and the expectation value of the number of particles in the final state works out to be the same as the classical result (Berger, 1974b). It thus appears that particle production, at least in this scheme, does not occur. The particles are always there (present in the initial conditions) but for early times their associated wavelengths were so much larger than the horizon that they were unnoticed. At late times they evolve like short wavelength radiation in an almost flat background. What is produced is not the particle, but the particle-like behavior. It is obvious that much work remains to be done on this problem, particularly in looking at models in which the particles being produced are not gravitons.

5.3. Singularities

Perhaps the most profound problem of principle confronting physics today is the existence and nature of a singularity in cosmology. Early hopes that the singularity of an isotropic model would disappear once the high symmetry of the model was relaxed proved wrong. The singularity theorems of the mid-1960's (see Hawking and Ellis, 1973) showed that symmetry was irrelevant in the proof of singularity. Yet the dedicated religionists who saw in the existence of a singularity the sign of a creator have also been disappointed. Although each cosmological model is singular by being incomplete, incompleteness does not necessarily involve an infinity of a physical variable, nor does it necessarily prevent extension of the model toward the past.

A manifold is incomplete if it contains inextendible geodesic segments of finite affine parameter. For example, Figure 42 illustrates geodesics in the mathematically non-singular vacuum Taub–NUT–Misner model (Misner and Taub, 1968). No

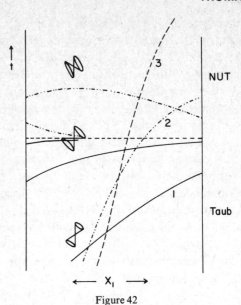

Figure 42

Timelike geodesics in T–NUT–M space. Geodesic 1 approaches the Misner border between Taub space (a type IX spatially homogeneous vacuum model) and NUT space (where the light cone has tipped over). Geodesic 1 has no unique limit point, since it wraps around the closed spacelike hypersurface in Taub space infinitely often. Geodesics 1 and 2 are incomplete, but geodesic 3 and any other geodesic leaving the model have infinite length. The T–NUT–M has a differentiable metric and is mathematically non-singular.

geodesic, such as number 3, can leave the model without becoming infinitely long, yet some geodesic segments cannot be extended because they have a continuum of limit points. Although geodesics 1 and 2 are incomplete, the metric itself shows no discontinuities when the manifold is extended beyond the region where geodesics end. Incompleteness, however, amounts at least to a physical trouble in the model. Any timelike or null geodesic may be the path of a real particle, and consequently it is far from sufficient to postulate that "real" matter may follow only certain geodesics which are complete.

Each general relativity cosmological model is incomplete if it obeys the following conditions (Hawking and Ellis, 1973):

1) $R_{\alpha\beta}k^{\alpha}k^{\beta} \geq 0$ for every non-spacelike vector k^{α}.

2) There exists a compact spacelike 3-surface S.

3) The unit normals to S are everywhere converging (or diverging) on S.

Condition 1 is the "energy condition", and is valid if the pressure is not too negative and if the speed of sound is always less than the velocity of light. Condition 3 basically recognizes the presently observed cosmic expansion. Condition 2 may be modified, of course, to some more appropriate condition in other circumstances, for example in a general spatially homogeneous model with a non-compact homogeneous space-like slice. Parameters of the real universe indicate that any general relativity model portraying the real universe must be incomplete (Hawking and Ellis, 1968).

The T–NUT–M model mentioned above is incomplete yet non-singular. Later work (Shepley, 1969; Ellis and King, 1974) has shown that if an infinite density singularity does occur in a model, this infinity (which represents an undoubted singularity) and incompleteness of geodesics need not have any relation. Ellis and King (1974) have described those circumstances in which incompleteness is not accompanied by a singularity in the matter which fills the model. The title of this important work *Was the Big Bang a Whimper?* seems to imply that incomplete geodesics in isolation are not important. It is more correct to say that incomplete geodesics signal a singularity which physically corresponds to a hypothetical test particle appearing at an infinity of limit points at an instant of time or else disappears from the universe in a finite time.

Thus when any matter is added to the empty T–NUT–M model (with the exception of quite special magnetic fields), the null hypersurface which acts as an interface between the "Taub" and "NUT" regions is replaced by a true singularity. Indeed, in some important classes of models it is possible to show that the density always does become infinite, so that incompleteness in isolation does not occur. The spatially homogeneous models with matter velocity normal to the spacelike invariant hypersurfaces are of such a class. So are all dust-filled type IX models, rotating or not. However, type V models with matter tilted with respect to the homogeneous hypersurfaces may have isolated incompleteness (Shepley, 1969; Collins, 1974).

What if anything breaks down simply because of incompleteness is unknown. The Schmidt (1971) method of defining the structure of a singular boundary could be of some help, since the singular points are mathematically well-described. The method assigns to a higher dimensional manifold, the bundle of orthonormal frames, B, a positive-definite metric (see Fig. 43). The bundle's boundary is in principle simply defined (though in practice the computation is quite difficult) by limit points of Cauchy sequences. The same method of going from bundle to manifold (using the Lorentz group action in the bundle) then is used to define a manifold boundary as equivalence sets of bundle boundary points. This method is different from an earlier and simpler technique of Geroch (1967a), but does correct certain deficiencies. Geroch's technique looked only at incomplete geodesics and thus could not consider all incomplete physical paths (such as segments of paths of bounded acceleration). Sachs (1973) (see Eardley, Liang and Sachs, 1972) has described a method similar to Schmidt's using the bundle of unit timelike vectors (the sphere bundle). This bundle is the smallest manifold on which a positive-definite metric can be defined from the original indefinite metric. Duncan (1973) (see Duncan and Shepley, 1974,

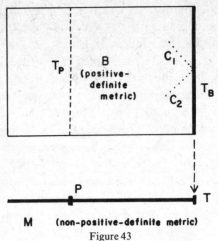

M   (non-positive-definite metric)

Figure 43

The Schmidt–Sachs method of defining sinular points of M is associated a higher-dimensional space B with positive-definite metric. Schmidt's technique uses for B the bundle of orthonormal frames. The association with M is particularly natural, since the association is through the action of the Lorentz group on B. C_1 and C_2, incomplete Cauchy sequences, define points on T_B the boundary of B. The boundary of M is defined by extending the bundle association to T_B. Boundary points are defined for each path in M which is geodesically incomplete or which is a finite timelike path of bounded acceleration.

1975, and also compare Hajicek and Schmidt, 1971) has shown that an equivalence relation can be defined which makes the Sachs and Schmidt techniques obviously equivalent, but the original Sachs method of taking equivalence classes of bundle boundary points was an equivalent technique. Possible non-uniqueness is this general method's principle bugaboo.

It would be inappropriate to go into further mathematical details. However, the sphere bundle is the structure used (although with an indefinite metric) in relativistic kinetic theory. There is therefore the possibility, as yet only a speculation, that the Schmidt–Sachs technique may directly yield physical information on the singularity which at best is almost impossible to discover by other methods.

More esoteric methods of treating incompleteness have also been devised. Miller and Kruskal (1973) have extended certain incomplete manifolds by dropping the Hausdorff criterion that two distinct points should be coverable to disjoint open sets. Figure 44 shows a manifold in coordinates u, v. The dashed lines are orbits of the isometry group of the metric, which is from a compact, incomplete torus given as an example by Misner (1963). When the subset of all points $\{m\pi/2, n\pi/2, m + n \text{ even}\}$ is deleted, the model is geodesically complete. When also points $\{m\pi/2, n\pi/2, m + n \text{ odd}\}$ are removed and identifications made, the result is a maximal analytic extension of Misner's torus. This extension, however, is non-Hausdorff, and the physical

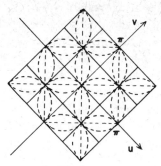

Figure 44

The Miller–Kruskal (1973) manifold, M, used to extend a compact incomplete torus T. T was given by Misner (1963) as a paradigm of an incomplete compact manifold. M minus points $\{m\pi/2, n\pi/2,$ such that $m + n$ is even$\}$ is geodesically complete. M minus $\{m\pi/2, n\pi/2,$ all integers $m, n\}$ with identifications is the Miller–Kruskal maximal extension of T. The extension is non-Hausdorff. The illustrated manifold, in which the dashed lines define orbits of the isometies defined from the metric of T, is an possible example of a structure needed to give a physical interpretation of incompleteness.

implications of the extension or of the manifold illustrated in the figure have not been given.

In most discussions of cosmic singularities it is assumed that incompleteness is accompanied by a singularity in a matter parameter, such as an infinity in the density. The nature of a region near such a singularity in a classical model, that is, the oscillatory or non-oscillatory approach to the singularity, is a subject best handled by the Hamiltonian techniques discussed elsewhere. The question of the horizon size is also best handled by those methods as is the important question of when the horizon is large enough to include all cosmic matter.

The physical effects of the singularity itself may be necessary to describe galaxy formation. The Eardley–Liang–Sachs technique (1972) in velocity-dominated models describes such structure in the approximation where the curvature of space-like slices is ignored. The initial singularity hypersurface may be assigned several structure functions, whose physical significance may only be suggested. For example, one such function describes an effective cosmic initial time which is position dependent. Of course, no approximation technique can give a definite answer to the singularity problem, but the method may be important in describing conditions just past singularity.

The question of what and when physics in addition to classical general relativity becomes important near a singularity is a vital one. In an isotropic model similar to the real universe the horizon distance becomes equal to the Compton radius of an elementary particle at about 10^{-23} seconds after the initial singularity. Before this time, presumably the quantum field theoretic properties of particles dominated classical properties. Parker and Fulling (1973) have shown that the quantum mechanical properties include an effective oscillating pressure. Figure 45 is a computed example by Parker and Fulling showing a turn-around in the cosmic radius $a(t)$ because of momentarily negative terms in the stress-energy tensor when a is less than the Compton length m^{-1} associated with the massive quantum field in their model. The singularity is thus exorcised by invoking an evanescent violation of the energy condition of the singularity theorems (see, also, Bekenstein, 1974). When a is large compared to the Compton length m^{-1}, the model is indistinguishable from a classical model. If such considerations are effective, the chaos of quantum gravitation, appearing as foam when curvature distance scales reach 10^{-33} cm, is avoided.

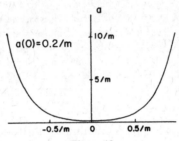

Figure 45

The Parker–Fulling (1973) non-singular model. The radius $a(t)$ is plotted for an isotropic model with classical metric filled with a quantum field of mass m. Because the quantum field gives rise to a rapidly oscillating pressure which can become negative, even violating the energy conditions in the singularity theorems, the radius need never become zero. These effects may become important when a is of the order of the Compton length m^{-1} associated with the field.

Finally, there is the question of what the real universe thinks about singularities. The blackbody radiation shows that any general relativity model of the real universe must be at least incomplete toward the past (Hawking and Ellis, 1968). Collins and Hawking (1973a) have shown, however, that present data do not give good limits on anisotropy near the singularity in many cases. (It is large anisotropy in a type V model which allows a separation between incompleteness and infinite density, where presumed quantum mechanical properties are important.) It is thus ever more important to understand the real cosmos. On the other hand, the real universe may be singular, it perhaps being necessary to treat cosmic time as an intrinsically positive quantity (Misner, 1969b).

5.4. Remnants—Mini-black holes

Although the subject is not directly connected with the cosmological problem, we note that Hawking (1974) has recently discussed the production of particles in the strong field region of black holes. He finds that low mass black holes very quickly dissipate their mass in the production of particle-antiparticle pairs. The lifetime decreases as the inverse third power of mass of the black hole. Hawking estimates that the smallest black hole which could have survived since the earliest days of the universe has mass 10^{15} grams. This means that if a substantial number of small black holes exist which are the relics of the big bang, they have mass exceeding this lower bound. Although the arguments of Gott et al. (1974) tend to exclude a sufficient density of mini-black holes to close the universe, they could have directly observable consequences if, for instance, near encounters or collisions between the Earth and such a hole occurred (Jackson and Ryan, 1973).

6. CONCLUSIONS

The explosion of activity in theories of cosmology due to the new data that became available during the 1960's has to an extent run its course. Cosmology has gone off into tangential areas—such as Quantum Cosmology—which is not so much a means of describing the universe as a model for quantizing General Relativity. We have seen one great influx of observation. Observations of other types, on effects at first glance divorced from cosmology—as was the deuterium question which turns out to put limits on the deceleration parameter—may soon produce another such burst. Or theoretical breakthroughs such as are possible in the quantized theory of gravity may be the stimulus for the next rush of work in cosmology. It is a fascinating, magnificient study and we are awed by its beauty and its simplicity that often hide behind seeming complexity.

Acknowledgments

We wish to thank all our colleagues who made material available for this work. Special thanks also to Susan Lytton, Eva Danks, and Kathy Hilburn for their help in completing the manuscipt and figures in a very short time. This work was supported in part by NSF grants GP-34639X-1 and GP-41655-X.

REFERENCES

ALLEN, C. W., *Astrophysical Quantities,* Oxford University Press, Inc., New York, 1955.
ALPHER, R. A., FOLLIN, J. W. and HERMAN, R. C., *Phys. Rev.* **92**, 1347 (1953).
ARP. H., *Ap. J. Suppl.* **14**, 1 (1966).
ARP, H., *Astron. Astrophys.* **3**, 418 (1969).
ARP, H., *Astrophys. Lett.* **5**, 257 (1970).
ARP, H., *Astrophys. Lett.* **7**, 221 (1971).

ARP, H., *Ap. J.* **183**, 411 (1973).

BAHCALL, J. N., MCKEE, C. F. and BAHCALL, N. A., *Astrophys. Lett.* **10**, 147 (1972).

BASCHEK, B., SARGENT, W. L. W. and SEARLE, L., *Ap. J.* **173**, 611 (1972).

BEKENSTEIN, J., *Ann. Phys.* **82**, 535 (1974).

BELINSKII, V. A., LIFSHITZ, E. M. and KHALATNIKOV, I. M., *Uspekhi Fiz. Nauk* **102**, 463 (1971) [Eng. Trans. *Sov. Phys.-Upsekhi* **13**, 745 (1971)].

BERGER, B. K., Ph.D. Thesis, University of Maryland, 1972.

BERGER, B. K., *Ann. Phys. (N.Y.)*. **83**, 458 (1974a).

BERGER, B. K., Preprint (1974b).

BERTOTTI, B., *Proc. Roy. Soc.* **A294**, 195 (1966).

BIANCHI, L., *Mem. Soc. It. Della. Sc.* (Dei XL) (3) 11, 267 (1897).

BLAIR, A. G., BEERY, J. G., EDESKUTY, F., HIEBERT, R. D., SHIPLEY, J. P. and WILLIAMSON, K. D., *Phys. Rev. Lett.* **27**, 1154 (1971).

BONDI, H., *Mon. Not. Roy. Astron. Soc.* **107,** 410 (1947).

BONDI, H., *Cosmology,* 2nd ed., Cambridge University Press, 1960.

BONNOR, W., *Mon. Not. Roy. Astron. Soc.* **159**, 261 (1972).

BORTOLOT, V. J., CLAUSER, J. F. and THADDEUS, P., *Phys. Rev. Lett.* **22**, 307 (1969).

BOUGHN, S. P., FRAM, D. M. and PARTRIDGE, R. B., *Ap. J.* **165**, 439 (1971)

BOYNTON, P. E. and PARTRIDGE, R. B., *Ap. J.* **181**, 243 (1973).

BOYNTON, P. E., STOKES, R. A. and WILKINSON, D. T., *Phys. Rev. Lett.* **21**, 462 (1968).

BRANS, C. and DICKE, R. H., *Phys. Rev.* **124**, 925 (1961).

BURBIDGE, E. M. and BURBIDGE, G. R., *Ap. J.* **134**, 244 (1961).

BURBIDGE, E. M., BURBIDGE, G. R., SOLOMAN, P. M. and STRITTMATTER, P. A., *Ap. J.* **170**, 233 (1971).

BURBIDGE, G. R., *Nature Physical Science* **246**, 17 (1973).

BURBIDGE, G. R. and BURBIDGE, E. M., *Nature* **222**, 735 (1969).

BURBIDGE, G. R. and O'DELL, S. L., *Ap. J. Lett.* **182**, L47 (1973).

BURBIDGE, G. R., O'DELL, S. L. and STRITTMATTER, P. A., *Ap. J.* **175**, 601 (1972).

CALLAN, C., Thesis, Princeton University, 1964.

CARLITZ, R., FRAUTSCHI, S. and NAHM, W., *Astron. and Astrophys.* **26**, 171 (1973).

CARPENTER, R. L., GULKIS, S. and SATO, T., *Ap. J. Lett.* **182**, L61 (1973).

CARSWELL, R. F., *Mon. Not. Roy. Astron. Soc.* **144**, 279 (1969).

CHERNIN, A. D., *Dokl. Akad. Nauk SSSR* **206**, 62 (1972a) [Eng. Trans. *Sov. Phys. Dokl.* **17**, 825 (1973)].

CHERNIN, A. D., *Astrophys. Lett.* **10**, 125 (1972b).

CHITRE, D., Thesis, University of Maryland, 1972a.

CHITRE, D., *Phys. Rev.* **D6**, 3390 (1972b).

COLGATE, S., *Ap. J. Lett.* **181**, L53 (1973).

COLLINS, C. B. Preprint "Tilting at Cosmological Singularities," 1974.

COLLINS, C. B. and HAWKING, S. W., *Mon. Not. Roy. Astron. Soc.* **162**, 307 (1973a).

COLLINS, C. B. and HAWKING, S. W., *Ap. J.* **180**, 317 (1973b).

COLLINS, C. B. and STEWART, J. M., *Mon. Not. Roy. Astron. Soc.* **153**, 419 (1971).

CONKLIN, E. K., *Nature* **222**, 971 (1969).

CONKLIN, E. K. and BRACEWELL, R. N., *Nature* **216**, 777 (1967).

DAUTCOURT, G., *Gen. Rel. Grav.* **2**, 97 (1971).

DESITTER, W., *Proc. Kon. Ned. Akad. Wet.* **19**, 121 (1917a).

DESITTER, W., *Proc. Kon. Ned. Akad. Wet.* **20**, 229 (1917b).

DEVAUCOULEURS, G., *Science* **167**, 1203 (1970).

DICKE, R. H., *Ap. J.* **152**, 1 (1968).

DICKE, R. H., PEEBLES, P. J. E., ROLL, P. G. and WILKINSON, D. T., *Ap. J.* **142**, 414 (1965).

DOROSHKEVICH, A. G., LUKASH, V. and NOVIKOV, I. D., *Zh. Eksp. Teor. Fiz.* **60**, 1201 (1971) [Eng. Trans. *Sov. Phys.–JETP* **33**, 649 (1971)].

DOROSHKEVICH, A. G., ZEL'DOVICH, YA. B. and NOVIKOV, I. D., *Zh. Eksp. Teor. Fiz. Pis'ma* **5**, 119 (1967a) [Eng. Trans. *JETP Lett.* **5**, 96 (1967)].

DOROSHKEVICH, A. G., ZEL'DOVICH, YA. B. and NOVIKOV, I. D., *Zh. Eksp. Teor. Fiz.* **53**, 644 (1967b) [Eng. Trans. *Sov. Phys.–JETP* **26**, 408 (1968)].

DOROSHKEVICH, A. G., ZEL'DOVICH, YA. B. and NOVIKOV, I. D., *Astron. Zh.* **44**, 295 (1967c) [Eng. Trans. *Sov. Astron.–AJ* **11**, 233 (1967)].

DOROSHKEVICH, A. G., ZEL'DOVICH, YA. B. and NOVIKOV, I. D., private communication, 1968.

DOROSHKEVICH, A. G., ZEL'DOVICH, YA. B. and NOVIKOV, I. D., *Astrofizika* **5**, 439 (1969).

DOROSHKEVICH, A. G., ZEL'DOVICH, YA. B. and NOVIKOV, I. D., *Zh. Eksp. Teor. Fiz.* **60**, 3 (1971) [Eng. Trans. *Sov. Phys.–JETP* **33**, 1 (1971)].

DUNCAN, D. P., Dissertation, University of Texas at Austin, 1973.

DUNCAN, D. P. and SHEPLEY, L. C., *Nuovo Cimento B,* **24**, 130 (1974).

DUNCAN, D. P. and SHEPLEY, L. C., *J. Math. Phys.* **16**, 485 (1975).

DYER, C. C. and ROEDER, R. C., *Ap. J.* **189**, 167 (1974).

EARDLEY, D., LIANG, E. P. T. and SACHS, R., *J. Math. Phys.* **13**, 99 (1972).

EDELEN, D., *Nuovo Cimento B* **55B**, 155 (1968).

EINSTEIN, A., *S.-B. Preuss Akad. Wiss., Berlin* 142, 1917.

EINSTEIN, A. and STRAUS, E. G., *Rev. Mod. Phys.* **18**, 148 (1946). Also in *Rev. Mod. Phys.* **17**, 120 (1945).

ELLIS, G. F. R., *Gen. Rel. Grav.* **2**, 7 (1971).

ELLIS, G. F. R. and KING, A. R., *Comm. Math. Phys.* **38**, 119 (1974).

ELLIS, G. F. R. and MACCALLUM, M. A. H., *Comm. Math. Phys.* **12**, 108 (1969).

EPSTEIN, E. E., *Ap. J. Lett.* **148**, L157 (1967).

EWING, M. S., BURKE, B. F. and STAELIN, D. H., *Phys. Rev. Lett.* **19**, 1251 (1967).

FIELD, G. B. in *Stars and Stellar Systems,* Vol. 9 (A. SANDAGE snd M. SANDAGE, eds.), University of Chicago Press, to appear (1975).

FIELD, G. B. and HITCHCOCK, J., *Ap. J.* **146**, 1 (1966).

GEROCH, R. P., Thesis, Princeton University, 1967a.

GEROCH, R. P., *J. Math. Phys.* **8**, 782 (1967b).

GEROCH, R. P., *Ann. Phys.* **48**, 526 (1968).

GOTT, J. F., GUNN, J. E., SCHRAMM, D. N. and TINSLEY, B. M., *Ap. J.* **194**, 543 (1974).

GOWDY, R. H., *Phys. Rev. Lett.* **27**, 826 (1971).

GOWDY, R. H., *Ann. Phys.* (*N.Y.*) **83**, 203 (1974).

GREENSTEIN, J. L., *Ap. J.* **144**, 496 (1966).

GURR, H. S., REINES, F. and SOBEL, H. W., *Phys. Rev. Lett.* **28**, 1406 (1972).

HAGEDORN, R., *Astron. and Astrophys.* **5**, 184 (1970).

HAJICEK, P. and SCHMIDT, B., *Comm. Math. Phys.* **23**, 285 (1971).

HARRISON, E. R., *Rev. Mod. Phys.* **39**, 862 (1967).

HARRISON, E. R., in *Cargese Lectures in Physics,* Vol. 6 (E. SCHATZMAN, ed.), Gordon and Breach, New York, 1973.

HAWKING, S. W., *Proc. Roy. Soc.* **A294**, 511 (1966*a*).

HAWKING, S. W., *Proc. Roy. Soc.* **A295**, 490 (1966*b*).

HAWKING, S. W., *Ap. J.* **145**, 544 (1966*c*).

HAWKING, S. W., *Mon .Not. Roy. Astron. Soc.* **142**, 129 (1969).

HAWKING, S. W., *Nature* **248**, 30 (1974).

HAWKING, S. W. and ELLIS, G. F. R., *Ap. J.* **152**, 25 (1968).

HAWKING, S. W. and ELLIS, G. F. R., *The Large Scale Structure of Space–Time,* Cambridge University Press, Cambridge, 1973.

HOUCK, J. R., SOIFER, B. T., HARWIT, M. O. and PIPHER, J. L., *Ap. J. Lett.* **178**, L29 (1972).

HOWELL, T. F. and SHAKESHAFT, J. R., *Nature* **210**, 1318 (1966).

HOWELL, T. F. and SHAKESHAFT, J. R., *Nature* **216**, 753 (1967).

HOYLE, F. and BURBIDGE, G. R., *Ap. J.* **144**, 534 (1966).

HOYLE, F. and FOWLER, W. A., *Nature* **197**, 533 (1963).

HU, B. L. and REGGE, T., *Phys. Rev. Lett.* **29**, 1616 (1972).

HUBBLE, E. P., *Proc. Nat. Acad. Sci. U.S.A.* **15**, 168 (1929).

HUGHSTON, L. P., *Ap. J.* **158**, 987 (1969).

HUGHSTON, L. P. and SHEPLEY, L. C., *Ap. J.* **160**, 333 (1970).

JACKSON, A. A. and RYAN, M., *Nature* **245**, 88 (1973).

JACOBS, K. C., *Nature* **215**, 1156 (1967).

JACOBS, K. C., *Ap. J.* **155**, 379 (1969).

JEANS, J. H., *Astronomy and Cosmogony,* Cambridge University Press, Cambridge, 1929.

JONES, J. and JONES, B., *Nature* **227**, 475 (1970).

KANTOWSKI, R., *Ap. J.* **155**, 89 (1969).

KRISTIAN, J., *Ap. J. Lett.* **179**, L61 (1973).

KRISTIAN, J. and SACHS, R., *Ap. J.* **143**, 379 (1966).

KUNDT, W., *Comm. Math. Phys.* **4**, 143 (1967).

LAYZER, D. and HIVELEY, R., *Ap. J.* **179**, 361 (1973).

LEMAITRE, A. G., *Ann. Soc. Sci. Bruxelles* **47A**, 49 (1927) [Eng. Trans. *Mon. Not. Roy. Astron. Soc.* **91**, 483 (1931)].

LEUNG, Y. C. and WANG, C. G., *Ap. J.* **181**, 895 (1973).

LIANG, E. P. T., Ph.D. Thesis, University of California, Berkeley, 1971.

LIANG, E. P. T., *J. Math. Phys.* **13**, 386 (1972).

LIANG, E. P. T., Preprint, University of Utah, 1975.

LIFSHITZ, E., *J. Phys. USSR* **10**, 116 (1946).

LIFSHITZ, E. and KHALATNIKOV, I. M., *Advan. Phys.* **12**, 185 (1963).

LONGAIR, M. S. and SUNYAEV, R. A., *Astrophys. Lett.* **4**, 65 (1969).

MACCALLUM, M. A. H. and ELLIS, G. F. R., *Comm. Math. Phys.* **19**, 31 (1970).

MACCALLUM, M. A. H., STEWART, J. M. and SCHMIDT, B. G., *Comm. Math. Phys.* **17**, 343 (1970).

MACCALLUM, M. A. H. and TAUB, A. H., *Comm. Math. Phys.* **25**, 173 (1972).

MATZNER, R. A., *Ap. J.* **154**, 1123 (1968).

MATZNER, R. A., *Ap. J.* **157**, 1085 (1969a).

MATZNER, R. A., *Astrophys. Space Science* **4**, 459 (1969b).

MATZNER, R. A., Talk at Fourteenth General Assembly I.A.U. Brighton (title only published in *Transactions,* Vol. XIVB, Reidel), 1970.

MATZNER, R. A., *Ann. Phys.* **65**, 438 (1971a).

MATZNER, R. A., *Ann. Phys.* **65**, 482 (1971b).

MATZNER, R. A., *Ap. J.* **171**, 433 (1972).

MATZNER, R. A. and CHITRE, D., *Comm. Math. Phys.* **22**, 173 (1971).

MATZNER, R. A. and MISNER, C. W., *Ap. J.* **171**, 415 (1972).

MATZNER, R. A., SHEPLEY, L. C. and WARREN, J. B., *Ann. Phys.* **57**, 401 (1970).

MILLEA, M. F., MCCOLL, M., PEDERSON, R. J. and VERNON, JR., F. L., *Phys. Rev. Lett.* **26**, 919 (1971).

MILLER, J. G. and KRUSKAL, M. D., *J. Math. Phys.* **14**, 484 (1973).

MISNER, C. W., *J. Math. Phys.* **4**, 924 (1963).

MISNER, C. W., *Phys. Rev. Lett.* **19**, 533 (1967a).

MISNER, C. W., *Nature* **214**, 40 (1967b).

MISNER, C. W., *Ap. J.* **151**, 431 (1968a).

MISNER, C. W., in *Battelle Recontres* (C. DEWITT and J. WHEELER, eds.), W. A. Benjamin, Inc., New York, 1968b.

MISNER, C. W., *Phys. Rev. Lett.* **22**, 1071 (1969a).

MISNER, C. W., *Phys. Rev.* **186**, 1328 (1969b).

MISNER, C. W., *Phys. Rev.* **186**, 1319 (1969c).

MISNER, C. W., in *Relativity* (M. CARMELI, S. I. FICKLER and L. WITTEN, eds.), Plenum Press, New York, 1970.

MISNER, C. W., "Minisuperspace," in *Magic Without Magic: John Archibald Wheeler* (J. KLAUDER, ed.), H. Freeman and Co., San Francisco, 1972.

MISNER, C. W., *Phys. Rev.* **D**, in press (1974).

MISNER, C. W. and TAUB, A. H., *Zh. Eksp. Teor. Fiz.* **55**, 233 (1968) [Eng. Trans. *Sov. Phys–JETP* **28**, 122 (1969)].

MOSER, A. R., MATZNER, R. A. and RYAN, Jr., M. P., *Ann. Phys.* (*N.Y.*) **79**, 558 (1973).

MUEHLNER, D. and WEISS, R., *Phys. Rev. Lett.* **24**, 742 (1970).

MUEHLNER, D. and WEISS, R., *Phys. Rev.* **D7**, 326 (1973a).

MUEHLNER, D. and WEISS, R., *Phys. Rev. Lett.* **30**, 757 (1973b).

NOVIKOV, I. D., *Astron. Zh.* **45**, 538 (1968) [Eng. Trans. *Sov. Astron.–AJ* **12**, 427 (1968)].

OMNES, R. L., *Phys. Rev. Lett.* **23**, 38 (1969a).

OMNES, R. L., *Nature* **223**, 1349 (1969b).

OZERNOI, L. M., *Astr. Zh.* **49**, 1148 (1972) [Eng. Trans. *Sov. Astron.–AJ* **16**, 938 (1973)].

PARIJSKIJ, Y. N., *Ap. J. Lett.* **180**, L47 (1973a).

PARIJSKIJ, Y. N., *Astron. Zh.* **50**, 453 (1973b) [Eng. Trans. *Sov. Astron–AJ* **17**, 289 (1973)].

PARIJSKIJ, Y. N. and PYATUNIA, T. B., *Astron. Zh.* **47**, 1337 (1970) [Eng. Trans. *Sov. Astron.–AJ* **14**, 1067 (1971)].

PARKER, L., *Phys. Rev.* **183**, 1057 (1969).

PARKER, L., *Phys. Rev. Lett.* **28**, 705 (1972).

PARKER, L. and FULLING, S. A., *Phys. Rev.* **D7**, 2357 (1973).

PARTRIDGE, R. B., *American Sci.* **57**, 37 (1969).

PARTRIDGE, R. B. and WILKINSON, D. T., *Phys. Rev. Lett.* **18**, 557 (1967).

PEEBLES, P. J. E., *Ap. J.* **153**, 1 (1968).

PEEBLES, P. J. E., *Physical Cosmology,* Princeton University Press, Princeton, 1971.

PEEBLES, P. J. E., in *Formation and Dynamics of Galaxies,* I.A.U. Symposium 58 (J. SHAKE-SHAFT, ed.), Reidel, Dordrecht, 1974.

PEEBLES, P. J. E. and DICKE, R. H., *Ap. J.* **154**, 891 (1968).

PEEBLES, P. J. E. and YU, J. T., *Ap. J.* **162**, 815 (1970).

PEIMBERT, M., *Bull. Obs. Tonantzintla* No. 30 (1968).

PELYUSHENKO, S. A. and STANKEVICH, K. S., *Astron. Zh.* **46**, 283 (1969) [Eng. Trans. *Sov. Astron.–AJ* **13**, 223 (1969)].

PENZIAS, A. A., SCHRAML, J. and WILSON, R. W., *Ap. J.* **157**, L49 (1969).

PENZIAS, A. A. and WILSON, R. W., *Ap. J.* **142**, 419 (1965).

PENZIAS, A. A. and WILSON, R. W., *Astr. J.* (*N.Y.*) **72**, 315 (1967).

PERKO, T. E., Thesis, University of Texas, 1971.

PERKO, T. E., MATZNER, R. A. and SHEPLEY, L. C., *Phys. Rev.* **D6**, 969 (1972).

PIPHER, J. L., HOUCK, J. R., JONES, B. W. and HARWIT, M. O., *Nature* **231**, 375 (1971).

PUZANOV, V. I., SALMONOVICH, A. E. and STANKEVICH, K. S., *Astron. Zh.* **44**, 1129 (1967) [Eng. Trans. *Sov. Astron.–AJ* **11**, 905 (1968)].

REES, M. J., *Mon. Not. Roy. Astron. Soc.* **154**, 187 (1971a).

REES, M. J., in *General Relativity and Cosmology* (R. K. SACHS, ed.), Academic Press, New York, 1971b.

REES, M. J. and SCIAMA, D. W., *Nature* **213**, 374 (1967).

REES, M. J. and SCIAMA, D. W., *Nature* **217**, 511 (1968).

REFSDAL, S., *Ap. J.* **159**, 357 (1970).

ROEDER, R. C., Preprint (1974).

ROLL, P. G. and WILKINSON, D. T., *Phys. Rev. Lett.* **16**, 405 (1966).

RYAN, M. P., *J. Math. Phys.* **10**, 1724 (1969).

RYAN, M. P., *Ann. Phys.* **65**, 506 (1971a).

RYAN, M. P., *Ann. Phys.* **68**, 541 (1971b).

RYAN, M. P., *Hamiltonian Cosmology,* Springer-Verlag, Heidelberg, 1972a.

RYAN, M. P., *Ann. Phys.* (*N.Y.*) **72**, 584 (1972b).

RYAN, M. P., *J. Math. Phys.* **15**, 812 (1974).

RYAN, M. P. and SHEPLEY, L. C., *Homogeneous Relativistic Cosmologies,* Princeton University Press, Princeton, 1975.

SACHS, R. K., *Comm. Math. Phys.* **33**, 215 (1973).

SACHS, R. K. and WOLFE, A. M., *Ap. J.* **147**, 73 (1967).

SANDAGE, A., *Ap. J.* **178**, 1 (1972a).

SANDAGE, A., *Ap. J.* **178**, 25 (1972b).

SANDAGE, A., *Ap. J.* **183**, 743 (1973).

SANDAGE, A. and HARDY, E., *Ap. J.* **183**, 743 (1973).

SANDAGE, A., TAMMANN, G. A. and HARDY, E., *Ap. J.* **172**, 253 (1972).

SARGENT, W. L. W., *Ap. J. Lett.* **153**, L135 (1968).

SARGENT, W. L. W. and SEARLE, L., *Ap. J.* **145**, 652 (1966).

SARGENT, W. L. W. and SEARLE, L., *Ap. J. Lett.* **150**, L33 (1967).

SCHMIDT. B., *Gen. Rel. Grav.* **1**, 269 (1971).

SCHMIDT. M., *Ap. J.* **176**, 273 (1972).

SCHWARTZ, D. A., *Ap. J.* **162**, 439 (1970).

SEARLE, L. and SARGENT, W. L. W., *Ap. J.* **173**, 25 (1972).

SHEPLEY, L. C., *Phys. Lett.* **28A**, 695 (1969).

SHIVANANDAN, K., HOUCK, J. R. and HARWIT, M. O., *Phys. Rev. Lett.* **21**, 1460 (1968).

SOLHEIM, J. E., *Nature* **217**, 41 (1968*a*).

SOLHEIM, J. E., *Nature* **219**, 45 (1968*b*).

STEIGMAN, G., in *Cargese Lectures in Physics,* Vol. 6 (E. SCHATZMAN, ed.), Gordon and Breach, New York, 1973.

STEWART, J. M., *Mon. Not. Roy. Astron. Soc.* **145**, 347 (1969).

STEWART, J. M., *Ap. J.* **176**, 323 (1972).

STOCKTON, A., *Nature Physical Science* **246**, 25 (1973).

STOKES, R. A., PARTRIDGE, R. B. and WILKINSON, D. T., *Phys. Rev. Lett.* **19**, 1199 (1967).

THORNE, K. S., *Ap. J.* **148**, 51 (1967).

TOLMAN, R., *Proc. Nat. Acad. Sci. U.S.A.* **20**, 169 (1934).

TRENDOWSKI, C. J., Senior Thesis, Princeton University. Quoted in Peebles (1971).

VAN DEN BERGH, S., *Z. Astrophys.* **53**, 219 (1961).

WAGONER, R. V., *Ap. J.* **179**, 343 (1973).

WAGONER, R. V., FOWLER, W. A. and HOYLE, F., *Ap. J.* **148**, 3 (1967).

WEINBERG, S., *Gravitation and Cosmology: Principles and Applications of the General Theory of Relativity,* Wiley, New York, 1972.

WELCH, W. J., KEACHIE, S., THORNTON, D. D. and WRIXON, G., *Phys. Rev. Lett.* **18**, 1068 (1967).

WEYL, H., *Philosophy of Mathematics and Natural Science,* Princeton University Press, Princeton, 1949.

WHEELER, J. A., *Geometrodynamics,* Academic Press, New York, 1962.

WHEELER, J. A., Private communication to L. C. SHEPLEY, 1973.

WILKINSON, D. T. and PARTRIDGE, R. B., *Nature* **215**, 719 (1967).

WOLF, J., *Spaces of Constant Curvature,* McGraw-Hill, New York, 1967.

ZEL'DOVICH, YA. B., *Zh. Eksp. Teor. Fiz. Pis'ma* **12**, 443 (1970) [Eng. Trans. *JETP Lett.* **12**, 307 (1970)].

ZEL'DOVICH, YA. B., *Comm. in Astrophys. and Space Phys.* **3**, 179 (1971).

ZEL'DOVICH, YA. B. and NOVIKOV, I. D., *Zh. Eksp. Teor. Fiz. Pis'ma* **6**, 772 (1967) [Eng. Trans. *JETP Lett.* **6**, 236 (1967)].

ZEL'DOVICH, YA. B. and STAROBINSKY, A. A., *Zh. Eksp. Teor. Fiz.* **61**, 2161 (1971) [Eng. Trans. *Sov. Phys.–JETP* **34**, 1159 (1972)].

Recent Progress in Exact Solutions

WILLIAM KINNERSLEY

California Institute of Technology
Pasadena, California, U.S.A.

1. INTRODUCTION

In solving a set of nonlinear partial differential equations such as Einstein's equations, we have basically three different approaches to take: exact solutions, approximation schemes, and numerical computation. Despite some personal preferences, most people would agree that these have been listed in the order of decreasing aesthetic appeal. I would like to argue that from a practical standpoint as well, seeking exact solutions offers the best promise over the next few years. Let's look at the alternatives.

Approximation schemes have been used to considerable advantage in Relativity in the past, and we have the weak-field, and slow-motion approximations, plus some others. A serious drawback has always been the lack of a rigorous analysis of the validity of these schemes. This is not merely a question of mathematical nit-picking either, because examples exist in related fields such as Fluid Mechanics of perturbation schemes which are singular or contain subtle surprises. Brill and

Deser [1] and Fischer and Marsden [2] have recently made some interesting contributions to this problem.

Nevertheless most realistic problems in Relativity lie clearly outside the domain of any approximation scheme. The large amplitude production of gravitational waves which accompanies the formation of a neutron star is a prime example. For problems such as these we can only choose between exact solutions and numerical analysis. In the long run the latter will probably be the winner, and Relativity will become an exercise as devoid of intuition as Nuclear Physics. Presently, however, the application of numerical methods to Einstein's equations is still in its infancy. "Regge Calculus" (which analysts in other fields know and love as the method of finite elements) shows greater promise following the work of Collins and Williams [3] and Sorkin [4], but is still not quite at the stage where we can use it for practical calculations. We therefore turn to exact solutions as the third alternative.

Unfortunately, the study of exact solutions has acquired a rather low reputation in the past, for which there are several explanations. Most of the known exact solutions describe situations which are frankly unphysical, and these do have a tendency to distract attention from the more useful ones. But the situation is also partially the fault of those of us who work in this field. We toss in null currents, macroscopic neutrino fields and tachyons for the sake of greater "generality"; we seem to take delight at the invention of confusing anti-intuitive notation; and when all is done we leave our newborn metric wobbling on its vierbein without any visible means of interpretation.

At this time, when the number of known solutions is rapidly growing, I feel that an overall survey would be of considerable help. Kramer, Neugebauer and Stephani [5] gave a lengthy review of solutions in 1972 in German, and a forthcoming paper by Krasinski [6] will discuss solutions containing perfect fluid. The last survey previous to this was the well-known work of Jordan, Ehlers and Kundt [7] in 1960, more than a decade ago. The present paper will not try to be as complete, but will have two objectives. One is to give a brief summary of *all* known vacuum solutions and their interrelationships, referring to the bibliography for further details. The other is to elaborate on a few of the most important recent developments. I hope the latter aim will focus more attention in these areas, while the former will help to reduce the duplication of effort which has been so prevalent in the past.

2. SOLUTIONS WITH KILLING VECTORS

Most of the recent work in exact solutions which has physical relevance falls into this category. We will consider primarily the case in which the spacetime has two commuting 2-forming Killing vectors. The metric can then be put into a canonical form introduced originally by Lewis:

$$ds^2 = f(dt + \omega\, d\varphi)^2 - f^{-1}[e^{2\gamma}(d\rho^2 + dz^2) + \rho^2\, d\varphi^2] \tag{1}$$

where f, ω, γ are functions only of ρ and z. We regard (ρ, z, φ) as cylindrical coordinates in a flat space with gradient operator ∇. The Einstein field equations reduce to

$$\nabla \cdot [f^{-1}\nabla f + \rho^{-2}f^2\omega\,\nabla\omega] = 0 \tag{2}$$

$$\nabla \cdot [\rho^{-2}f^2\,\nabla\omega] = 0 \tag{3}$$

and

$$\begin{aligned}
\gamma_\rho &= \tfrac{1}{4}\rho f^{-2}(f_\rho^2 - f_z^2) - \tfrac{1}{4}\rho^{-1}f^2(\omega_\rho^2 - \omega_z^2) \\
\gamma_z &= \tfrac{1}{2}\rho f^{-2}f_\rho f_z - \tfrac{1}{2}\rho^{-1}f^2\omega_\rho\omega_z
\end{aligned} \tag{4}$$

The function γ (Thorne's "C-energy") is determined up to a constant by quadratures once f, ω are known. Should f become negative, we must use this constant to maintain the correct Lorentz signature, by replacing

$$\gamma \to \gamma + \tfrac{1}{2}i\pi$$

Then φ becomes the time coordinate. Otherwise γ may be ignored in the process of finding solutions.

Eq. (3) is the integrability condition for the existence of a "twist potential" Ω, defined by

$$\nabla\Omega = \rho^{-1}f^2\mathbf{e}_\varphi \times \nabla\omega$$

where $\mathbf{e}_\varphi$ is a unit vector in the φ direction. We can eliminate ω in favor of Ω, to obtain an alternative pair of field equations equivalent to Eqs. (2), (3):

$$\begin{aligned}
\nabla \cdot [f^{-2}(f\nabla f + \Omega\,\nabla\Omega)] &= 0 \\
\nabla \cdot [f^{-2}\,\nabla\Omega] &= 0
\end{aligned} \tag{5}$$

2.1. Wave solutions

Gravitational waves with two Killing vectors can be obtained from this metric by making a complex coordinate transformation. If we let

$$t = i\tilde{z}$$

$$z = i\tilde{t}$$

$$\omega = i\tilde{\omega}$$

the line element becomes

$$ds^2 = f^{-1}e^{2\gamma}(d\tilde{t}^2 - d\rho^2) - f(d\tilde{z} + \tilde{\omega}\, d\varphi)^2 - f^{-1}\rho^2\, d\varphi^2 \tag{6}$$

The same field equations now describe Jordan–Ehlers waves [7], which are cylindrical waves with two degrees of freedom, corresponding to two available wave polarizations. The more familiar Einstein–Rosen waves are included in them as the subcase $\tilde{\omega} = 0$. In that case, with

$$f = e^{2\psi}$$

we find that ψ is a solution of the ordinary cylindrical wave equation. Note that stationary *solutions* and Jordan–Ehlers *solutions* do *not* map into one another (unless an analytic continuation can be made) since both ω, $\tilde{\omega}$ are normally required to be real.

There is also a second complex coordinate transformation possible. If we let

$$t = i\hat{\rho}$$

$$\rho = i\hat{t}$$

$$\varphi = i\hat{\varphi}$$

the metric becomes

$$ds^2 = f^{-1}e^{2\gamma}(d\hat{t}^2 - dz^2) - f(d\hat{\rho} + \omega\,d\hat{\varphi})^2 - f^{-1}\hat{t}^2\,d\hat{\varphi}^2 \tag{7}$$

The waves are now independent of $\hat{\rho}$, $\hat{\varphi}$ and propagate along the z-axis. This second possibility had been noted by Ehlers [7] and Thorne [8]. It has also recently received attention by P. Szekeres [9], who interprets Eq. (7) as the interaction region of two colliding plane waves.

To see this we regard $\hat{\rho}$, $\hat{\varphi}$ as Cartesian coordinates and consider the case $\omega = 0$ (which amounts to giving all the waves the same linear polarization). Again letting

$$f = e^{2\psi}$$

we get a time-dependent analog of the cylindrical-wave equation,

$$\psi_{zz} - \psi_{\hat{t}\hat{t}} - \frac{1}{\hat{t}}\psi_{\hat{t}} = 0 \tag{8}$$

Szekeres shows that it is possible to smoothly join this solution to *pp* waves along the null boundaries $\hat{t} \pm z = \text{const}$ (see Fig. 1).

The importance of this result lies in the apparent singularity which occurs at $\hat{t} = 0$. Szekeres asserts that the singularity is physical, and furthermore suggests that this phenomenon is inevitable for *any* wave collision treated in the full Einstein theory! (It does not appear in linearized theory because the time required for the singularity to develop is inversely proportional to the product of the two wave amplitudes, which is second order.) Of course if that is true it would have drastic consequences for the radiation produced by astrophysical sources.

The question as to whether the singularity is physical is slightly obscured by a simultaneous coordinate singularity at $\hat{t} = 0$, which persists even in the limit of

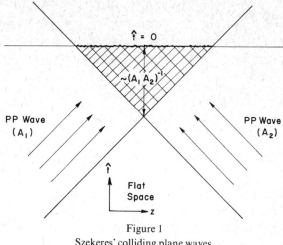

Figure 1

Szekeres' colliding plane waves.

flat space. In that case $\hat{\varphi}$ is a pseudoangle (see Fig. 2). The surfaces $\hat{t} = $ const bifurcate at $\hat{t} = 0$, and continuation beyond is possible, with $\hat{t}$ becoming spacelike. When waves are present, however, we expect the solution of Eq. (8) to have a logarithmic singularity for small $\hat{t}$,

$$\psi \sim A(z) \ln \hat{t}$$

For fixed z, the solution will locally approximate a Kasner metric [10], the well-known example of gravitational collapse which is truly singular.

The other question to be resolved is how general the phenomenon really is. One suspects that while the assumption of linear polarization is not crucial to the argument, the assumption of perfectly plane incident waves *is*. The collision of slowly diverging waves should be examined to see if the singularity can in that way be "defocused."

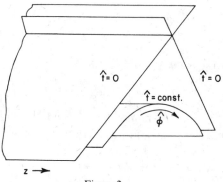

Figure 2

2.2. Tomimatsu–Sato solutions

The most important stationary solution to be recently discovered is the Tomimatsu–Sato (TS) family of metrics [11, 12]. The TS metrics are asymptotically flat with leading multipole moments

$$M = m$$

$$J = ma$$

$$Q = \left(\frac{2n^2 + 1}{3n^2} \right) ma^2 + \left(\frac{n^2 - 1}{3n^2} \right) m^3$$

where m, a are arbitrary real parameters and n is a positive integer. The case $n = 1$ is identical to the Kerr metric.

To best describe them we use the Ernst potential ξ [13, 14], related to the previous variables by

$$f + i\Omega = \frac{1 - \xi}{1 + \xi} \tag{9}$$

It is a complexified nonlinear version of the Newtonian potential and obeys the field equation

$$(\xi\xi^* - 1)\nabla^2\xi = 2\xi^* \nabla\xi \cdot \nabla\xi \tag{10}$$

The Kerr metric for $a < m$ can be described by the simple linear solution,

$$\xi_1^{-1} = -px + iqy \tag{11}$$

where

$$q = a/m$$
$$p^2 + q^2 = 1 \tag{12}$$

and x, y are prolate spheroidal coordinates.

With this as motivation, we look for solutions of the form

$$\xi_n^{-1} = \frac{w_n(x, y, p, q)}{u_n(x, y, p, q)} \tag{13}$$

where w_n, u_n are polynomials. We only require

$$\deg(w_n) = \deg(u_n) + 1$$

to insure asymptotic flatness. Tomimatsu and Sato have given what they believe to be an infinite sequence of such solutions. The next one after Kerr is

$$\xi_2^{-1} = \frac{p^2(x^4 - 1) - 2ipqxy(x^2 - y^2) - q^2(1 - y^4)}{2px(x^2 - 1) - 2iqy(1 - y^2)} \tag{14}$$

The complexity of their solutions increases so rapidly for larger values of n, that complete results have only been given up to $n = 4$.

The near-field structure of the TS metrics has been analyzed by Glass [15], Gibbons et al. [16], and Abramowicz et al. [17]. The function $f = g_{00}$ is of the form

$$f = A/B$$

where A, B are polynomials. The solution contains a sequence of ergospheres or surfaces of infinite redshift where $A = 0$, and a sequence of concentric ring singularities where $B = 0$. There are also certain points at which A and B vanish simultaneously. These points are directional singularities, similar to the ones occurring in Weyl metrics and studied by Gautreau and Anderson [18], Szekeres and Morgan [19], and others.

The surface $x = +1$ where $g_{tt}g_{\varphi\varphi} - (g_{t\varphi})^2$ vanishes was at first thought to be an event horizon. Gibbons and Russell-Clark showed this not to be the case for $n = 2$, because the degenerate metric induced on the surface is Lorentzian. Therefore, it cannot be a null surface. Tomimatsu and Sato [20] have pointed out, however, that this happens only when n is even. The surface is singularity-free only for $n = 1$.

Kinnersley and Kelley [21] discussed the weak-field limit of the TS metrics in terms of disc models, similar to those previously given by Israel [22] for the Kerr metric. They also considered the various limits which may be taken as $a \to m$. They obtained in this way a new family of exact solutions, which bear the same relationship to TS that Bardeen's "throat metric" [74] does to Kerr (see Sec. 3.1).

2.3. Transformation theorems

The first theorem by which one exact solution of the field equations may be used to generate others was discussed by Ehlers [23] in 1957. Since then his methods have been generalized by many authors. In the past few years transformation theorems have become a most effective tool for producing new solutions and for understanding the relationships that exist between old ones. The amount of hidden symmetry contained in the field equations is startling, and no one yet fully understands the reason for it.

We will restrict ourselves to the axially symmetric stationary case, although for many of the results only one Killing vector is required. As we stated previously, the vacuum field equations can be reduced to two equivalent forms:

$$\nabla \cdot [f^{-1} \nabla f + \rho^{-2} f^2 \omega \nabla \omega] = 0 \tag{15}$$

$$\nabla \cdot [\rho^{-2} f^2 \nabla \omega] = 0 \tag{16}$$

or

$$\nabla \cdot [f^{-2}(f \nabla f + \Omega \nabla \Omega)] = 0 \tag{17}$$

$$\nabla \cdot [f^{-2} \nabla \Omega] = 0 \tag{18}$$

To begin with, we have a symmetry group $\mathscr{G}$ of coordinate transformations which replace (t, φ) by linear combinations of themselves. $\mathscr{G}$ is isomorphic to $SL(2, R)$ and also to $SO(2, 1)$ and $SU(1, 1)$. It must be possible to rewrite the field equations in a way which is manifestly covariant under $\mathscr{G}$. The simplest covariant object available to us is

$$(f^1, f^2, f^3) \equiv \rho^{-1}(g_{tt}, g_{t\varphi}, g_{\varphi\varphi}) =$$
$$= (\rho^{-1}f, \rho^{-1}f\omega, \rho^{-1}f\omega^2 - \rho f^{-1}) \tag{19}$$

which transforms like an $SO(2, 1)$ vector with norm

$$f^1 f^3 - (f^2)^2 = -1$$

Eq. (16) can be rewritten as

$$\nabla \cdot [f^1 \nabla f^2 - f^2 \nabla f^1] = 0$$

We therefore define another 3-vector,

$$\mathbf{g}_i = \varepsilon_{ijk} f^j \nabla f^k \tag{20}$$

and observe that when $\mathscr{G}$ is applied, Eq. (16) becomes just one of *three* conservation laws,

$$\nabla \cdot \mathbf{g}_i = 0, \qquad i = 1, 2, 3 \tag{21}$$

Explicitly, the conserved quantities are

$$\mathbf{g}_1 = 2\omega f^{-1} \nabla f + \nabla \omega + \rho^{-2} f^2 \omega^2 \nabla \omega - 2\rho^{-1}\omega \nabla \rho$$
$$\mathbf{g}_2 = -2f^{-1} \nabla f - 2\rho^{-2} f\omega \nabla \omega + 2\rho^{-1} \nabla \rho$$
$$\mathbf{g}_3 = \rho^{-2} f^2 \nabla \omega$$

Eq. (21) for $i = 2, 3$ reproduces the original field equations, and for $i = 1$ it gives a new law which follows from them. Invariance under $\mathscr{G}$ was studied by Matzner et al. [24, 25].

The second version of the field equations, Eqs. (17), (18), has an invariance group $\mathscr{H}$ which is different from $\mathscr{G}$ but also isomorphic to $SL(2, R)$. One way of generating $\mathscr{H}$ is to use the following two particular transformations: a gage transformation on Ω,

$$f \to f$$
$$\Omega \to \Omega + a \tag{22}$$

and the Ehlers transformation [23],

$$f \to \frac{f}{(1 - b\Omega)^2 + b^2 f^2}$$

$$\Omega \to \frac{\Omega - b(f^2 + \Omega^2)}{(1 - b\Omega)^2 + b^2 f^2}$$

(23)

Physically, the latter is a "duality rotation" for gravity. It sends Schwarzschild mass into magnetic mass, and when applied to any stationary vacuum solution it leads to a one-parameter family of physically different solutions.

Although the action of $\mathscr{H}$ on f, Ω is nonlinear, it can be linearized by defining a new set of variables

$$(F^1, F^2, F^3) \equiv (f^{-1}, f^{-1}\Omega, f^{-1}(f^2 + \Omega^2))$$

(24)

One can check easily that F^a transforms like an $SO(2, 1)$ vector under $\mathscr{H}$, with norm

$$F^1 F^3 - (F^2)^2 = +1$$

Then in complete analogy with the treatment for $\mathscr{G}$, we make the field equations manifestly covariant under $\mathscr{H}$ by defining

$$\mathbf{G}_a = \varepsilon_{abc} F^b \nabla F^c$$

(25)

and obtain

$$\nabla \cdot \mathbf{G}_a = 0, \qquad a = 1, 2, 3$$

(26)

Explicitly,

$$\mathbf{G}_1 = f^{-2}(2f\Omega \nabla f + (\Omega^2 - f^2) \nabla \Omega)$$

$$\mathbf{G}_2 = f^{-2}(-2f \nabla f - 2\Omega \nabla \Omega)$$

$$\mathbf{G}_3 = f^{-2} \nabla \Omega$$

The original field equations, Eqs. (17), (18), are equivalent to Eq. (26) for $a = 2, 3$. For $a = 1$ we get a new law (*not* equivalent to Eq. (21) with $i = 1$). This reformulation of the problem was discussed by Neugebauer and Kramer [26].

The close parallel between the two formulations is not an accident. In fact there is a mapping

$$f \to \rho/f$$

$$\omega = i\Omega$$

(27)

which directly transforms one pair of field equations into the other, and maps $\mathscr{G}$ onto $\mathscr{H}$. Eq. (27) itself is nontrivial, and may be used to generate new solutions provided we can do the necessary analytic continuation from real ω to real Ω!

An alternative method of linearizing the action of $\mathcal{H}$ was given by Kinnersley [27]. By analogy with the Ernst potential we let

$$f + i\Omega = \frac{u - w}{u + w} \tag{28}$$

Then we find that the complex 2-spinor

$$\Gamma_\alpha = (u, w)$$

transforms as an $SU(1, 1)$ spinor under $\mathcal{H}$. The two sets of linearizing variables F^i and Γ_α are closely related:

$$F^i = \frac{\overline{\Gamma}_\alpha \sigma^i_{\alpha\beta} \Gamma_\beta}{\overline{\Gamma}_\gamma \Gamma_\gamma} \tag{29}$$

where $\sigma^i_{\alpha\beta}$ are the Pauli matrices.

The next logical step is to study the combined actions of $\mathcal{G}$ and $\mathcal{H}$ together. In a beautiful analysis, Geroch [28, 29] has shown that this leads to an infinite-dimensional symmetry group, acting on an infinite hierarchy of potentials and generating an infinite number of conservation laws. Although he was not able to write out the general transformation in a closed form, he speculated that the full group might suffice to generate *all* stationary axially symmetric vacuum metrics, starting with only one!

We will just sketch the results for the case in which electromagnetism is present. We must then also deal with a complex scalar potential Φ for the Maxwell field. The field equations are more complicated than Eqs. (15)–(18) but may still be written entirely in conservation form. They have an additional symmetry, the Harrison transformation [30], which can be most simply written as

$$\Phi \to \frac{\Phi + b\mathscr{E}}{1 - 2b^*\Phi - cc^*\mathscr{E}}$$

$$\mathscr{E} \to \frac{\mathscr{E}}{1 - 2b^*\Phi - cc^*\mathscr{E}} \tag{30}$$

$$\mathscr{E} \equiv f - \Phi\Phi^* + i\Omega$$

and which maps vaccum fields into charged ones. With all of the commutators included, the enlarged symmetry group $\mathcal{K}$ is now an eight-parameter group, isomorphic to $SU(2, 1)$. The sets of linearizing variables must also be enlarged, to $\Gamma_\alpha = (u, v, w)$ which forms a triplet representation of $SU(2, 1)$, or to F^i, $i = 1, \dots, 8$ which is an octet.

Three of the eight parameters in $\mathcal{K}$ are only gage transformations. The remaining five provide us with an automatic procedure for generating five-parameter *"families"* of stationary Einstein–Maxwell solutions from each solution which is known.

We *still* have not exhausted the amazing structure hidden in these field equations! There is a discrete transformation, due originally to Bonnor [31], which maps stationary vacuum fields into static Einstein–Maxwell ones. Bonnor's transformation can be used in either of two forms,

$$\hat{f} = -f^2$$
$$\hat{\Phi} = \psi \tag{31}$$

or else

$$\hat{f} = +f^2$$
$$\hat{\Phi} = i\psi \tag{32}$$

Eq. (32) lets us keep t as the time coordinate, but requires once more an analytic continuation for success. Bonnor [31], Perjés [32], and Misra et al. [33] have all independently applied the transformation to the Kerr metric, where one must analytically continue the Kerr parameter, $a \rightarrow i\alpha$.

In terms of the linearizing variables, Bonnor's transformation is *quadratic*:

$$\hat{u} = (w^*u + u^*w) = \tfrac{1}{2}(F^1 - F^3)$$
$$\hat{v} = i(w^*u - u^*w) = F^2 \tag{33}$$
$$\hat{w} = (u^*u + w^*w) = \tfrac{1}{2}(F^1 + F^3)$$

In other words, it takes advantage of a unique quadratic mapping of the group $SU(2, 1)$ into itself via the vector representation.

One other type of transformation should be mentioned, which is quite appealing but unfortunately has had only limited success. In certain cases the field equations can be reduced to a form which is translation invariant and (more importantly) analytic. A *complex* translation is then possible and leads to some new metrics. I know, however, of only two cases in which the method has succeeded (see Sec. 2.4).

2.4. Survey of stationary solutions

In Figs. 3, 4 we have given a capsule summary of the known stationary Einstein–Maxwell solutions, together with indications of how they are related to one another under generalizations, specializations, and transformations. They are all axially symmetric except for the ones which have been underlined.

The familiar static Weyl solutions arise from the substitutions

$$f = e^{2\psi}$$
$$\Omega = 0 \tag{34}$$

and the resulting field equation

$$\nabla^2\psi = 0$$

Static Vacuum Solutions

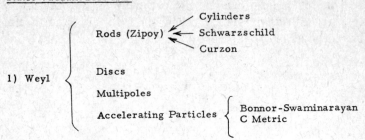

1) Weyl

    Rods (Zipoy) ← Cylinders
             ← Schwarzschild
             ← Curzon

    Discs

    Multipoles

    Accelerating Particles { Bonnor-Swaminarayan / C Metric

2) <u>Harrison</u>

3) <u>Islam</u>

Stationary Vacuum Solutions

1) Papapetrou
2) van Stockum
3) Lewis
4) C-NUT ← K-NUT ← NUT
5) <u>RRZ</u> ← <u>KS</u> ← KD ← Kerr
6) Tomimatsu-Sato ←
7) Kinnersley-Kelley ←   "Extreme" TS
8) Marek
9) NT
10) Kota-Perjés

Static Einstein-Maxwell Solutions

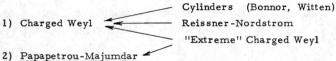

1) Charged Weyl ← Cylinders (Bonnor, Witten)
               ← Reissner-Nordstrom
               ← "Extreme" Charged Weyl

2) <u>Papapetrou-Majumdar</u> ←
3) Gautreau-Hoffman
4) Bonnor, Perjés, Misra
5) <u>Harrison</u>

Stationary Einstein-Maxwell Solutions

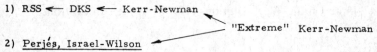

1) RSS ← DKS ← Kerr-Newman
                         "Extreme" Kerr-Newman

2) <u>Perjés</u>, Israel-Wilson ←
3) Cylinder (Datta, Raychaudhuri)

Figure 3

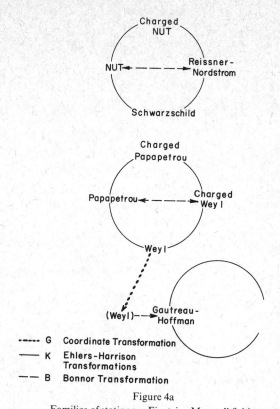

Figure 4a

Families of stationary Einstein–Maxwell fields.

Application of the internal symmetry group $\mathscr{K}$ leads to fields in which f, Ω, Φ are functionally dependent. The Ehlers transformation leads to the stationary vacuum Papapetrou solutions [34], while the Harrison transformation leads to the static "charged Weyl" solutions [35]. The general transformation of $\mathscr{K}$ is a combination of the two, and leads to a "charged Papapetrou" class of solutions that have recently been discussed by Bonnor [36]. Particular examples with whole-cylinder symmetry are the well-known Levi–Civita static cylinders and the charged cylinders of Bonnor [37] and Witten [38].

We could next look for solutions in which f, ω, Φ are the functionally dependent variables. We let

$$f = \rho g(\psi)$$

$$\omega = Kg'/g \tag{35}$$

$$\Phi = 0$$

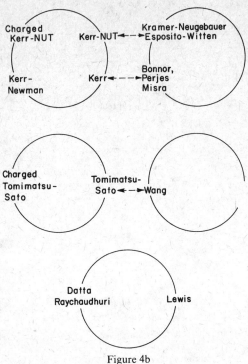

Figure 4b

Families of stationary Einstein–Maxwell fields.

and obtain reduced field equations

$$gg'' - (g')^2 = -K^{-2}$$
$$\nabla^2\psi = 0$$

(36)

Thus ψ again is an arbitrary harmonic function, and the solution for $g(\psi)$ has one of three possible forms:

$$g(\psi) = a \sinh \psi + b \cosh \psi, \qquad a \geq b$$

(37)

$$g(\psi) = a\psi + b$$

(38)

$$g(\psi) = a \sin \psi + b \cos \psi$$

(39)

In Eq. (37) the restriction on a, b is necessary to keep K real.

Could we not have simply obtained these solutions from the Papapetrou fields by means of the Kramer–Neugebauer transformation, Eq. (27), which turns Ω into ω? No, we cannot, for two reasons. In the first place this transformation turns the internal group $\mathscr{H}$ into the coordinate group $\mathscr{G}$. Any fields obtained in that way from Papapetrou must therefore be *equivalent* to Weyl by a change of coordinates. The second obstacle is the necessary continuation $K \to iK$. This is possible only for Eq. (37), by letting b exceed a. Hence Eq. (37) is equivalent to Weyl and the others are not. Eq. (38) was discovered by van Stockum [39] and Eq. (39) by Lewis [40].

Van Stockum's solution written out completely is

$$ds^2 = \rho\psi \, dt^2 - 2\rho \, d\varphi \, dt - \rho^{-1/2}(d\rho^2 + dz^2) \tag{40}$$

and has several points of interest. Note for this metric that t is actually a *null* coordinate. For $\psi = 0$ it reduces, not to flat space, but to the type D static B-metric (see Sec. 3.1). The van Stockum metric is a nondiverging type II solution, with double principal null vector lying in the t direction. It was more recently rediscovered by Tiwari and Misra [41]. Hoffman [42] gives a full discussion of all three solutions of this type. Transformations from $\mathscr{H}$ could be applied to the above metrics to get further solutions, but to my knowledge this has never been carried out. Some particular charged solutions in which f, ω, Φ are functionally dependent have been given by Datta and Raychaudhuri [43].

The next step is to consider the effect of the Bonnor transformation, Eq. (32). This can be applied to metrics that are either vacuum or static, and will almost always lead to solutions which belong to a different "family" under $\mathscr{H}$. The Weyl family is unique, in that it contains *both* stationary (Papapetrou) solutions *and* static (charged Weyl) solutions. What comes as a surprise is that these two solutions are *Bonnor transforms of each other.*

The Bonnor transformation does *not* commute with the coordinate group $\mathscr{G}$, so even though Eq. (37) above is equivalent to Weyl, it has a nontrivial Bonnor transform. It results in the static electromagnetic solutions discovered by Gautreau and Hoffman [44]. To my knowledge, the transform has not been applied to the other two cases.

Now let us see what solutions could be obtained by starting from Kerr. The Harrison transformation leads to the charged Kerr–Newman solution, and the Ehlers transformation to K–NUT space, completing a "family." The Bonnor transform leads out of the family to another static Einstein–Maxwell solution of Bonnor [31], Perjés [32], and Misra [33], mentioned earlier. The five-parameter family of *this* solution has also been explored, by Kramer and Neugebauer [45], and by Esposito and Witten [46]! Tomimatsu–Sato solutions have been charged [47] and "Bonnorized" [48].

Finally, there is a circumstance we should point out that occurs for certain "extreme" values of the parameters, say $e = m$ or $a = m$. Consider the Reissner–Nordstrom solution as a familiar example. The Harrison transformation acts on the two-dimensional parameter space (e, m) of these solutions, and maps them into each other. However, it does so hyperbolically. That is, if the new solution produced is $(\tilde{e}, \tilde{m})$, then

$$\tilde{m}^2 - \tilde{e}^2 = m^2 - e^2$$

What this means is that the "extreme" solutions form an invariant subfamily under $\mathscr{K}$. Analytic continuation is the only way that class can be entered or exited.

The "extreme" case seems to lead to simpler field equations. Extreme charged Weyl and extreme Kerr–Newman have thereby been successfully generalized to yield classes of non-axially symmetric solutions [49–52], and extreme TS has been generalized to allow noninteger values of n [21].

Most of the remaining solutions [53–58] shown in Fig. 3 stand by themselves with regard to the symmetry transformations that are presently known. Several authors have attempted to give classifications for the stationary solutions by various approaches [59–61].

2.5. Brans–Dicke solutions

Although this paper is devoted to the exact solutions of Einstein's theory, some Brans–Dicke solutions have been found recently that are so interesting that they must be mentioned. It is well known that Brans–Dicke theory is conformally related to an Einstein theory, in which gravitation is coupled to a massless scalar meson. The coupling constant is

$$\kappa = \omega + \tfrac{3}{2}$$

where ω is the Brans–Dicke parameter. Given an Einstein-scalar solution $(g_{\mu\nu}, \phi)$, the Brans–Dicke solution is

$$\bar{g}_{\mu\nu} = e^{-\phi} g_{\mu\nu} \tag{42}$$

Penney [62] showed that any vacuum Weyl metric $f = e^{2\psi}$ leads to a one-parameter family of axially-symmetric Einstein-scalar solutions,

$$\psi' = (1 + \kappa A^2)^{-1/2} \psi$$
$$\phi = A(1 + \kappa A^2)^{-1/2} \psi \tag{43}$$

Buchdahl [63] showed that a similar procedure could be carried out on *any* static vacuum solution, and he emphasized the relevance to Brans–Dicke theory. Sneddon and McIntosh [64] then examined the possibility of Ehlers–Harrison transformations in the presence of this scalar field. They found that ϕ did not enter the field equations for f, Ω at all, only the quadrature for γ. Operations from the group $\mathscr{K}$ may therefore be performed on Brans–Dicke solutions also. They have given as an example the Brans–Dicke–NUT metric.

Of course the Bonnor transformation is also unaffected by ϕ, and we can use it to generate a very interesting result. Starting from the Bonner, Perjés, Misra static solution, we add a scalar field via Eq. (43). We conformally transform this to get a Brans–Dicke solution. A Bonnor transformation is then all that is required to produce for us a Brans–Dicke–Kerr metric! This important result also was obtained by McIntosh [65]. If the Einstein theory is ever brought tumbling down by the experimentalists, those of us in exact solutions will have at least one place of refuge.

3. ALGEBRAICALLY SPECIAL FIELDS

Many exact solutions have been found by assuming that the Weyl tensor is algebraically special. The three reduced field equations which result from this assumption were already given by Kerr [66], in the original letter announcing his spinning mass solution. The present status of the algebraically special vacuum solutions is shown in Fig. 5, where they are listed by Petrov type, and by whether or not their rays have divergence and rotation. Solutions which are type I with geodesic rays have been included, since similar techniques have been used to derive them.

3.1. Nondiverging rays

This class of fields was first studied systematically by Kundt [67]. He gave exhaustive solutions for types III and N, which together comprise the gravitational waves with plane wave-surfaces. For type N the wave-surfaces must be geodetic, and the solutions are then known as "plane-fronted" waves. All of Kundt's waves have two independent modes of polarization. Correspondingly they contain two amplitude functions whose time dependence is arbitrary, and whose spatial dependence along the front is harmonic. This means that in general the waves are inhomogeneous, and both metric and Riemann tensor may become singular somewhere on the front. This feature is supposed to be analogous to the electromagnetic *TEM* waves in a waveguide, which become singular on the axis, and are therefore realistic solutions only after a central conductor has been inserted.

The more general of these solutions also contain "twist" τ (in Newman–Penrose

Petrov Type	Nondiverging $\rho = 0$	Nonrotating $\rho = \bar{\rho}$	Rotating $\rho \neq \bar{\rho}$
N	Complete (1961)	Complete (1967)	Equations Only
III	Complete (1961)	Complete (1967)	Equations Only
D	Complete (1968)	Complete (1968)	Complete (1968)
II	Partial	Complete (1962)	Nonradiative Only
I, $\kappa = 0$	(Ruled Out)	Complete (1962)	Stationary Only

Figure 5
Status of algebraically special solutions.

notation) which causes the direction of the propagation vector to vary steadily in time, like the beam emitted by a panning searchlight. The type N waves without twist are the "pp waves." When a bounded source emits gravitational radiation, the far-field limit as would be received by a detector is a homogeneous pp wave. Alternative forms for the pp metric have been given by Rosen [68], by Bondi, Pirani and Robinson [69], and by Misner, Thorne and Wheeler [70], and a detailed discussion is given in Ehlers and Kundt [71].

The nondiverging type D fields are also completely known. The general solution contains just two arbitrary parameters, and can be written as:

$$ds^2 = 2\,du\,d\Sigma - f^{-1}\,dx^2 - f\,dy^2 \tag{44}$$

where

$$f(x) = \frac{2amx + l(a^2 - x^2)}{2a(x^2 + a^2)}$$

$$d\Sigma = \left(\frac{r^2 l}{2a(x^2 + a^2)} - \frac{2r^2 a^2 f}{(x^2 + a^2)^2} \right) du + dr - \tag{45}$$

$$- \frac{2rx}{x^2 + a^2}dx + \frac{arf}{x^2 + a^2}dy$$

Here the parameter a is redundant, and may be rescaled to unity. The metric was first given by Carter [72] and later shown by Kinnersley [73] to exhaust the class. Bardeen [74] has discussed the role played by this solution as a limiting metric which is valid in the immediate vicinity of an extreme Kerr throat. Included in this class of solutions is a trio of metrics which can be obtained from Eq. (44) by performing the singular limit $a, l \to 0$ in various ways. They are known as the static B-metrics, and listed in Jordan, Ehlers and Kundt [7].

Nondiverging type II fields have not yet been studied in detail. I know of only one solution, the stationary van Stockum metric (see Sec. 2). It is an example also of a metric of the generalized Kerr–Schild [91] type, with a static B-metric as its background space.

3.2. Nonrotating rays

The case in which the rays have divergence but no rotation was treated by Robinson and Trautman [75] (RT). All solutions of this class are "known," although some are much better known than others! The form of the metric is

$$ds^2 = H\,du^2 + 2\,du\,dr - r^2\,d\sigma^2$$

$$H = -\frac{2m}{r} + K + \frac{2r}{3m}\nabla^2 K \tag{46}$$

$$\dot{m} + 3m\dot{K} + 2K\,\nabla^2 K + \nabla^4 K = 0$$

where $d\sigma^2$ is the line element of an evolving 2-space Σ with Laplacian operator ∇^2 and Gaussian curvature K, and $m(u)$ denotes the mass.

To obtain type N, we put $m = 0$, $K = +1$. Then Σ is isometric to a sphere, and the solutions are determined by the set of all isometric mappings of a sphere into itself, with arbitrary time dependence. Unfortunately one-to-one mappings yield a metric which is flat, and consequently for non-trivial solutions the embedding of Σ must have at least one singular point. This angular singularity prevents the waves from representing realistic spherical radiation. (However, it should be emphasized that this statement applies only to type N, and many RT solutions exist which do *not* have angular singularities.)

Type III is much like type N, but with the relaxed condition

$$\nabla^2 K = 0$$

This equation has only one solution which can be written down exactly, and it leads to an infinite (but not exhaustive) class of type III RT solutions discussed by Foster and Newman [76], Brans [77], and Cahen [78]. To find the remaining solutions would require numerical analysis which no one has done; however, the equation makes clear that K itself must have singular points, and hence these fields also cannot represent realistic waves.

The only type D RT solutions are Schwarzschild, and its two-parameter generalization: the "static C metric" listed in Jordan, Ehlers and Kundt [7]:

$$ds^2 = (x + y)^{-2} (F \, dt^2 - F^{-1} \, dy^2 - G^{-1} \, dx^2 - G \, dz^2)$$

$$G(x) = 1 - x^2 - 2mAx^3 - e^2 A^2 x^4 \qquad (47)$$

$$F(y) = - G(-y)$$

Kinnersley and Walker [79] discussed its interpretation as the field of a uniformly accelerating point mass. In general it does have an angular singularity; however, for the "extreme" charged case $e = m$ it *does not*. The solution contains both incoming and outgoing electromagnetic and gravitational radiation. While runaway charges are not to be found in the real world, this metric still represents one of the most realistic radiating solutions we have to date. It also suggests that charged versions of some other RT solutions may be less singular than the vacuum version.

Type II RT solutions are the most general case. Foster and Newman [76] analyze those which are perturbations of Schwarzschild. They are smooth in the angular variables, and when analyzed into spherical harmonics, each mode is found to have an exponentially damped time-dependence. The damping time is of the order of the unperturbed Schwarzschild mass. This sounds like quite desirable behavior for a realistic radiating solution, despite the very special time-dependence. However, several objections have been raised. One is that the exponentials blow up in the infinite past and there is no way of deciding on the presence of incoming radiation.

Another is the essential singularity that the perturbation exhibits near the Schwarzschild throat. For both of these reasons the RT solutions lie outside the class of "regular Schwarzschild perturbations" as studied by Regge and Wheeler [80] and others [81–83].

3.3. Rotating rays

In case the rays have rotation our results are much less complete. To my knowledge, no exact type III or type N rotating solutions have ever been found! The problem has been reduced to a very simple set of equations in each case. Kerr's equations [66] for type III are

$$\partial_u D^* D^* \Omega + (\partial_u D\Omega)(\partial_u D^* \Omega^*) = 0$$

$$D^* D^* D\Omega - DDD^* \Omega^* = 0 \tag{48}$$

$$D = \partial_\zeta - \Omega\, \partial_u$$

where Ω and ζ are complex. For type N, Exton [84] has obtained a similar set of lower order,

$$D^*\, \partial_u D\Omega = 0$$

$$D^* D\Omega - DD^* \Omega^* = 0 \tag{49}$$

Type D was completely solved by Kinnersley [73], and the results are shown in Fig. 6. The most general type D vacuum metric, called C–NUT, is stationary and axially symmetric. It is determined by four arbitrary parameters representing mass, "magnetic" mass, spin, and acceleration. The type D metrics previously obtained:

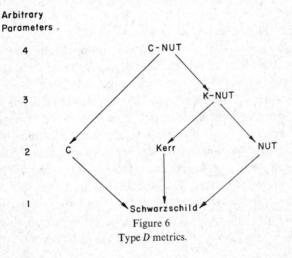

Figure 6

Type D metrics.

Kerr, NUT, K–NUT and the C metric, may be gotten by setting one or another of the parameters to zero. Debever [85] and now Plebanski and Demiański [86] have presented the solution in progressively simpler forms. With a slight change of notation, the Plebanski–Demiański version is

$$ds^2 = (x + y)^{-2} \{F(dt - x^2 \, dz)^2 - F^{-1} \, dy^2 - G^{-1} \, dx^2 - G(dz + y^2 \, dt)^2\}$$

$$F = (1 + x^2y^2)^{-1} \{(g^2 - \gamma - \tfrac{1}{6}\Lambda) + 2ly + \varepsilon y^2 + 2my^3 + (e^2 + \gamma - \tfrac{1}{6}\Lambda) \, y^4\} \quad (50)$$

$$G = (1 + x^2y^2)^{-1} \{-(g^2 - \gamma - \tfrac{1}{6}\Lambda) + 2lx - \varepsilon x^2 + 2mx^3 - (e^2 + \gamma - \tfrac{1}{6}\Lambda) \, x^4\}$$

This includes the electric and magnetic charge e, g and the cosmological constant Λ. The mass parameters are m and l, and ε, γ are related to the spin and acceleration. Upon specialization it reduces directly to either the C metric, Eq. (47), or to the form of K–NUT space given by Carter [72]. We should also mention that Hughston and Sommers [87] have recently given a short and elegant explanation of the fact that type D metrics always have two Killing vectors. Type D metrics have also been discussed by Cahen and Sengier [88].

Rotating type II solutions have been vigorously pursued by Robinson and his co-workers, and the results to date are shown in Fig. 7 (co-workers fell by the wayside as the generality increased!). Starting off is the Kerr–Debney [89] solution, a stationary three-parameter generalization of Kerr. Demiański [90] has independently discovered it, and studied some of its curious properties. For example, its complex divergence has the form

$$\rho = -\left(r + iaP_1(\theta) + ibQ_1(\theta)\right)^{-1} \quad (51)$$

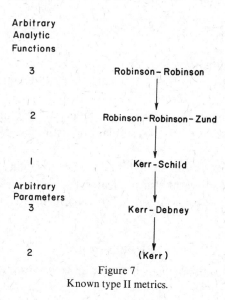

Figure 7
Known type II metrics.

where P_1 and Q_1 are the two kinds of Legendre function. Instead of the familiar Kerr disc, we find the Kerr–Debney singularity extending along the symmetry axis, in the form of an infinite spiral staircase. The fact that it is type II is also rather unusual, because a stationary metric would "normally" be either type I or type D.

The Kerr–Schild (KS) solutions [91] are those whose metric has the form

$$g_{\mu\nu} = \eta_{\mu\nu} + H l_\mu l_\nu \tag{52}$$

where l_μ is a null vector tangent to a shear free geodesic. They are all stationary, and have angular singularities due to the presence of one harmonic function of solid angle. In this case the function is the rotation itself. The only case in which the singularity does not appear is the Kerr metric. A very restricted class of KS metrics has been studied from a different viewpoint by Schiffer et al. [92].

The Robinson–Robinson–Zund (RRZ) solutions [93–95] are stationary solutions which generalize Kerr–Schild. They have two harmonic functions of angle, the added one being the mass aspect which was constant for KS. Further generalizations were made by the Robinsons to a nonstationary class of metrics [96–98]. The key assumption which leads to RR metrics is that the Riemann tensor must fall off asymptotically as r^{-3}. Thus they are all nonradiative. Indeed, they have the polynomial time-dependence which is typical of nonradiative solutions.

Finally, the type I fields with shearing but geodesic rays have been studied by similar techniques. Newman and Tamburino [99] found a two-parameter static solution in this category as well as a nonstatic one. Unti and Rorrence [100] showed that all others must be asymptotically cylindrical. Kota and Perjés [101] found the only stationary solution to be a metric with three parameters.

Notes Added in Proof

1. A third accelerating-particle solution was discussed in W. Israel and K. Khan, "Collinear Particles and Bondi Dipoles in General Relativity," *Nuovo Cim.* **33**, 331 (1964).

2. Professor Robinson informs me that a twisting type III solution may be obtained from [96] Eq. (5.18) by setting $v = 0$. A similar but distinct one is reported in a recent preprint by A. Held.

3. A twisting type N solution, something that has long eluded us, has now been found by I. Hauser and submitted for publication.

4. The Islam three-variable solution [57] has been shown to be reducible to the Schwarzschild solution.

REFERENCES

1. D. BRILL and S. DESER, "Instability of Closed Spaces in General Relativity," *Commun. Math. Phys.* **31**, 291 (1973).

2. A. FISCHER and J. MARSDEN, "Linearization Stability of the Einstein Equations," *Bull. Am. Math. Soc.* **79**, 997 (1973).

3. P. COLLINS and R. WILLIAMS, "Dynamics of the Friedmann Universe Using Regge Calculus," *Phys. Rev.* **D7**, 965 (1973).

4. R. SORKIN, "Development of Simplectic Methods for the Metrical and Electromagnetic Fields," Ph.D. Thesis, California Institute of Technology, 1974 (unpublished).

5. D. KRAMER, G. NEUGEBAUER and H. STEPHANI, "Konstruktion und Charakterisierung von Gravitationsfeldern," *Fortschritte der Physik* **20**, 1 (1972).

6. A. KRASINSKI, "Some Solutions of the Einstein Field Equations for a Rotating Perfect Fluid," *J. Math. Phys.* **16**, 125 (1975).

7. P. JORDAN, J. EHLERS and W. KUNDT, *Akad. Wiss. Mainz, Abh. Math. Nat. Kl. Jahg.* 1960, No. 2.

8. K. THORNE, "Geometrodynamics of Cylindrical Systems," Ph.D. Thesis, Princeton University, 1965 (unpublished).

9. P. SZEKERES, "Colliding Plane Gravitational Waves," *J. Math. Phys.* **13**, 286 (1972).

10. E. KASNER, "Solutions of the Einstein Equations Involving Functions of Only One Variable," *Trans. Am. Math. Soc.* **27**, 155 (1925).

11. A. TOMIMATSU and H. SATO, "New Exact Solution for the Gravitational Field of a Spinning Mass," *Phys. Rev. Lett.* **29**, 1344 (1972).

12. A. TOMIMATSU and H. SATO, "New Series of Exact Solutions for Gravitational Fields of Spinning Masses," *Prog. Theor. Phys. (Kyoto)* **50**, 95 (1973).

13. F. ERNST, "New Formulation of the Axially Symmetric Gravitational Field Problem. I," *Phys. Rev.* **167**, 1175 (1968).

14. F. ERNST, "New Formulation of the Axially Symmetric Gravitational Field Problem. II," *Phys. Rev.* **168**, 1415 (1968).

15. E. GLASS, "Structure of the Tomimatsu-Sato Gravitational Field," *Phys. Rev.* **D7**, 3127 (1973).

16. G. GIBBONS and R. RUSSELL-CLARK, "Note on the Sato-Tomimatsu Solution of Einstein's Equations," *Phys. Rev. Lett.* **30**, 398 (1973).

17. M. ABRAMOWICZ, J. LASOTA and M. DEMIAŃSKI, "A Note on Tomimatsu-Sato Metric," Preprint No. 26, Polish Academy of Sciences, 1973.

18 R. GAUTREAU and J. ANDERSON, "Directional Singularities in Weyl Gravitational Fields," *Phys. Lett.* **25A**, 291 (1967).

19. P. SZEKERES and F. MORGAN, "Extensions of the Curzon Metric," *Commun. Math. Phys.* **32**, 313 (1973).

20. A. TOMIMATSU and H. SATO, "On the Event Horizon of the TS Metrics," Preprint RIFP-177, Kyoto University, 1973.

21. W. KINNERSLEY and E. KELLEY, "Limits of the Tomimatsu-Sato Gravitational Field," *J. Math. Phys.* **15**, 2121 (1974).

22. W. ISRAEL, "Source of the Kerr Metric," *Phys. Rev.* **D2**, 641 (1970).

23. J. EHLERS, "Konstruktionen und Charakterisierung der Einsteinschen Gravitationsfeldgleichungen," Dissertation, Hamburg, 1957.

24. R. MATZNER and C. MISNER, "Gravitational Field Equations for Sources with Axial Symmetry and Angular Momentum," *Phys. Rev.* **154**, 1229 (1967).

25. R. MATZNER and Y. NUTKU, "On Stationary Axisymmetric Solutions of the Einstein Field Equations," *Ap. J.* **167**, 149 (1971).

26. G. NEUGEBAUER and D. KRAMER, "Eine Methode zur Konstruktion Stationärer Einstein-Maxwell Felder," *Ann. Physik* **24**, 62 (1969).

27. W. KINNERSLEY, "Generation of Stationary Einstein-Maxwell Fields," *J. Math. Phys.* **14**, 651 (1973).

28. R. GEROCH, "A Method for Generating New Solutions of Einstein's Equations. I," *J. Math. Phys.* **12**, 918 (1971).

29. R. GEROCH, "A Method for Generating New Solutions of Einstein's Equations. II," *J. Math. Phys.* **13**, 394 (1972).

30. B. K. HARRISON, "New Solutions of the Einstein-Maxwell Field Equations from Old," *J. Math. Phys.* **9**, 1744 (1968).

31. W. BONNOR, "An Exact Solution of the Einstein-Maxwell Equations Referring to a Magnetic Dipole," *Z. Phys.* **190**, 444 (1966).

32. Z. PERJÉS, "A Method for Constructing Certain Axially Symmetric Einstein-Maxwell Fields," *Nuovo Cim.* **55B**, 600 (1968).

33. R. MISRA, D. PANDEY, D. SRIVASTAVA and S. TRIPATHI, "New Class of Solutions of the Einstein-Maxwell Fields," *Phys. Rev.* **D7**, 1587 (1973). Erratum: *Phys. Rev.* **D8**, 1942 (1973).

34. A. PAPAPETROU, "Eine Rotationssymmetrische Lösung in der Allgemeinen Relativitätstheorie," *Ann. Physik* (6) **12**, 309 (1953).

35. H. WEYL, "Zur Gravitationstheorie," *Ann. Physik* **54**, 117 (1917).

36. W. BONNOR, "A Class of Stationary Solutions of the Einstein-Maxwell Equations," *Commun. Math. Phys.* **34**, 77 (1973).

37. W. BONNOR, "Certain Exact Solutions of the Equations of General Relativity with an Electrostatic Field," *Proc. Phys. Soc.* **66A**, 145 (1953).

38. L. WITTEN, "A Geometric Theory of the Electromagnetic and Gravitational Fields," in *Gravitation: An Introduction to Current Research* (L. WITTEN, ed.), Wiley, 1962.

39. W. VAN STOCKUM, "The Gravitational Field of a Distribution of Particles Rotating about an Axis of Symmetry," *Proc. Roy. Soc. Edinburgh* **A57**, 135 (1937).

40. T. LEWIS, "Some Special Solutions of the Equations of Axially Symmetric Gravitational Fields," *Proc. Roy. Soc. (London)* **A136**, 176 (1932).

41. R. TIWARI and M. MISRA, "The Gravitational and Electromagnetic Fields Due to Rotating Bodies in General Relativity Theory," *Proc. Nat. Inst. Sci. India* **A28**, 771 (1962).

42. R. HOFFMAN, "Stationary Axially Symmetric Generalizations of the Weyl Solutions in General Relativity," *Phys. Rev.* **182**, 1361 (1969).

43. B. DATTA and A. RAYCHAUDHURI, "Stationary Electromagnetic Fields in General Relativity," *J. Math. Phys.* **9**, 1715 (1968).

44. R. GAUTREAU and R. HOFFMAN, "Class of Exact Solutions of the Einstein-Maxwell Equations," *Phys. Rev.* **D2**, 271 (1970).

45. D. KRAMER and G. NEUGEBAUER, "Eine Exakte Stationäre Lösung der Einstein-Maxwell Gleichungen," *Ann. Physik* **24**, 59 (1969).

46. P. ESPOSITO and L. WITTEN, "A Five-Parameter Exterior Solution of the Einstein-Maxwell Equations," *Phys. Rev.* **D8**, 3302 (1973).

47. F. ERNST, "Charged Version of Tomimatsu–Sato Spinning Mass Field," *Phys. Rev.* **D7**, 2520 (1973).

48. M. Wang, "Class of Solutions of Axially-Symmetric Einstein-Maxwell Equations," *Phys. Rev.* **D9**, 1835 (1974).

49. A. Papapetrou, "A Static Solution of the Equations of the Gravitational Field for an Arbitrary Charge Distribution," *Proc. Roy. Irish Acad.* **A51**, 191 (1947).

50. S. Majumdar, "A Class of Exact Solutions of Einstein's Field Equations," *Phys. Rev.* **72**, 390 (1947).

51. Z. Perjés, "Solutions of the Coupled Einstein–Maxwell Equations Representing the Fields of Spinning Sources," *Phys. Rev. Lett.* **27**, 1688 (1971).

52. W. Israel and G. Wilson, "A Class of Stationary Electromagnetic Vacuum Fields," *J. Math. Phys.* **13**, 865 (1972).

53. W. Bonnor and N. Swaminarayan, "An Exact Solution for Uniformly Accelerated Particles in General Relativity," *Z. Physik* **177**, 240 (1964).

54. D. Zipoy, "Topology of Some Spheroidal Metrics," *J. Math. Phys.* **7**, 1137 (1966).

55. J. Marek, "Some Solutions of Einstein's Equations in General Relativity Theory," *Proc. Camb. Phil. Soc.* **64**, 167 (1968).

56. B. K. Harrison, "Exact Three-Variable Solutions of the Field Equations of General Relativity," *Phys. Rev.* **116**, 1285 (1959).

57. J. Islam, "An Exact Three-Variable Solution of Einstein's Vacuum Field Equations," Preprint, University of Washington, Seattle, 1973. (See added notes.)

58. B. K. Harrison, "Electromagnetic Solutions of the Field Equations of General Relativity," *Phys. Rev.* **138**, B488 (1965).

59. R. D'Inverno and R. Russell-Clark, "Classification of the Harrison Metrics," *J. Math. Phys.* **12**, 1258 (1971).

60. H. Levy, "Classification of Stationary Axisymmetric Gravitational Fields," *Nuovo Cim.* **56B**, 253 (1968).

61. Z. Perjés, "Three-Dimensional Relativity for Axisymmetric Stationary Space-Times," *Commun. Math. Phys.* **12**, 275 (1969).

62. R. Penney, "Axially Symmetric Zero-Mass Meson Solutions of Einstein's Equations," *Phys. Rev.* **174**, 1578 (1968).

63. H. Buchdahl, "Static Sources in the Brans–Dicke Theory," *Gen. Rel. and Grav. J.* **4**, 319 (1973).

64. G. Sneddon and C. McIntosh, "Generation of Solutions of Brans–Dicke Equations," Preprint, Monash University, Clayton, Australia, 1973.

65. C. McIntosh, "A Family of Brans–Dicke–Kerr Solutions," Preprint, Monash University, Clayton, Australia, 1973.

66. R. Kerr, "Gravitational Field of a Spinning Mass as an Example of Algebraically Special Metrics," *Phys. Rev. Lett.* **11**, 237 (1963).

67. W. Kundt, "The Plane-Fronted Gravitational Waves," *Z. Physik* **163**, 77 (1961).

68. N. Rosen, "Plane Polarized Waves in the General Theory of Relativity," *Phys. Z. Sowjetunion* **12**, 366 (1937).

69. H. Bondi, F. Pirani and I. Robinson, "Gravitational Waves in General Relativity. III. Exact Plane Waves," *Proc. Roy. Soc. (London)* **A251**, 519 (1959).

70. C. Misner, K. Thorne and J. Wheeler, *Gravitation*, Freeman, 1973.

71. J. Ehlers and W. Kundt, "Exact Solutions of the Gravitational Field Equations," in *Gravitation: An Introduction to Current Research* (L. Witten, ed.), Wiley, 1962.

72. B. CARTER, "A New Family of Einstein Spaces," *Phys. Lett.* **26A**, 399 (1968).

73. W. KINNERSLEY, "Type D Vacuum Metrics," *J. Math. Phys.* **10**, 1195 (1969).

74. J. BARDEEN, letter to C. MISNER, 1972 (unpublished).

75. I. ROBINSON and A. TRAUTMAN, "Some Spherical Gravitational Waves in General Relativity," *Proc. Roy. Soc. (London)* **A265**, 463 (1962).

76. J. FOSTER and E. T. NEWMAN, "Note on the Robinson–Trautman Solutions," *J. Math. Phys.* **8**, 189 (1967).

77. C. BRANS, "Complex Two-Form Representation of the Einstein Equations: The Petrov Type III Solutions," *J. Math. Phys.* **12**, 1616 (1971).

78. M. CAHEN and J. SPELKENS, "Espaces de Type III Solutions des Equations de Maxwell–Einstein," *Bull. Acad. R. Belg.* (5)**53**, 817 (1967).

79. W. KINNERSLEY and M. WALKER, "Uniformly Accelerating Charged Mass in General Relativity," *Phys. Rev.* **D2**, 1359 (1970).

80. T. REGGE and J. WHEELER, "Stability of a Schwarzschild Singularity," *Phys. Rev.* **108**, 1063 (1957).

81. C. VISHVESHWARA, "Stability of the Schwarzschild Metric," *Phys. Rev.* **D1**, 2870 (1970).

82. F. ZERILLI, "Effective Potential for Even Parity Regge–Wheeler Gravitational Perturbation Equations," *Phys. Rev. Lett.* **24**, 737 (1970).

83. R. PRICE, "Nonspherical Perturbations of Relativistic Gravitational Collapse. I. Scalar and Gravitational Perturbations," *Phys. Rev.* **D5**, 2419 (1972).

84. A. EXTON, private communication 1969.

85. R. DEBEVER, "On Type *D* Expanding Solutions of Einstein–Maxwell Equations," *Bull. Soc. Math. de Belg.* **23**, 360 (1971).

86. J. PLEBANSKI and M. DEMIANSKI, "Rotating, Charged and Uniformly Accelerating Mass in General Relativity," submitted to *J. Math. Phys.*, and *Lett. al Nuovo Cim.* **9**, 202 (1974).

87. L. HUGHSTON and P. SOMMERS, "Spacetimes with Killing Tensors," *Commun. Math. Phys.* **32**, 147 (1973).

88. M. CAHEN and J. SENGIER, "Espaces de Classe D Admettant un Champ Electromagnetique," *Bull. Acad. R. Belg.* (5)**53**, 801 (1967).

89. R. KERR and G. DEBNEY, "Einstein Spaces with Symmetry Groups," *J. Math. Phys.* **11**, 2807 (1970).

90. M. DEMIANSKI, "New Kerr-Like Space-Time," *Phys. Lett.* **42A**, 157 (1972).

91. G. DEBNEY, R. KERR and A. SCHILD, "Solutions of the Einstein and Einstein–Maxwell Equations," *J. Math. Phys.* **10**, 1842 (1969).

92. M. SCHIFFER, R. ADLER, J. MARK and C. SHEFFIELD, "Kerr Geometry as Complexified Schwarzschild Geometry," *J. Math. Phys.* **14**, 52 (1973).

93. I. ROBINSON, J. ROBINSON and J. ZUND, "Degenerate Gravitational Fields with Twisting Rays," *J. Math. Mech.* **18**, 881 (1969).

94. J. ZUND, "Degenerate Gravitational Fields with Twisting Rays, II," *Lett. al Nuovo Cim.* **4**, 879 (1972).

95. J. ZUND, "Degenerate Gravitational Fields with Twisting Rays, III," *Lett. al Nuovo Cim.* **7**, 233 (1973).

96. I. ROBINSON and J. ROBINSON, "Vacuum Metrics without Symmetry," *Int. J. Theor. Phys.* **2**, 231 (1969).

97. D. TRIM and J. WAINWRIGHT, "Nonradiative Algebraically Special Space-Times," *J. Math. Phys.* **15**, 535 (1974).

98. I. ROBINSON, A. SCHILD and H. STRAUSS, "A Generalized Reissner–Nordstrom Solution," *Int. J. Theor. Phys.* **2**, 243 (1969).

99. E. T. NEWMAN and L. TAMBURINO, "Empty-Space Metrics Containing Hypersurface-Orthogonal Shearing Rays," *J. Math. Phys.* **3**, 902 (1962).

100. T. UNTI and R. TORRENCE, "Theorem on Gravitational Fields with Geodesic Rays," *J. Math. Phys.* **7**, 535 (1966).

101. J. KOTA and Z. PERJÉS, "All Stationary Vacuum Metrics with Shearing Geodesic Eigenrays," *J. Math. Phys.* **13**, 1695 (1972).

The Bondi–Metzner–Sachs Group:
Its Complexification and Some Related
Curious Consequences

E. T. NEWMAN

University of Pittsburgh, Pittsburgh, Pennsylvania, U.S.A.

The Bondi–Metzner–Sachs (BMS) group is (to my knowledge) the only group in use in general relativity, which can be considered as a symmetry group in some sense, and still be applicable to a wide class of physical situations, namely to all asymptotically flat Einstein or Einstein–Maxwell space-times. Due to the fact that the BMS group is so similar in structure [1, 2] to the Poincaré group (both being the semi-direct product of the homogeneous Lorentz group with an abelian group, the abelian part being four parameter for the Poincaré group and infinite parameter for the BMS group) there has been considerable hope that it would play an important role, either via its representation theory [1, 3, 4] or through its reduction to the Poincare group [5, 6] in a quantized version of general relativity or even in particle physics.

The BMS group first arose [7, 8] as the set of coordinate transformations which preserved some natural appearing coordinate conditions in the neighborhood of future null infinity; shortly afterwards it was recognized [1, 9] that the group could be understood as, in some sense, an approximate symmetry group and satisfied an asymptotic version of the Killing equation; finally Penrose [10, 11] and from a slightly different point of view, Winicour [12] gave a precise and clear statement of the geometric origin of the BMS group and its relation to symmetries.

We will now give a brief and intuitive review of the Penrose version of the BMS group. (The interested reader should see [11] for the formal statements and proofs.)

The basic entity which we must deal with is future null infinity (or if we had desired, past null infinity) referred to as $\mathscr{I}^+$. $\mathscr{I}^+$ is a three dimensional manifold $(S^2 \times R)$ which is the boundary of the conformally compactified asymptotically flat space-time. Intuitively we can think of $\mathscr{I}^+$ as the points added to the physical space-time by taking the future directed limit along each null generator of an arbitrary family of retarded null surfaces.

$\mathscr{I}^+$ possesses the following properties which can be derived [11] from the assumption that we are dealing with an asymptotically flat solution to the vacuum Einstein (or Einstein–Maxwell) equation.

(a) $\mathscr{I}^+$ possesses a *degenerate* conformal metric with signature $(0, +, +)$ such that there exists a choice of conformal factor and coordinate system in which the metric takes the form

$$dl^2 = o \cdot du^2 + \frac{4d\rho \, d\bar{\rho}}{(1 + \rho\bar{\rho})^2} \tag{1}$$

with ρ and $\bar{\rho}$ being the complex stereographic coordinates of the sphere. Intuitively (1) could be understood as the limit of the physical metric ds in a null coordinate system divided by the affine length r (along the null rays) as $r \to \infty$, i.e.,

$$dl^2 = \lim_{r \to \infty} \frac{ds^2}{r^2} \tag{2}$$

If we had chosen a different null coordinate system in (2) this would have induced a conformal rescaling of $\mathscr{I}^+$.

(b) There further exists, for each choice of conformal factor, a preferred "length" u along each (null) generator of $\mathscr{I}^+$, i.e., along the rays ρ and $\bar{\rho}$ const, such that the ratio of du/dl is invariant under conformal rescalings. By du here we mean the following; consider two displacements on $\mathscr{I}^+$ both leading from a point on one generator to a second generator, i.e., they both have the same dl but have different positions on the second generator, the difference being du.

We henceforth adopt this preferred u as the coordinate which labels cuts of $\mathscr{I}^+$. This u when used with the conformal factor and coordinate system associated with (1) leads to what is referred to as a Bondi-type coordinate system.

The BMS group is now defined as those mappings of $\mathscr{I}^+ \to \mathscr{I}^+$ which preserve both angles defined by the degenerate metric (1) and the ratio du/dl, sometimes [11] referred to as a null angle. In a Bondi frame one obtains for the mapping

$$u \to u' = K\big(u + \alpha(\rho, \bar{\rho})\big) \tag{3}$$

$$\rho \to \rho' = \frac{\alpha\rho + \beta}{\gamma\rho + \delta}, \qquad \alpha\delta - \beta\gamma = 1 \tag{4}$$

where $\alpha(\rho, \bar{\rho})$ is an arbitrary real regular function on the sphere and K is defined from

$$\frac{d\rho' d\bar{\rho}'}{(1 + \rho'\bar{\rho}')^2} = K^2 \frac{d\rho \, d\bar{\rho}}{(1 + \rho\bar{\rho})^2} \tag{5}$$

The homogeneous Lorentz transformations are given by (4) while the infinite parameter abelian group is described by the $\alpha(\rho, \bar{\rho})$ and referred to as a super-translation. (If the α is chosen to have only $l = 0$ and $l = 1$ spherical harmonics the BMS group reduces to the Poincaré group. In general there is no canonical way to make this reduction [5, 6].)

An additional property of $\mathscr{I}^+$ which is important for us is the following:

(c) Using a Bondi type coordinate system, there exists (associated with the slicing) a complex function $\sigma^0(u, \rho, \bar{\rho})$ on $\mathscr{I}^+$, which from the physical space point of view is a measure of the asymptotic shear of the null cones which intersect $\mathscr{I}^+$ at $u = $ const ($\dot{\sigma}^0$ is the Bondi news function). Furthermore under the super-translation subgroup of the BMS group, namely $\rho' = \rho$, $u' = u + \alpha(\rho, \bar{\rho})$ to the new cuts $u' = $ const, the asymptotic shear associated with the new cuts becomes

$$\sigma'^0(u', \rho, \bar{\rho}) = \sigma^\circ(u' - \alpha, \rho, \bar{\rho}) + \partial^2\alpha \qquad (6)$$

Two important questions for us are, under what circumstances can we find α's or new cuts $u' = $ const such that $\sigma'^0 = 0$ and how many such α's exist.

In the case of stationary space-times one can show rather easily that there exists a four parameter family of α's, or equivalently a four parameter family of null surfaces, such that their asymptotic shear vanishes. We refer to each one of these surfaces or its intersection with $\mathscr{I}^+$ as a "good" cone or a "good" cut. The set or space of good cuts is a four dimensional manifold and will be referred to as $\mathscr{H}_s$ or stationary heaven. We remark that in the case of stationary space-times the BMS group can be reduced to the Poincaré group by asking for the subgroup of the BMS transformations which map the good cuts into themselves. It is thus convenient to think of the $\mathscr{H}_s$ (for stationary space-times) as being isometric to Minkowski space. In fact one can show that for the proper choice of coordinates (x^μ) on $\mathscr{H}_s$, starting from two good cuts $\alpha_1 = \alpha(x_1^\mu, \rho, \bar{\rho})$ and $\alpha_2 = \alpha(x_2^\mu, \rho, \bar{\rho})$ one has the rather surprising result

$$\eta_{\mu\nu}(x_2^\mu - x_1^\mu)(x_2^\nu - x_1^\nu) = \left[\frac{1}{16\pi i} \int \frac{d\Omega}{(\alpha_1 - \alpha_2)^2} \right]^{-1} \qquad (7)$$

where $d\Omega$ is the surface area element on the sphere.

If the stationary space-time were specialized further to Minkowski space then a "good" cone can be shown to be simply the ordinary light cone from an interior point and one really has a one-to-one relation between points of $\mathscr{H}_s$ and Minkowski space points. In other cases $\mathscr{H}_s$ is unrelated to actual space-time points.

When we are no longer dealing with stationary metrics but with general asymptotically flat spaces the situation is much more difficult. First of all there exist, in general, *no cuts* of $\mathscr{I}^+$ such that $\sigma'^0(u', \rho, \bar{\rho}) = 0$, i.e., there exist no good cones. Though that appears to end the search for good cones, there is an unusual way out that leads to some curious results.

We consider the analytic extension of $\mathscr{I}^+$ to a three complex-dimensional manifold $\mathscr{I}_c^+$ with the obvious generalizations of the properties (a), (b) and (c).

Working with a Bondi-type coordinate system and conformal factor we have; (a′) there exists a degenerate conformal metric (which is complex)

$$dl^2 = o \cdot du^2 + \frac{4d\rho\, d\tilde{\rho}}{(1 + \rho\tilde{\rho})^2} \tag{8}$$

where u, ρ, and $\tilde{\rho}$ are three *independent* complex variables ($\tilde{\rho}$ is *not* the complex conjugate of ρ).

(b′) There exists a preferred complex u along each generator, ρ and $\tilde{\rho}$ constant.

(c′) There exists a complex function $\sigma^0(u, \rho, \tilde{\rho})$ on $\mathscr{I}_c^+$ which is the analytic extension of $\sigma^0(u, \rho, \bar{\rho})$ on $\mathscr{I}^+$ and which can be viewed as the asymptotic shear of the null cones (of the analytically extended physical space-time) which intersect $\mathscr{I}_c^+$ at $u = \text{const}$ (complex).

From (a′) and (b′) we obtain the complex version of the BMS group as the transformations which preserve complex angles and the ratio du/dl, namely

$$u' = K\big(u + \alpha(\rho, \tilde{\rho})\big) \tag{9}$$

$$\rho' = \frac{\alpha\rho + \beta}{\gamma\rho + \delta}, \qquad \tilde{\rho}' = \frac{\tilde{\alpha}\tilde{\rho} + \tilde{\beta}}{\tilde{\gamma}\tilde{\rho} + \tilde{\delta}} \tag{10}$$

$$\alpha\delta - \beta\gamma = \tilde{\alpha}\tilde{\delta} - \tilde{\beta}\tilde{\gamma} = 1$$

where the variables with tilde are independent of the variables without tilde and where K is defined from

$$\frac{d\rho'\, d\tilde{\rho}'}{(1 + \rho'\tilde{\rho}')^2} = K^2 \frac{d\rho\, d\tilde{\rho}}{(1 + \rho\tilde{\rho})^2} \tag{11}$$

and $\alpha(\rho, \tilde{\rho})$ is an arbitrary regular *complex* function on the analytically extended sphere. (By regular we mean it is expandable in spherical harmonics when $\tilde{\rho} = \bar{\rho}$)

One then has the generalization of (6) to

$$\sigma'^0(u, \rho, \tilde{\rho}) = \sigma^0(u - \alpha, \rho, \tilde{\rho}) + \partial^2\alpha(\rho, \tilde{\rho}) \tag{12}$$

where α is now complex.

Though we do not have a rigorous proof, there appears to be little doubt that there exists a *four-complex parameter* family of cuts of $\mathscr{I}_c^+$ with $\sigma'^0 = 0$, i.e., that there exists a four-complex dimensional manifold $\mathscr{H}$ (heaven) of good cones or good cuts, which is a generalization of the $\mathscr{H}_s$ defined earlier for stationary space-times.

We state without proof the following curious and possibly startling properties of $\mathscr{H}$:

(i) If one takes two neighboring good cuts $\alpha_1 = \alpha(z^\mu, \rho, \tilde{\rho})$ and $\alpha_2 = \alpha(z^\mu + dz^\mu, \rho, \tilde{\rho})$ (with z^μ being the complex labels of the good cuts or points of $\mathscr{H}$) and their difference $d\alpha = \alpha_2 - \alpha_1$, then there exists the generalization of (7), namely

$$g_{\mu\nu}\, dz^\mu\, dz^\nu = \left[\frac{1}{16\pi i} \int \frac{d\Omega}{(d\alpha)^2} \right]^{-1} \tag{13}$$

i.e., we have induced on $\mathcal{H}$ a complex metric (which is non-degenerate) and thus $\mathcal{H}$ becomes a complex Riemannian manifold.

(ii) The metric is conformal to a (complex) vacuum solution of the Einstein equations [13], i.e., $g_{\mu\nu} = \theta^2 \mathring{g}_{\mu\nu}$ where $\mathring{g}_{\mu\nu}$ satisfies $R_{\mu\nu}(\mathring{g}_{\mu\nu}) = 0$. It is conceivable that it will eventually be proven that $\theta = 1$.

(iii) There does not exist [13] any real four dimensional subspaces of $\mathcal{H}$ given by $z^\mu = x^\mu$ (x^μ real) such that the induced metric (from (13)) is real, except when $\mathcal{H}$ is flat. When there is radiation in the physical space-time, the Weyl tensor of $\mathcal{H}$ is non-vanishing.

(iv) $\mathcal{H}$ lends itself to a natural description of the Penrose theory of asymptotic twistors [14].

Though we have as yet no clear physical interpretation for $\mathcal{H}$ and it might be nothing more than a mathematical curiosity, there are nevertheless several strong indications that $\mathcal{H}$ does have physical significance.

First of all if one believes that twistor theory is of physical importance then due to the intimate connection between twistors and $\mathcal{H}$, it would be highly unlikely that $\mathcal{H}$ could be without importance.

Of more immediate interest is the observation that some structures in $\mathcal{H}$ have direct geometric meaning in the physical space-time. In particular for every (complex) curve in $\mathcal{H}$ there exists in the space-time a null geodesic congruence, *which is asymptotically shear free.* For many asymptotically flat space-times (e.g., algebraically special ones) and probably for most, there exists a unique curve in $\mathcal{H}$ (which in some sense can be considered as the complex center of mass [15]) and thus in the physical space a unique asymptotic shear free congruence. The twist (or curl) of this congruence can be considered to be a direct measure of the spin angular momentum of the source. If, in addition, there is an asymptotically vanishing Maxwell field present, it appears likely that a second curve exists in $\mathcal{H}$ which can be considered as the complex center of charge [13]. Though it is not proven, *except for the charged Kerr metric*, it seems that if the two complex lines coincide then the gyromagnetic ratio of the source will have the Dirac value, i.e., e/mc.

There are several other indications of the possible physical use of $\mathcal{H}$ and there even exists a possible physical interpretation for it as a space for asymptotic observations. They are however still sufficiently tentative and nebulous that it is best to await their future development before describing them.

The details and proofs of the material presented here will be given in a paper being prepared with R. Penrose.

Note Added in Proof

The metric $g_{\mu\nu}$ of $\mathcal{H}$ does satisfy the vacuum Einstein equations.

Acknowledgment

The author would like to thank Prof. J. Ehlers, the Max Planck Institut für Physik und Astrophysik, and the Akademie der Wissenschaften und der Literatur in Mainz for their aid and encouragement, while this work was being completed. This work was supported by the NSF Grant No. 22789.

REFERENCES

1. R. K. SACHS, *Phys. Rev.* **128**, 2851 (1962).
2. R. GEROCH and E. T. NEWMAN, *J. Math. Phys.* **12**, 314 (1971.
3. V. J. CANTONI, *J. Math. Phys.* **7**, 1361 (1966); **8**, 1700 (1967).
4. P. J. MCCARTHY, *Proc. Roy. Soc. (London)* **A333**, 317 (1973).
5. E. T. NEWMAN and R. PENROSE, *J. Math. Phys.* **7**, 863 (1966).
6. B. SCHMIDT. Gravity Prize Essay, 1974.
7. H. BONDI, A. W. K. METZNER and M. J. G. VAN DER BURG, *Proc. Roy. Soc. (London)* **A269**, 21 (1962).
8. R. K. SACHS, *Proc. Roy. Soc. (London)* **A270**, 103 (1962).
9. E. T. NEWMAN and T. UNTI, *J. Math. Phys.* **3**, 891 (1962).
10. R. PENROSE, in *Relativity, Groups and Topology* (C. M. DE WITT and B. S. DE WITT, eds.), Les Houches Lectures 1963, Blackie & Son, London, 1964.
11. R. PENROSE, "Relativistic Symmetry Groups," in *Group Theory in Non-Linear Problems*, NATO Advanced Study Institutes Series C; Math & Phys. Sciences (A. O. BARUT, ed.), D. Reidel Publ. Co., Dordrecht, Holland, 1974. This should be considered as a major reference on symmetries in general relativity.
12. J. WINICOUR, *J. Math. Phys.* **9**, 861 (1968).
13. R. PENROSE, private communication.
14. R. PENROSE and M. MACCALLUM, Phys. Reports Sec. C of *Phys. Lett.* **6**, (1972).
15. E. T. NEWMAN and J. WINICOUR, *J. Math. Phys.,* to be published.

Numerical Integration of Exact Time-Dependent Einstein Equations with Axial Symmetry*

J. PACHNER

Department of Physics and Astronomy, University of Regina
Regina, Saskatchewan, Canada

ABSTRACT

Numerical integration of exact Einstein equations with axial symmetry is not an easy problem; therefore the reasons that justify such a difficult and sometimes tedious research are discussed first.

The solution of the problem of the numerical integration of exact time-dependent Einstein equations with axial symmetry may be divided into two parts: 1. Analytical part, in which the equations describing the behavior of a perfect fluid and the corresponding exterior field are reduced to the form suitable for the computer. 2. Development of the computer programs.

The extent of programming and of running the programs was limited by a very modest research grant. Therefore only two simple numerical examples illustrate the integration method. Further programming is needed before the method may be applied for solving astrophysical problems.

The paper is concluded by a brief discussion of an interesting relativistic effect.

INTRODUCTION

Numerical integration of exact Einstein equations with axial symmetry is not an easy problem; it is therefore appropriate first to mention the reasons that justify such a difficult and sometimes tedious research.

In the actual Universe one can hardly imagine an astronomical object which exhibits no rotation at all with respect to the background cosmic field. Because of the conservation of angular momentum, the angular velocity of a first slowly rotating and collapsing object steadily increases during the contraction and may reach at its late stages such a high value that the centrifugal acceleration can stop the radial contraction and revert it to a new expansion (a well-known Newtonian effect).

* Supported in part by the National Research Council of Canada.

143

However, since the Einstein field equations are of hyperbolic type, this high radial acceleration generates also a gravitational radiation that may decisively influence its axial contraction (a relativistic effect caused not by the non-linearity of Einstein equations, but by their hyperbolic character; remember the radiation generated in Maxwell electrodynamics by an accelerated motion of an electric charge).

It is obvious that in such a situation the deviations from spherical symmetry cannot be considered as small. The applicability of the Penrose theorem [1] on the inevitable occurrence of a gravitational collapse is thus seriously restricted in the actual Universe, because one of the assumptions upon which the theorem is based is "that the deviations from the above situation" (i.e., from the spherical symmetry) "are not too great". The assumption tacitly implies that the self-interaction of a collapsing object with its own gravitational radiation may be neglected.

The question whether the gravitational wave generated by the radial acceleration supports the axial contraction or acts in the opposite direction and how strong is this self-interaction may be decided by analytical methods, but the problem of the time evolution from a given initial situation can be solved only by a numerical integration of exact time-dependent Einstein equations with axial symmetry. The tensorial perturbation calculus [2] presumes a slow rotation which generates merely a weak gravitational radiation; it cannot be therefore applied in the latter stages of contraction when the rotation has become fast.

The solution of the problem of the numerical integration of Einstein equations may be divided into two parts: 1. Analytical part, in which the complete system of equations describing the behavior of a perfect fluid and the corresponding exterior field are reduced to the form suitable for the computer. 2. Development of the computer programs.

The integration method is thoroughly described in two larger papers [3, 4]; in the present communication only its main ideas can be expounded and its applicability illustrated by two simple numerical examples.

1. ANALYTICAL PART

Part 1 may be subdivided into the following steps:

1.1. The complete system of 13 equations describing the behavior of a perfect fluid under assumption that all the thermodynamical processes are adiabatic and no nuclear energy is being released is reduced in a particular system of comoving coordinates to six Einstein field equations, one equation of continuity, and four Lichnerowicz initial conditions.

1.2. Without loss of generality the choice of the comoving coordinate system is restricted by further conditions so that one physical situation with its own past gravitational history corresponds only to one set of initial data.

1.3. The field equations are written out in their explicit form.

1.4. The junction conditions and the field equations for the exterior field are derived.

1.5. Cauchy initial problem is properly formulated.

1.6. Approximate equations for the distance weak field zone are deduced.

1.7. The problem of Lichnerowicz initial conditions is solved.

Each of the steps has its own problems that are to be solved, but the most important of all is the proper formulation of the Cauchy initial data.

Definitions. The metric is assumed to have the signature $+2$ and a system of units is used in which the velocity of light c and the Newtonian constant of gravitation G equal to 1.

A comma indicates a partial derivative, but where there is no danger of confusion the comma is omitted. The Riemannian derivative is denoted by a semicolon. Greek indices run from 1 to 4, Latin indices from 1 to 3.

Let p indicate the pressure, ρ the proper rest mass density, and ε the proper internal energy per unit mass. The four-velocity will be denoted by $u^\mu = dx^\mu/ds$, and the usual substitution

$$g = \det g_{\alpha\beta}$$

applied throughout.

The components of the energy momentum tensor of the perfect fluid are defined by the expression

$$T_{\mu\nu} = e^\chi \rho u_\mu u_\nu + p g_{\mu\nu} \tag{1}$$

in which

$$e^\chi = 1 + \varepsilon + p/\rho \tag{2}$$

represents the proper enthalpy per unit mass.

1.1. The complete system of equations

The complete system consists of one equation of state (of any given form in which the pressure vanishes when the mass density equals to zero)

$$p = p(\rho, \varepsilon) \tag{3}$$

of one conservation law of baryon number, which reduces in the case under consideration to the equation of continuity

$$(-g)^{-1/2} \left[\rho u^\mu (-g)^{1/2} \right]_{,\mu} = 0 \tag{4}$$

of one normalization equation for the four-velocity

$$g_{\mu\nu} u^\mu u^\nu = -1 \tag{5}$$

of $(1 + 3)$ equations of motion

$$T^{\nu}_{4\,;\nu} = 0 \tag{6}$$

$$T^{\nu}_{i\,;\nu} = 0 \tag{7}$$

and of six independent Einstein field equations

$$R_{ik} = 8\pi(T_{ik} - \tfrac{1}{2}Tg_{ik}) \tag{8}$$

The initial conditions for the metric tensor cannot be chosen quite arbitrarily, but they must satisfy four Lichnerowicz initial conditions [5]

$$\tfrac{1}{2}I^4_\mu \equiv R^4_\mu - \tfrac{1}{2}R\delta^4_\mu - 8\pi T^4_\mu = 0 \tag{9}$$

Since the system of 13 equations has to be satisfied by 17 functions, i.e., by 3 quantities of state, 4 components of the four-velocity, and 10 components of the metric tensor, four coordinate conditions must now be added. The author succeeded to reduce the preceding equations to six Einstein equations (8) for six unknown components g_{ik} and to one algebraic equation

$$\rho = \Psi(x^j)\,(g_{44}/g)^{1/2} \tag{10}$$

for the unknown mass density ρ only in a particular system of comoving coordinates defined by the $(3 + 1)$ conditions [6, 7]

$$u^i = 0 \tag{11a}$$

$$(\partial/\partial x^i)\ln\left[e^x(-g_{44})^{1/2}\right] = 0 \tag{11b}$$

The internal energy ε may be considered as a known function, since it is given by the equation (following from (6))

$$d\varepsilon/d\rho = p/\rho^2 \tag{12}$$

expressing the conservation of energy in a perfect fluid at constant entropy, and four components $g_{\mu4}$ are determined by the formula (resulting from (7))

$$\left.\begin{array}{l} g_{\mu4} = -(A_\mu e^{-x})(A_4 e^{-x}) = -u_\mu u_4 \\ A_i = A_i(x^j), \qquad A_4 = A_4(x^4) \end{array}\right\} \tag{13}$$

The function $\Psi(x^j)$ is given by the initial distribution of the mass density. Under the coordinate transformation that preserve the coordinate conditions (11a, b),

$$\bar{x}^i = \bar{x}^i(x^1, x^2, x^3), \qquad \bar{x}^4 = \bar{x}^4(x^4) \tag{14}$$

the A_μ are components of a four-vector whose A_i components are determined by the initial distribution of vorticity.

1.2. Choice of the coordinate system

With the help of the coordinate transformations (14) it is always possible to reduce the given coordinate system to a new one in which the direction of the x^1-axis coincides at any point with the direction of the vorticity vector Ω^μ defined by the formula [8]

$$\Omega^\mu = \tfrac{1}{2}(-g)^{-1/2}\, \varepsilon^{\mu\alpha\beta\gamma} u_\alpha u_{\beta,\gamma}, \qquad \varepsilon^{\mu\alpha\beta\gamma} = \pm 1, 0 \tag{15}$$

so that

$$\Omega^1 \neq 0, \qquad \Omega^2 = \Omega^3 = \Omega^4 = 0 \tag{16}$$

and in which

$$A_1 = A_2 = 0, \qquad A_3 = A(x^2, x^3), \qquad |A_4| = 1 \tag{17}$$

at any moment, and at the initial moment only

$$g_{12} = g_{13} = 0 \quad \text{everywhere,} \tag{18a}$$

$$g_{23} = 0 \qquad \text{at } x^1 = 0 \tag{18b}$$

These requirements restrict the coordinate transformations to

$$\bar{x}^1 = \bar{x}^1(x^1), \quad \bar{x}^2 = \bar{x}^2(x^2), \quad \bar{x}^3 = \bar{x}^3(x^3), \quad \bar{x}^4 = \pm x^4 + \text{const} \tag{19}$$

In the numerical integration the unknown functions g_{ik} are computed at equally spaced grid points. The exterior field at large distances from the body, or from an insular system of bodies, generating the field varies in space as well as in time more smoothly than the interior field. With the help of the transformations (19) the geometrical distances between the grid points may be thus chosen at $t = 0$ far greater at the periphery of the integration domain than inside the bodies. This requirement and the condition that the metric has to be Euclidean in the infinitesimal neighborhood of the x^1-axis together with the initial distribution of the mass density and of the vorticity filaments determine thus uniquely and in the most natural way the comoving coordinate system.

1.3. Axially symmetric field equations

The components A_1 and A_2 reduce to zero only in the cylindrical coordinates (z, r, ϕ, t). It is advantageous to write the metric in the form

$$ds^2 = e^{2\alpha}\, dz^2 + e^{2\beta}\, dr^2 + (e^{2\eta} - A^2 e^{-2\chi})\, d\phi^2 - e^{-2\chi}\, dt^2 + 2U\, dz\, dr +$$

$$+ 2V\, dz\, d\phi + 2W\, dr\, d\phi - 2A e^{-2\chi}\, d\phi\, dt \tag{20}$$

The unknown functions $\alpha, \beta, \eta, U, V, W$ depend on z, r, t, and, in agreement with (17),

$$A = A(r) \tag{21}$$

It is very important to take $g_{33} = (e^{2\eta} - A^2 e^{-2\chi})$, because then Eq. (10) determines ρ as an algebraic function of $e^{2\alpha}$, $e^{2\beta}$, $e^{2\eta}$, U, V, W, but not of χ, which depends on the equation of state and may thus be a complicated function of ρ.

The metric (20) is Euclidean in the infinitesimal neighborhood around the z-axis if

$$\eta(z, r, t) = \sigma(z, r, t) + \ln r \tag{22a}$$

and

$$\beta = \sigma \quad \text{at } r = 0 \tag{22b}$$

The vorticity vector vector Ω^1 is finite at the z-axis if

$$A = r^2 a(r) = r^2 a^* e^{2v(r)}, \qquad a^* = \text{const}, \qquad v(0) = 0 \tag{23}$$

Its square, given by the relation

$$|\Omega|^2 = g_{\alpha\beta}\Omega^\alpha\Omega^\beta \tag{24}$$

is identical with the square of the angular velocity measured ($c = 1$) by a local observer.

It has been shown [9] that the field is regular at the z-axis at least at the initial moment and at its infinitesimal past and future only if

$$U = ru, \qquad V = r^2 v, \qquad W = r^3 w \tag{25}$$

and if the functions α, β, σ, u, v, w, ρ, v are even functions in r. If the mass exhibits also a reflection symmetry with respect to the hyperplane $z = 0$, the functions α, β, σ, w, ρ are even functions in z, while u and v are odd functions in z.

The six independent Einstein field equations (8) may be now expressed as follows

$$\alpha_{44} = \frac{4\pi\rho(e^\chi - 2p/\rho) + P_{11}}{e^{2\chi}k^{44}} \tag{26a}$$

$$\beta_{44} = \frac{4\pi\rho(e^\chi - 2p/\rho) + P_{22}}{e^{2\chi}k^{44}} \tag{26b}$$

$$\sigma_{44} = \frac{4\pi\rho[(e^\chi - 2p/\rho) + a^2 r^2 e^{-2\sigma - 2\chi}(e^\chi + 2p/\rho)] + P_{33}}{e^{2\chi}k^{44}} - e^{-2\sigma - 2\chi}r^2 a^2 \chi_{44} \tag{26c}$$

$$u_{44} = \frac{8\pi\rho(e^\chi - 2p/\rho)u + P_{12}}{e^{2\chi}k^{44}} \tag{26d}$$

$$v_{44} = \frac{8\pi\rho(e^\chi - 2p/\rho)v + P_{13}}{e^{2\chi}k^{44}} \tag{26e}$$

$$w_{44} = \frac{8\pi\rho(e^\chi - 2p/\rho)w + P_{23}}{e^{2\chi}k^{44}} \tag{26f}$$

The subscripts at the functions α, β, σ, u, v, w, ρ, χ, ν indicate henceforth the partial derivatives; the comma is omitted. The functions $k^{\mu\nu}$ occurring in (26a–f) and in the functions $P_{\mu\nu}$ are defined in Appendix A of [3]. They are not tensors; their superscripts just indicate in which relation they stand to the contravariant components of the metric tensors. The functions $P_{\mu\nu}$ are defined in Appendix B of [3] together with the substitutions introduced to simplify the formulas for $P_{\mu\nu}$. The $P_{\mu\nu}$ are not tensors; their subscripts just indicate in which relation they stand to the covariant components of Ricci tensor.

Since the components g_{ik} $(i \neq k)$ vanish at the z-axis as given by Eq. (25), the field is here regular. In spite of it the functions $P_{\mu\nu}$ contains terms that have an indeterminate form $0/0$ or $\infty - \infty$. Therefore two sets of the functions $P_{\mu\nu}$ must be used: one set for the space with $r > 0$, and the other for the z-axis and denoted by an asterisk, $P^*_{\mu\nu}$, in which the indeterminacy is analytically evaluated.

The equation continuity (10) may be reduced to the form

$$\rho = \overline{\Psi} K^{1/2} e^{-\alpha-\beta-\sigma} \tag{27}$$

The function K is given by the formula in Appendix A of [3]. The function $\overline{\Psi}$ depending on the spatial coordinates only is computed by equation (27) from the initial data of the metric and mass density.

The functions $P_{\mu\nu}$ and $P^*_{\mu\nu}$ and the field equation (26c) contain the derivatives of the function χ. With the help of Eqs. (2), (12), (27), and the field equation (8) with $\mu = \nu = 4$ they are reduced to expressions containing the derivatives of the mass density, but neither (ρ_4/ρ) nor any second derivative with respect to the time-like coordinate [3].

Once the equation of state (3) is given and Eq. (12) integrated, the enthalpy e^χ, the pressure, as well as the functions E and F occurring in the formulas for the derivatives of the function χ, may be considered as known functions of the mass density ρ. The derivatives of any unknown function occurring in $P_{\mu\nu}$ with respect to spatial coordinates may be expressed, using Lagrange formulas for numerical differentiation, by the function itself at the given point and its neighborhood [10, 4]. If the cross section $\phi = \text{const}$ of the integration domain is divided into a two-dimensional grid of n equally spaced points where the unknown functions α, β, σ, u, v, w, and ρ are to be calculated, the set of six partial differential equations (26a–f) may be now considered as a set of $6n$ simultaneous ordinary differential equations and integrated numerically, using, for instance, the fourth-order Runge–Kutta method. The algebraic equation (27) determines at each point the seventh unknown function ρ. However, in each computation of the right-hand sides of Eqs. (26a–f) all the k^{ik} and all the derivatives must be evaluated anew and the integration must be carried out for all $6n$ functions simultaneously.

The finite number n of the grid points is a source of truncation errors in the computation of spatial derivatives. Another source of truncation errors is the integration method for ordinary differential equations. The general relativity yields, however, a sensitive indicator for estimating the total amount of all the errors of the numerical calculation: If the four components of I_μ^4 given by Eq. (9) vanish at the initial moment, then they must vanish also at any moment $t \gtrless 0$ [5]. Due to the numerical errors the I_μ^4 will differ from zero at $t \gtrless 0$ and this difference may be used as a criterion of the total amount of errors, for it is highly improbable that the errors could partially cancel each other in the I_μ^4 in such a way that the order of magnitude of the I_μ^4 would be smaller than the order of magnitude of all the errors.

1.4. The junction conditions and the exterior field

Let Σ be a smooth hypersurface separating the interior field from the exterior field of the empty space. The hyperspace itself is a part of both subdomains. In the co-moving coordinates of the interior metric the hypersurface is given by the equation

$$S(z, r) = 0 \tag{28}$$

According to Lichnerowicz [5], the junction conditions require the continuity of the metric tensor and of its normal derivatives across the hypersurface Σ if the metric is expressed in the admissible coordinates. However, Synge [11] has shown that even the non-admissible coordinates may be used on both sides of the hyper-surface Σ, providing that they are obtained from the admissible ones by a C^1 transformation and that the following four junction conditions are satisfied

$$(G_\mu^\nu S_{,\nu})_{\text{interior}} = (G_\mu^\nu S_{,\nu})_{\text{exterior}} \quad \text{at } \Sigma \tag{29}$$

G_μ^ν being the Einstein tensor. The components $g_{\mu\nu}$ are still continuous across Σ (since their transformation law involves only the first derivatives $\partial x^\alpha / \partial \bar{x}^\beta$), but their first normal derivatives $g_{\mu\nu,\varepsilon}$ may now be discontinuous.

If the interior metric is expressed in comoving coordinates, four junction conditions reduce to one condition [3]

$$p = 0 \quad \text{at } \Sigma \tag{30}$$

The mass density ρ varies in time according to Eq. (27). If condition (30) has to be satisfied at any moment, the pressure in the equation of state (3) must vanish when the mass density vanishes. Condition (30) reduces then to

$$\bar{\Psi}(z, r) = 0 \quad \text{at } \Sigma \tag{31}$$

In the exterior domain the harmonic coordinates, defined by the four coordinate conditions [12]

$$(-\bar{g})^{-1/2} (\partial / \partial \bar{x}^\nu) [\bar{g}^{\mu\nu} (-\bar{g})^{1/2}] = 0 \tag{32}$$

must be used if the Einstein field equations

$$R^{\mu\nu} = 0 \tag{33}$$

have to reduce in the distant weak field to the homogeneous wave equations

$$\Box \bar{h}^{\mu\nu} = 0 \tag{34}$$

in which

$$\bar{h}^{\mu\nu} = \bar{g}^{\mu\nu} - \bar{\eta}^{\mu\nu} \tag{35}$$

denote the deviations of $\bar{g}^{\mu\nu}$ from the Minkowskian values $\bar{\eta}^{\mu\nu}$. From Eq. (34) it is obvious that the harmonic coordinates guarantee the metric generated by one or more bodies forming an insular system to become automatically Minkowskian at infinity.

The transformation relations from the interior comoving coordinates x^{μ} to the exterior harmonic coordinates $\bar{x}^{\mu}$ follow from the equation [12]

$$(-g)^{-1/2} (\partial/\partial x^{\alpha}) [(-g)^{1/2} g^{\alpha\beta} (\partial\bar{x}^{\mu}/\partial x^{\beta})] = 0 \tag{36}$$

The six boundary values $\bar{g}^{ik}$

$$\bar{g}^{ik} = g^{\alpha\beta} (\partial\bar{x}^{i}/\partial x^{\alpha}) (\partial\bar{x}^{k}/\partial x^{\beta}) \tag{37}$$

are the source of the gravitational field generated by the object inside the hypersurface Σ. The remaining components $\bar{g}^{\mu 4}$ are given by the coordinate conditions (32).

When one or more bodies forming an insular system are surrounded by empty space, Eq. (36) and the boundary conditions (37) must be satisfied at the hypersurface Σ_i of each body if the back-scattering of radiation and the gravitational interaction (including the gravitational waves) are to be taken into account. It is obvious that such a difficult boundary value problem hardly can be either solved analytically or programmed for a computer.

After a thorough examination the author does not see any other approach practicable than the following one: Let Λ be a spherical hypersurface where the exterior field is so weak that the exact field equations (33) may be replaced with a sufficient accuracy by the approximate ones (34). The body, or the insular system of bodies moving along the z-axis to preserve the supposed axial symmetry, are assumed not to be surrounded by the empty space-time, but the whole domain inside Λ is now considered as the interior field where the mass density and the metric tensor are C^3 continuous and expressed in the above introduced particular system of comoving coordinates. The mass density takes very low values outside the bodies and equals to zero at Λ. In this way the many body problem is reduced to one body problem and the boundary between the interior field and the exterior field of the empty space is shifted to the region where the interior metric differs from the Minkowskian metric by very small quantities $\bar{h}^{\mu\nu}$ so that Eq. (36) with the background metric reduces here to the wave equation of Euclidean space with the particular solution

$$\left.\begin{array}{ll} \bar{x}^1 \equiv \bar{z} = z, & \bar{x}^2 \equiv \bar{x} = r \cos \phi \\ \bar{x}^3 \equiv \bar{y} = r \sin \phi, & \bar{x}^4 \equiv \bar{t} = t \end{array}\right\} \tag{38}$$

The harmonic coordinates outside Λ are thus identical with the Cartesian coordinates. The transformation from the interior metric to the exterior metric by Eq. (37) becomes now trivial.

1.5. Cauchy initial data

Because of the transient character of the generation of gravitational waves, the Cauchy initial data at the entire hypersurface $\bar{t} = $ const of the empty space never can be known, for in them the whole past gravitational history of the objects generating the waves is contained. The $\bar{g}^{\mu\nu}$ must, of course, satisfy the Lichnerowicz initial conditions (9), but this is a minor restriction. An improper choice of the initial values $\bar{g}^{\mu\nu}$ might imply the presence of gravitational waves generated not by the bodies, but somewhere at infinity.

However, the Einstein equations are quasilinear hyperbolic differential equations of second order for the integration of which the Cauchy data at the initial hypersurface $t = \bar{t} = 0$ must be given. Fortunately, in the weak field zone the field equations reduce to the homogeneous wave equations for which the classical Huyghens principle [13, 14] holds. The principle asserts that sharp signals are transmitted in three-dimensional space as sharp signals, that is that the solution of the wave equations describing the propagation of signals emitted at $t = 0$ depends upon the data at the boundary of the conoid of dependence, not upon the data inside. The principle implies that the signals are transmitted only in the direction of the propagation of waves, but not in the opposite direction, for the propagation towards the source of radiation would cause reverberation and make the transmission of sharp signals impossible (this occurs in the space of even number of dimensions).

Consequently, if the space outside Λ is assumed to be empty and with no gravitational waves incoming from infinity, then the initial data inside Λ, satisfying the initial conditions (9), represent the Cauchy initial data which determine the whole past and future gravitational history of all the bodies inside Λ. The gravitational waves propagating towards infinity are to be computed either with the help of the Huyghens principle from the field at a wave front Ξ lying in the vicinity of Λ or with the help of the Fourier transform from the field at Λ. There exists no back-scattering at the wave front Ξ and, consequently, no back-scattering at the hypersurface Λ, because the background metric is here Minkowskian.

If a wave incoming from infinity should be present, then the solution of the field equations (26a–f) for the interior domain would be determined not only by Cauchy initial data inside Λ, but also by a boundary condition at Λ during the time interval when the incoming wave was crossing the hypersurface Λ. If the integral of Eqs. (26a–f) is determined by Cauchy initial data inside Λ only, then this fact implies the absence of whatever incoming radiation.

1.6. Weak field zone

The field $\bar{h}^{\mu\nu}_\Lambda$ at Λ may be expanded into a series of spherical harmonics [3]

$$\bar{h}^{\mu\nu}_\Lambda(\theta, t) = \sum_{n=0}^{\infty} A^{\mu\nu}_n(t)\, P_n(\cos\theta) \tag{39}$$

It can be proved with the help of Fourier transform [3] that the field outside Λ may be computed by the equation

$$\bar{h}^{\mu\nu}(r, \theta, t) \cong (r_\Lambda/r) \sum_{n=0}^{\infty} A^{\mu\nu}_n(t - [r - r_\Lambda])\, P_n(\cos\theta) =$$
$$= (r_\Lambda/r)\, \bar{h}^{\mu\nu}_\Lambda(\theta, t - [r - r_\Lambda]) \tag{40}$$

This approximate relation may be used only for those wavelengths of Fourier spectrum which are much shorter than the radius r_Λ of the hypersurface Λ, i.e., when

$$2\pi r_\Lambda/\lambda_c \gg 1 \tag{41}$$

λ_c being the longest wavelength to which the detector of gravitational waves is sensitive. The weak radiation field (or, at least, its observable part) thus propagates, in the terminology of Courant and Hilbert [13], as a "relatively undistorted" progressive wave.

1.7. Lichnerowicz initial conditions

A general solution of this problem is far more difficult than the time evolution problem of Sec. 1.3. It has been solved analytically for the following three cases:

(i) An ideal fluid with a non-vanishing pressure is rotating. At $t = 0$, the functions α, β, σ, and the mass density ρ have just reached their extremum values:

$$\alpha_4 = \beta_4 = \sigma_4 = \rho_4 = 0 \tag{42}$$

In addition to $u = v = 0$ (which represent no restriction of generality) also $w = 0$ must vanish everywhere as a consequence of $\rho_4 = 0$; we thus have

$$u = v = w = 0 \tag{43}$$

The unique solution of Eqs. (9) with $\mu = 1, 2$ is

$$u_4 = 0 \tag{44}$$

The functions v_4 and w_4 have been chosen as unknown functions that are determined by Eqs. (9) with $\mu = 3, 4$ [3].

(ii) An incoherent (i.e., pressureless) matter is rotating. The initial conditions are given by Eqs. (42), as a consequence of which Eqs. (43) and (44) still hold, but the functions w_4 and ρ have been chosen as unknown functions that are again determined by Eqs. (9) with $\mu = 3, 4$ [9]. This solution is far simpler than the preceding case.

(iii) An ideal fluid with a non-vanishing pressure does not rotate. At $t = 0$ the body, or an insular system of bodies, explode or implode in the radial direction so that

$$v_4 = w_4 = 0 \tag{45}$$

and either α_4 or β_4 is given. The initial conditions (45) together with (43) satisfy not only the initial condition (9) with $\mu = 3$ but also the field equations (26e–f) at any moment. The functions β_4, u_4 and σ_4 or α_4, u_4 and σ_4, respectively, have been chosen as unknown functions that are determined by Eqs. (9) with $\mu = 1, 2, 4$ (an unpublished study).

2. DEVELOPMENT OF COMPUTER PROGRAMS

The whole computer program is divided into a time evolution program and an initial data program. Since the Lichnerowicz initial conditions do not determine the initial data uniquely but only restrict their choice, the problem of the admissible initial data can be solved in different ways. Therefore it is appropriate to keep both programs apart, not to merge them into one program.

In order to eliminate the occurrence of numbers that are too big or too small a dimensionless system of units is introduced first.

A part of both programs depends immediately upon the equation of state. This subroutine can easily be removed from the programs and replaced by a new one according to the choice of the equation of state.

2.1. Reduction to a dimensionless system of units

The dimensionless system of units is defined by the formulas

$$\left.\begin{array}{llll} z = N^{-1}\rho_i^{-1/2}\tilde{z}, & r = N^{-1}\rho_i^{-1/2}\tilde{r}, & \phi = \phi \\ t = (Nc)^{-1}\rho_i^{-1/2}\tilde{t}, & \rho = \rho_i\tilde{e}, & a^* = N\rho_i^{1/2}\tilde{a}^* \end{array}\right\} \tag{46}$$

The proper rest mass densities ρ and ρ_i, the angular velocity ω_i related to a^* by the formula (following from Eqs. (24, (15), and (23))

$$\omega_i = ca^*e^{-2\beta_i - \chi_i} \tag{47}$$

and the coordinates (z, r, ϕ, t) are measured in (46) and (47) in the c.g.s. system of units.

The subscript i indicates the values at the origin of coordinates at $t = 0$. The quantities measured in the dimensionless system of units are denoted by a tilde. The reduction coefficient N and its inverse are given by the formulas

$$N = (4\pi G/c^2)^{1/2} = 3.055 \times 10^{-14}\,(\text{cm/g})^{1/2}, \qquad N^{-1} = 3.274 \times 10^{13}\,(\text{g/cm})^{1/2}$$

$$(48\text{a})$$

Hence

$$Nc = 0.9157 \times 10^{-3}\,(\text{cm}^{3/2}/\text{g}^{1/2}\text{sec}), \qquad (Nc)^{-1} = 1.092 \times 10^{3}\,(\text{g}^{1/2}\text{sec}/\text{cm}^{3/2})$$

$$(48\text{b})$$

In the case of incoherent matter the results of the numerical computation cover the whole range of densities. In an ideal fluid the numerical results refer only to the central density ρ_i at $t = 0$ which appears in the equation of state. Nevertheless it is very advantageous to use the dimensionless system of units also in this case.

The reduction of the field equations from Part 1 to the equations with dimensionless quantities can be carried out very easily. After having divided the former by the constant $4\pi\rho_i$ we find that the reduction consists in replacing $4\pi\rho$ by $\tilde{\rho}$ and considering then the equations as written throughout in dimensionless quantities. The results of the numerical computation are converted to the c.g.s. system of units using the reduction formulas (46) and (47).

2.2. Equation of state

The equation of state is assumed to be given analytically in the form (3). If it is represented by a set of equations each covering a limited range of the mass density, the equations must be C^3 continuous also at the end points of their intervals of validity, because only then the metric tensor remains C^3 continuous [5] (for the proper enthalpy per unit mass e^χ defined by (2) is related to $g_{\mu 4}$ by equation (13)).

After the equation of state has been substituted into Eq. (12) and the integration carried out, one obtains the relation

$$\varepsilon = \varepsilon(\rho) \tag{49}$$

and with its help

$$p/\rho = \text{function of } \rho \tag{50a}$$

and

$$e^\chi = 1 + \varepsilon + p/\rho = \text{function of } \rho \tag{50b}$$

$$E(\rho) = e^{-\chi}(dp/d\rho) \tag{50c}$$

$$F(\rho) = e^{-\chi}[\rho(d^2p/d\rho^2) - (1 + E)(dp/d\rho)] \tag{50d}$$

$$F'(\rho) = \rho^2(d^2E/d\rho^2) - 2F \tag{50e}$$

In Eqs. (50a–e) the independent variable ρ measured in the c.g.s. system of units is to be replaced by $\tilde{e}$ according to Eq. (46). Then five equations constitute now the exchangeable subroutine of the equation of state. The functions $E(\rho)$, $F(\rho)$, and $F'(\rho)$ appear in the formulas reducing the derivatives of χ to the derivatives of the mass density.

2.3. Time evolution program

Since the field is axially symmetric around the z-axis and reflection symmetric to the hyperplane $z = 0$, the integration domain may be restricted to one quadrant of the hyperplane $\phi = $ const. The integration domain is covered by a two-dimensional rectangular grid of equally spaced points where the functions α, β, σ, u, v, w, and ρ are to be computed as functions of the time-like coordinate t.

The input data are furnished by the initial data program. Since the functions $P_{\mu\nu}$ contain the k^{ik} and k^{44} and certain repeatedly occurring products of two or more quantities, three sets of substitutions are introduced in order to save the computer time (at the expense of a greater extent of its memory).

Besides the exchangeable subroutine of the equation of state, the time evolution program consists of

1) the unchangeable set of substitutions I and formulas for the functions k^{ik} and k^{44};

2) the Lagrange differentiation formulas which depend upon the chosen number and arrangement of the grid points in the integration domain;

3) the unchangeable set of substitutions II and III and formulas for the functions $P_{ik}, P_{44}, D, P_{\rho}, \chi_{44}$;

4) the suitably chosen integration method for the system of simultaneous ordinary differential equations;

5) the unchangeable set of formulas for the functions P_{i4} and I_{μ}^{4} which are computed only at the moments when the results of integration are printed;

6) the formulas depending upon the chosen number and arrangement of the grid points which determine the time dependent coefficients in the series of spherical harmonics for the exterior field.

Once the number and arrangement of the grid points in the integration domain, the Lagrange differentiation formulas, and the method of the numerical integration have been set, the time evolution program is quite general and may easily be adjusted for any equation of state.

A detailed description of the program, including all the formulas and substitutions, is given in [4].

2.4. Initial data program

A detailed description of the programs (i) and (ii) of Sec. 1.7 is also given in [4].

The program (i) consists of a large set of input data, of a set of definitions, of an algebraic part, of the integration of one quasi-linear partial differential equation of the first order, and of the exchangeable subroutine of the equation of state.

The program (ii) is far simpler, but it can be used only for incoherent matter. It consists of a small set of input data, of a small algebraic part, and of the integration of one ordinary differential equation of the first order.

The program (iii) of Sec. 1.7 has not yet been written out, but from its equations one may expect that it will be more complicated than the program (ii), but certainly simpler than the program (i).

3. TWO NUMERICAL EXAMPLES

The extent of programming and of running the programs was limited by a very modest research grant. Therefore only two simple numerical examples are given here. They show that the applied method really works and how it works, but nevertheless they also illustrate an interesting relativistic effect. However, further programming is needed before the method may be used for solving astrophysical problems.

The first computer programs, based upon the integration method described in [9] and [3], but restricted to the special case of zero pressure in the equation of state, were elaborated during a few weeks in 1972 by R. Teshima [10]. In these programs the number of the grid points was very low, $n = 15$. Figures 1 and 2 show how the proper rest mass density varies in time in the equatorial plane (solid curves) and along the axis of rotation (dotted curves). The completely different behavior of the mass density in Figs. 1 and 2 is caused by the different values of the dimensionless parameter κ^2, which equals to 0.36 and 1.44, respectively. The parameter is defined by the formula

$$\kappa^2 = \frac{|\Omega_i|^2}{4\pi G \rho_i} \tag{51}$$

in which $|\Omega_i|$ and ρ_i denote the square of the angular velocity and the proper rest mass density, respectively, at the origin of coordinates at $t = 0$. The discussion of this phenomenon is postponed to Part 4.

In order to estimate the stability of the integration method, the integration was run with increasing time and then run in the opposite direction with decreasing time. The discrepancies begin to appear in Fig. 1 when $\tilde{t}$ has decreased to 0.125, while in Fig. 2 no discrepancies appear at all in the integration results with increasing and decreasing time until $\tilde{t} = 0$. The applied Runge–Kutta integration method may be

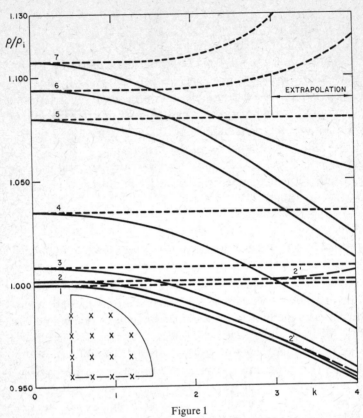

Figure 1

Relative mass density ρ/ρ_i in a spherical object of the radius $\tilde{R} = 0.5$. Initial values: $\tilde{a} = -0.2\tilde{r}^2$, $\tilde{\beta} = 0$, $\sigma = -0.02\tilde{r}^2$, $\tilde{w} = \tilde{\alpha}_4 = \tilde{\beta}_4 = \tilde{\sigma}_4 = \tilde{u}_4 = \tilde{v}_4 = 0$, $\tilde{a} = \tilde{a}^* = 0.6$, $\kappa^2 = 0.36$ (at $\tilde{t} = 0$ all the vorticity filaments are parallel to the $\tilde{z}$-axis), $n = 15$. Solid curves: density in the equatorial plane against $k = 8\tilde{r}$, increasing time: 1) $\tilde{t} = 0.000$; 2) $\tilde{t} = 0.1250$; 3) $\tilde{t} = 0.2500$; 4) $\tilde{t} = 0.5000$; 5) $\tilde{t} = 0.7500$; 6) $\tilde{t} = 0.8125$; 7) $\tilde{t} = 0.8671875$ (stop because $\rho \to \infty$ at the point with $\tilde{z} = 0.375$, $\tilde{r} = 0.250$). Dotted curves: corresponding densities at the axis of rotation against $k = 8\tilde{z}$, increasing time. Broken curves: time decreasing from $\tilde{t} = 0.8125$ to 0.1250; no discrepancies in curves 6–3. Computer time: 715 sec (time increasing from $\tilde{t} = 0.0$ to $\tilde{t} = 0.8125$ and decreasing to 0.1250). Inset: distribution of the grid points across the integration domain.

thus considered as stable during the whole integration. On the other hand, the computer stopped in both cases because the mass density started to increase fast at one point at the periphery of the integration domain. This point will be discussed in connection with the new programs.

Unfortunately, in 1973, R. Teshima was not available to modify his programs. The new programmer, recommended to the author, used Teshima's time evolution program as a basis of his programming in which he had (i) to increase the number of grid points from 15 to 43, (ii) to rewrite the field equations containing new terms

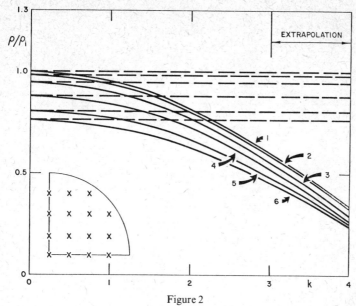

Figure 2

Relative mass density ρ/ρ_i in a spherical object of the radius $\tilde{R} = 0.5$. Initial values: $\tilde{\alpha} = -0.1\tilde{r}^2$, $\tilde{\beta} = 0$, $\sigma = (1.36/3)\tilde{r}^2$, $\tilde{w} = \tilde{\alpha}_4 = \tilde{\beta}_4 = \tilde{\sigma}_4 = \tilde{u}_4 = \tilde{v}_4 = 0$, $\tilde{a} = \tilde{a}^*e^{2\tilde{v}(\tilde{r})} = 1.2\exp(0.2\tilde{r}^2)$, $\kappa^2 = 1.44$ (at $\tilde{t} = 0$ all the vorticity filaments are parallel to the $\tilde{z}$ axis), $n = 15$. Solid curves: density in the equatorial plane against $k = 8\tilde{r}$, increasing time: 1) $\tilde{t} = 0.000$; 2) $\tilde{t} = 0.125$; 3) $\tilde{t} = 0.250$; 4) $\tilde{t} = 0.375$; 5) $\tilde{t} = 0.500$; 6) $\tilde{t} = 0.5625$ (stop because $\rho \rightarrow \infty$ at the point with $\tilde{z} = 0.250$, $\tilde{r} = 0.375$). Dotted curves: corresponding densities at the axis of rotation against $k = 8\tilde{z}$, increasing time. No discrepancies at all when time decreasing. Computer time 361 sec (time increasing from $\tilde{t} = 0.0$ to $\tilde{t} = 0.5625$ and decreasing to 0.0). Inset: distribution of the grid points across the integration domain.

involving the pressure and its derivatives, and reduced by the author to a more compact form by introducing the substitutions II and III, and (iii) to enlarge the program by introducing the computation of the functions I_μ^4. The new time evolution program was elaborated during three months and tested for the special case of zero pressure with the help of Teshima's initial data program (the program (ii) of Sec. 1.7) which required merely a very slight modification.

The distribution of grid points across the integration domain is shown in Fig. 3. The integration was run for the same initial data as in Figs. 1 and 2. Its results are demonstrated in Tables 1a, 1b and 2a, 2b. Tables 1a and 2a show that the initial data program yields the required accuracy: the error bound of the fourth-order Runge–Kutta integration method was chosen 10^{-8}; the functions I_3^4 and I_4^4 never exceed this order of magnitude. Tables 1b and 2b show that the functions I_μ^4 remain at a sufficiently low level at and around the center of the integration domain, while at its periphery they took inadmissibly high values already after a few steps of integra-

TABLE 1a (for key see p. 164)

0   T =   0.0

ALPHA

0.0	-0.781250D-03	-0.312500D-02	-0.703125D-02	-0.125000D-01	-0.195312D-01	-0.281250D-01
0.0	-0.781250D-03	-0.312500D-02	-0.703125D-02	-0.125000D-01	-0.195312D-01	-0.281250D-01
0.0	-0.781250D-03	-0.312500D-02	-0.703125D-02	-0.125000D-01	-0.195312D-01	-0.281250D-01
0.0	-0.781250D-03	-0.312500D-02	-0.703125D-02	-0.125000D-01	-0.195312D-01	-0.281250D-01
0.0	-0.781250D-03	-0.312500D-02	-0.703125D-02	-0.125000D-01	-0.195312D-01	
0.0	-0.781250D-03	-0.312500D-02	-0.703125D-02	-0.125000D-01		
0.0	-0.781250D-03	-0.312500D-02	-0.703125D-02			

BETA

0.0	0.0	0.0	0.0	0.0	0.0	0.0
0.0	0.0	0.0	0.0	0.0	0.0	0.0
0.0	0.0	0.0	0.0	0.0	0.0	0.0
0.0	0.0	0.0	0.0	0.0	0.0	0.0

0.0	0.0	0.0	0.0	0.0	0.0
0.0	0.0	0.0	0.0	0.0	
0.0	0.0	0.0	0.0		

SIGMA

0.0	-0.781250D-04	-0.312500D-03	-0.703125D-03	-0.125000D-02	-0.195312D-02	-0.281250D-02
0.0	-0.781250D-04	-0.312500D-03	-0.703125D-03	-0.125000D-02	-0.195312D-02	-0.281250D-02
0.0	-0.781250D-04	-0.312500D-03	-0.703125D-03	-0.125000D-02	-0.195312D-02	-0.281250D-02
0.0	-0.781250D-04	-0.312500D-03	-0.703125D-03	-0.125000D-02	-0.195312D-02	
0.0	-0.781250D-04	-0.312500D-03	-0.703125D-03	-0.125000D-02		
0.0	-0.781250D-04	-0.312500D-03	-0.703125D-03			

U

0.0	0.0	0.0	0.0	0.0	0.0	0.0
0.0	0.0	0.0	0.0	0.0	0.0	0.0
0.0	0.0	0.0	0.0	0.0	0.0	0.0
0.0	0.0	0.0	0.0	0.0	0.0	0.0
0.0	0.0	0.0	0.0	0.0	0.0	
0.0	0.0	0.0	0.0	0.0		
0.0	0.0	0.0	0.0			

V

0.0	0.0	0.0	0.0	0.0	0.0	0.0
0.0	0.0	0.0	0.0	0.0	0.0	0.0
0.0	0.0	0.0	0.0	0.0	0.0	0.0
0.0	0.0	0.0	0.0	0.0	0.0	0.0
0.0	0.0	0.0	0.0	0.0	0.0	
0.0	0.0	0.0	0.0	0.0		
0.0	0.0	0.0	0.0			

W

0.0	0.0	0.0	0.0	0.0	0.0	0.0
0.0	0.0	0.0	0.0	0.0	0.0	0.0
0.0	0.0	0.0	0.0	0.0	0.0	0.0
0.0	0.0	0.0	0.0	0.0	0.0	0.0
0.0	0.0	0.0	0.0	0.0	0.0	
0.0	0.0	0.0	0.0	0.0		
0.0	0.0	0.0	0.0			

RHO

1.00000	0.999203	0.996815	0.992841	0.987290	0.980177	0.971520
1.00000	0.999203	0.996815	0.992841	0.987290	0.980177	0.971520
1.00000	0.999203	0.996815	0.992841	0.987290	0.980177	0.971520
1.00000	0.999203	0.996815	0.992841	0.987290	0.980177	0.971520
1.00000	0.999203	0.996815	0.992841	0.987290	0.980177	
1.00000	0.999203	0.996815	0.992841	0.987290		
1.00000	0.999203	0.996815	0.992841			

I41

0.0	0.0	0.0	0.0	0.0	0.0	0.0
0.0	0.0	0.0	0.0	0.0	0.0	0.0
0.0	0.0	0.0	0.0	0.0	0.0	0.0
0.0	0.0	0.0	0.0	0.0	0.0	0.0
0.0	0.0	0.0	0.0	0.0	0.0	
0.0	0.0	0.0	0.0	0.0		
0.0	0.0	0.0	0.0			

I42

0.0	0.0	0.0	0.0	0.0	0.0	0.0

0.0	0.0	0.0	0.0	0.0	0.0	0.0
0.0	0.0	0.0	0.0	0.0	0.0	0.0
0.0	0.0	0.0	0.0	0.0	0.0	0.0
0.0	0.0	0.0	0.0	0.0	0.0	
0.0	0.0	0.0	0.0	0.0		
0.0	0.0	0.0	0.0			

I43

0.0	0.431024D-12	0.363877D-11	-0.178199D-12	0.176059D-10	-0.413723D-11	0.102398D-08
0.0	0.431024D-12	0.363877D-11	-0.178189D-12	0.176059D-10	-0.413723D-11	0.102398D-08
0.0	0.431024D-12	0.363877D-11	-0.178189D-12	0.176059D-10	-0.413723D-11	0.102398D-08
0.0	0.431024D-12	0.363877D-11	-0.178199D-12	0.176059D-10	-0.413723D-11	0.102398D-08
0.0	0.430137D-12	0.366659D-11	-0.551749D-12	0.228634D-10	-0.130824D-09	
0.0	0.428807D-12	0.370546D-11	-0.123407D-11	0.399435D-10		
0.0	0.427260D-12	0.377105D-11	-0.321382D-11			

I44

0.155431D-14	-0.334399D-12	-0.217826D-11	0.914824D-13	-0.108320D-10	0.257705D-11	-0.651130D-09
0.155431D-14	-0.334399D-12	-0.217826D-11	0.919265D-13	-0.108322D-10	0.257705D-11	-0.651130D-09
0.155431D-14	-0.334399D-12	-0.217826D-11	0.919265D-13	-0.108322D-10	0.257705D-11	-0.651130D-09
0.155431D-14	-0.334399D-12	-0.217826D-11	0.914824D-13	-0.108322D-10	0.257705D-11	-0.651133D-09
0.888178D-15	-0.333955D-12	-0.219513D-11	0.320188D-12	-0.140723D-10	0.816573D-11	
0.155431D-14	-0.332845D-12	-0.221845D-11	0.732969D-12	-0.245621D-10		
0.133227D-14	-0.329736D-12	-0.225642D-11	0.191136D-11			

TABLE 1b (for key see p. 164)

4   T =   0.18750

ALPHA

0.352509D-02	0.274685D-02	0.412131D-03	-0.347906D-02	-0.892674D-02	-0.159314D-01	-0.245077D-01
0.352508D-02	0.274685D-02	0.412131D-03	-0.347904D-02	-0.892666D-02	-0.159309D-01	-0.244999D-01
0.352510D-02	0.274685D-02	0.412133D-03	-0.347910D-02	-0.892679D-02	-0.159318D-01	-0.244881D-01
0.352513D-02	0.274686D-02	0.412132D-03	-0.347898D-02	-0.892661D-02	-0.159332D-01	-0.244400D-01
0.352509D-02	0.274685D-02	0.412121D-03	-0.347986D-02	-0.893068D-02	-0.159210D-01	
0.352550D-02	0.274700D-02	0.411487D-03	-0.350687D-02	-0.898505D-02		
0.352730D-02	0.274816D-02	0.400381D-03	-0.385475D-02			

BETA

-0.422084D-02	-0.421682D-02	-0.420473D-02	-0.418448D-02	-0.415601D-02	-0.412061D-02	-0.413435D-02

-0.422085D-02	-0.421682D-02	-0.420473D-02	-0.418449D-02	-0.415602D-02	-0.412231D-02	-0.415773D-02
-0.422086D-02	-0.421682D-02	-0.420473D-02	-0.418446D-02	-0.415614D-02	-0.412106D-02	-0.414389D-02
-0.422091D-02	-0.421682D-02	-0.420474D-02	-0.418449D-02	-0.415734D-02	-0.411813D-02	-0.334207D-02
-0.422084D-02	-0.421682D-02	-0.420473D-02	-0.418483D-02	-0.416196D-02	-0.422662D-02	
-0.422073D-02	-0.421671D-02	-0.420650D-02	-0.422144D-02	-0.401617D-02		
-0.422096D-02	-0.421579D-02	-0.422851D-02	-0.499782D-02			

SIGMA

-0.422084D-02	-0.428998D-02	-0.449741D-02	-0.484327D-02	-0.532773D-02	-0.595159D-02	-0.672539D-02
-0.422085D-02	-0.428998D-02	-0.449741D-02	-0.484327D-02	-0.532773D-02	-0.595254D-02	-0.674184D-02
-0.422086D-02	-0.428998D-02	-0.449742D-02	-0.484326D-02	-0.532776D-02	-0.595481D-02	-0.677927D-02
-0.422091D-02	-0.428998D-02	-0.449742D-02	-0.484327D-02	-0.532793D-02	-0.595469D-02	-0.670966D-02
-0.422084D-02	-0.428998D-02	-0.449741D-02	-0.484336D-02	-0.532788D-02	-0.594969D-02	
-0.422073D-02	-0.428989D-02	-0.449796D-02	-0.465250D-02	-0.530281D-02		
-0.422096D-02	-0.428918D-02	-0.450632D-02	-0.504540D-02			

U

0.0	0.0	0.0	0.0	0.0	0.0	0.0
0.365092D-06	0.132199D-06	-0.359990D-07	0.204616D-07	0.131157D-07	0.538623D-05	0.120334D-03
0.710846D-06	0.261020D-06	-0.756215D-07	0.119665D-06	-0.173732D-06	0.126617D-04	0.350451D-03
-0.413093D-06	-0.103849D-06	-0.159890D-07	-0.846994D-07	-0.256230D-06	-0.452710D-04	-0.982071D-04
-0.101642D-05	-0.478596D-06	0.172545D-06	0.477124D-06	0.175382D-04	0.148200D-04	
-0.776192D-03	-0.278559D-06	-0.453568D-05	-0.747056D-04	0.609983D-04		
-0.685705D-04	-0.747615D-05	0.578509D-05	-0.227626D-03			

V

0.0	0.0	0.0	0.0	0.0	0.0	0.0
0.357956D-08	-0.123498D-07	0.248835D-07	-0.615088D-09	0.392290D-06	0.428959D-04	0.769879D-03
0.515271D-08	-0.158999D-07	-0.114843D-07	-0.167595D-06	0.847342D-06	0.609435D-04	0.133067D-03
-0.133362D-07	-0.367000D-07	0.234878D-06	-0.367805D-06	-0.145623D-04	-0.180486D-03	-0.552690D-02
0.314360D-06	0.170700D-06	-0.490721D-06	0.582761D-05	-0.108952D-03	0.135307D-02	
0.133507D-05	-0.358470D-05	0.689466D-04	0.298508D-03	-0.339166D-02		
0.375754D-04	-0.163334D-04	0.882689D-03	0.160814D-01			

W

0.223563D-01	0.223497D-01	0.223249D-01	0.222838D-01	0.222208D-01	0.221249D-01	0.213706D-01
0.223574D-01	0.223502D-01	0.223241D-01	0.222840D-01	0.222241D-01	0.220910D-01	0.212366D-01
0.223557D-01	0.223497D-01	0.223253D-01	0.222858D-01	0.222186D-01	0.221891D-01	0.224061D-01
0.223531D-01	0.223519D-01	0.223213D-01	0.222743D-01	0.221674D-01	0.221796D-01	0.360351D-01
0.223585D-01	0.223481D-01	0.223238D-01	0.222950D-01	0.220452D-01	0.185964D-01	
0.223320D-01	0.224043D-01	0.219494D-01	0.190800D-01	0.326057D-01		
0.219462D-01	0.223334D-01	0.201276D-01	-0.423807D-01			

RHO

1.00493	1.00411	1.00166	0.997591	0.991903	0.984619	0.975838
1.00493	1.00411	1.00166	0.997591	0.991903	0.984621	0.975869
1.00493	1.00411	1.00166	0.997591	0.991903	0.984623	0.975881
1.00493	1.00411	1.00166	0.997591	0.991905	0.984622	0.974994
1.00493	1.00411	1.00166	0.997592	0.991913	0.984711	
1.00493	1.00411	1.00167	0.997664	0.991799		
1.00493	1.00411	1.00171	0.998985			

I41

0.0	0.0	0.0	0.0	0.0	0.0	0.0
*-0.153819D-04	0.385509D-06	-0.594786D-06	-0.250724D-04	-0.180293D-03	-0.342901D-02	-0.369555D-01
-0.137802D-03	0.774382D-06	-0.778409D-05	0.124378D-03	-0.122749D-04	0.629071D-02	0.242119
-0.118332D-03	-0.195003D-04	0.586826D-04	-0.700750D-03	-0.431533D-02	-0.368062D-01	1.81083
0.766414D-03	0.743976D-04	-0.451792D-03	-0.517016D-02	0.424886D-01	-0.339328	
-0.162333D-02	0.198446D-03	-0.119040D-01	-0.285257	0.407684		
0.221201D-01	0.366952D-03	-0.121000	-3.44873			

I42

0.0	0.322013D-05	-0.399835D-05	0.499860D-04	-0.104832D-03	-0.983538D-02	-0.174180
0.0	0.450820D-05	0.466807D-06	-0.366148D-04	-0.528942D-04	-0.180822D-01	-0.237408
0.0	0.215944D-04	-0.257865D-04	0.157936D-03	-0.920748D-03	-0.100923D-01	-0.521854D-01
0.0	0.940891D-04	-0.985245D-04	-0.128870D-03	-0.712253D-02	0.416891D-01	2.54498
0.0	0.349528D-02	-0.494241D-03	-0.107405D-02	0.571930D-02	-0.454632	
0.0	0.130592D-02	-0.228780D-01	-0.117186	1.16543		
0.0	0.230397D-02	-0.252682	-3.17936			

I43

0.0	0.336424D-06	-0.554120D-05	0.136487D-03	0.517135D-03	0.770049D-03	0.312356D-01
0.0	-0.430395D-06	0.771818D-05	-0.162406D-03	-0.607985D-03	-0.126288D-02	-0.366984D-01
0.0	0.428624D-06	-0.103102D-04	0.275816D-03	0.119040D-02	0.364558D-02	0.793870D-01
0.0	-0.252862D-05	0.389738D-04	-0.790731D-03	-0.291462D-02	-0.326519D-01	-0.248039
0.0	0.287831D-05	-0.714718D-04	0.243530D-02	0.322987D-01	-0.690181D-01	
0.0	-0.398475D-04	0.930315D-03	-0.274754D-01	-0.252298		
0.0	-0.112060D-03	0.295061D-02	-0.879288D-02			

I44

-0.518414D-05	-0.908572D-06	0.190213D-05	-0.476722D-04	-0.443935D-03	-0.121146D-01	-0.214324
-0.446252D-05	-0.560632D-06	-0.666209D-05	0.727245D-04	-0.163382D-03	-0.175743D-01	-0.229727
-0.576722D-04	0.185658D-06	-0.637051D-06	-0.460561D-04	-0.165408D-02	-0.136372D-01	0.518602D-02
-0.167320D-03	-0.156296D-04	-0.271719D-04	-0.326034D-04	-0.587959D-02	0.322749D-01	3.22642
0.168958D-03	0.664915D-04	-0.518314D-03	-0.769858D-02	0.615476D-02	-0.484918	
-0.120612D-02	0.847510D-03	-0.235460D-01	-0.268245	1.13816		
0.176772D-01	0.201808D-02	-0.284510	-4.61344			

0   T =   0.0

ALPHA

0.0	-0.390625D-03	-0.156250D-02	-0.351562D-02	-0.625000D-02	-0.976562D-02	-0.140625D-01
0.0	-0.390625D-03	-0.156250D-02	-0.351562D-02	-0.625000D-02	-0.976562D-02	-0.140625D-01
0.0	-0.390625D-03	-0.156250D-02	-0.351562D-02	-0.625000D-02	-0.976562D-02	-0.140625D-01
0.0	-0.390625D-03	-0.156250D-02	-0.351562D-02	-0.625000D-02	-0.976562D-02	-0.140625D-01
0.0	-0.390625D-03	-0.156250D-02	-0.351562D-02	-0.625000D-02	-0.976562D-02	
0.0	-0.390625D-03	-0.156250D-02	-0.351562D-02	-0.625000D-02		
0.0	-0.390625D-03	-0.156250D-02	-0.351562D-02			

BETA

0.0	0.0	0.0	0.0	0.0	0.0	0.0
0.0	0.0	0.0	0.0	0.0	0.0	0.0
0.0	0.0	0.0	0.0	0.0	0.0	0.0
0.0	0.0	0.0	0.0	0.0	0.0	0.0

0.0	0.0	0.0	0.0	0.0	0.0
0.0	0.0	0.0	0.0	0.0	
0.0	0.0	0.0	0.0		

SIGMA

0.0	0.177083D-02	0.708333D-02	0.159375D-01	0.283333D-01	0.442708D-01	0.637500D-01
0.0	0.177083D-02	0.708333D-02	0.159375D-01	0.283333D-01	0.442708D-01	0.637500D-01
0.0	0.177083D-02	0.708333D-02	0.159375D-01	0.283333D-01	0.442708D-01	0.637500D-01
0.0	0.177083D-02	0.708333D-02	0.159375D-01	0.283333D-01	0.442708D-01	0.637500D-01
0.0	0.177083D-02	0.708333D-02	0.159375D-01	0.283333D-01	0.442708D-01	
0.0	0.177083D-02	0.708333D-02	0.159375D-01	0.283333D-01		
0.0	0.177083D-02	0.708333D-02	0.159375D-01			

U

0.0	0.0	0.0	0.0	0.0	0.0	0.0
0.0	0.0	0.0	0.0	0.0	0.0	0.0
0.0	0.0	0.0	0.0	0.0	0.0	0.0
0.0	0.0	0.0	0.0	0.0	0.0	0.0
0.0	0.0	0.0	0.0	0.0	0.0	
0.0	0.0	0.0	0.0	0.0		
0.0	0.0	0.0	0.0	0.0		

V

0.0	0.0	0.0	0.0	0.0	0.0	0.0
0.0	0.0	0.0	0.0	0.0	0.0	0.0
0.0	0.0	0.0	0.0	0.0	0.0	0.0
0.0	0.0	0.0	0.0	0.0	0.0	0.0
0.0	0.0	0.0	0.0	0.0	0.0	
0.0	0.0	0.0	0.0	0.0		
0.0	0.0	0.0	0.0	0.0		

W

0.0	0.0	0.0	0.0	0.0	0.0	0.0
0.0	0.0	0.0	0.0	0.0	0.0	0.0
0.0	0.0	0.0	0.0	0.0	0.0	0.0
0.0	0.0	0.0	0.0	0.0	0.0	0.0
0.0	0.0	0.0	0.0	0.0	0.0	
0.0	0.0	0.0	0.0	0.0		
0.0	0.0	0.0	0.0	0.0		

RHO

1.00000	0.999527	0.998039	0.995327	0.991043	0.984703	0.975687
1.00000	0.999527	0.998039	0.995327	0.991043	0.984703	0.975687
1.00000	0.999527	0.998039	0.995327	0.991043	0.984703	0.975687
1.00000	0.999527	0.998039	0.995327	0.991043	0.984703	0.975687
1.00000	0.999527	0.998039	0.995327	0.991043	0.984703	
1.00000	0.999527	0.998039	0.995327	0.991043		
1.00000	0.999527	0.998039	0.995327			

I41

0.0	0.0	0.0	0.0	0.0	0.0	0.0
0.0	0.0	0.0	0.0	0.0	0.0	0.0
0.0	0.0	0.0	0.0	0.0	0.0	0.0
0.0	0.0	0.0	0.0	0.0	0.0	0.0
0.0	0.0	0.0	0.0	0.0	0.0	
0.0	0.0	0.0	0.0	0.0		
0.0	0.0	0.0	0.0	0.0		

I42

0.0	0.0	0.0	0.0	0.0	0.0	0.0

0.0	0.0	0.0	0.0	0.0	0.0	0.0
0.0	0.0	0.0	0.0	0.0	0.0	0.0
0.0	0.0	0.0	0.0	0.0	0.0	0.0
0.0	0.0	0.0	0.0	0.0	0.0	
0.0	0.0	0.0	0.0	0.0		
0.0	0.0	0.0	0.0	0.0		

I43

0.0	0.183666D-11	0.123996D-11	0.496928D-10	-0.866347D-10	0.989677D-09	-0.153714D-07
0.0	0.183666D-11	0.123997D-11	0.496927D-10	-0.866349D-10	0.989677D-09	-0.153714D-07
0.0	0.183666D-11	0.123997D-11	0.496928D-10	-0.866347D-10	0.989677D-09	-0.153714D-07
0.0	0.183667D-11	0.123997D-11	0.496928D-10	-0.866349D-10	0.989677D-09	-0.153714D-07
0.0	0.185477D-11	0.740624D-12	0.562403D-10	-0.176092D-09	0.306208D-08	
0.0	0.188214D-11	-0.117902D-12	0.710240D-10	-0.535139D-09		
0.0	0.202462D-11	-0.572050D-11	0.236418D-09			

I44

-0.138112D-12	-0.226774D-11	-0.156186D-11	-0.618199D-10	0.108279D-09	-0.127713D-08	0.205579D-07
-0.138112D-12	-0.227740D-11	-0.156142D-11	-0.618208D-10	0.108277D-09	-0.127713D-08	0.205578D-07
-0.138112D-12	-0.226774D-11	-0.156097D-11	-0.618199D-10	0.108279D-09	-0.127713D-08	0.205579D-07
-0.138112D-12	-0.226752D-11	-0.156164D-11	-0.618194D-10	0.108277D-09	-0.127713D-08	0.205579D-07
-0.136557D-12	-0.228950D-11	-0.959899D-12	-0.698706D-10	0.220628D-09	-0.395426D-08	
-0.140554D-12	-0.232436D-11	0.741629D-13	-0.680989D-10	0.672496D-09		
-0.136557D-12	-0.252931D-11	0.690226D-11	-0.292669D-09			

4   T =   0.12500

ALPHA

0.467525D-02	0.430293D-02	0.318625D-02	0.132593D-02	-0.127709D-02	-0.462085D-02	-0.864276D-02
0.467534D-02	0.430293D-02	0.318625D-02	0.132579D-02	-0.127712D-02	-0.462216D-02	-0.870252D-02
0.467504D-02	0.430293D-02	0.318624D-02	0.132601D-02	-0.127724D-02	-0.461763D-02	-0.867842D-02
0.467525D-02	0.430294D-02	0.318624D-02	0.132563D-02	-0.127748D-02	-0.460774D-02	-0.893488D-02
0.467570D-02	0.430294D-02	0.318629D-02	0.132636D-02	-0.126312D-02	-0.469285D-02	
0.467158D-02	0.430295D-02	0.318533D-02	0.134432D-02	-0.118875D-02		
0.477674D-02	0.430648D-02	0.317655D-02	0.143266D-02			

BETA

0.496651D-02	0.496517D-02	0.496081D-02	0.495213D-02	0.493787D-02	0.491194D-02	0.525289D-02
0.496655D-02	0.496517D-02	0.496084D-02	0.495204D-02	0.493840D-02	0.490961D-02	0.543500D-02
0.496678D-02	0.496517D-02	0.496093D-02	0.495153D-02	0.494123D-02	0.488164D-02	0.576349D-02
0.496637D-02	0.496514D-02	0.496099D-02	0.495149D-02	0.494307D-02	3.491333D-02	0.209465D-02
0.496667D-02	0.496518D-02	0.496083D-02	0.495059D-02	0.497874D-02	0.547680D-02	
0.496507D-02	0.496516D-02	0.496051D-02	0.501036D-02	0.519190D-02		
0.500871D-02	0.496985D-02	0.492684D-02	0.656803D-02			

SIGMA

0.496651D-02	0.671489D-02	0.119608D-01	0.207066D-01	0.329563D-01	0.487150D-01	0.680050D-01
0.496655D-02	0.671489D-02	0.119608D-01	0.207066D-01	0.329563D-01	0.487151D-01	0.680329D-01
0.496678D-02	0.671489D-02	0.119609D-01	0.207065D-01	0.329565D-01	0.487146D-01	0.681064D-01
0.496637D-02	0.671488D-02	0.119608D-01	0.207066D-01	0.329562D-01	0.487177D-01	0.677476D-01
0.496667D-02	0.671489D-02	0.119608D-01	0.207065D-01	0.329578D-01	0.487250D-01	
0.496507D-02	0.671485D-02	0.119606D-01	0.207108D-01	0.329575D-01		
0.500871D-02	0.671761D-02	0.119607D-01	0.207430D-01			

U

0.0	0.0	0.0	0.0	0.0	0.0	0.0
-0.269651D-05	-0.931866D-06	0.287824D-06	-0.192539D-06	0.277198D-05	-0.171500D-04	-0.161887D-02
-0.253781D-05	-0.763094D-06	0.145727D-06	-0.371911D-07	-0.100388D-06	-0.368920D-04	-0.403222D-02
0.894022D-05	0.331211D-05	-0.124244D-05	0.179341D-05	-0.634962D-05	0.131351D-03	-0.212524D-02
-0.145914D-04	-0.630249D-05	0.353291D-05	-0.943708D-05	0.199029D-04	-0.214787D-02	
0.706664D-04	0.300648D-04	-0.198399D-04	0.114314D-03	-0.101270D-02		
-0.964333D-03	-0.496073D-03	0.268327D-03	-0.449950D-02			

V

0.0	0.0	0.0	0.0	0.0	0.0	0.0
-0.141159D-06	0.346117D-06	-0.492291D-06	0.993987D-06	-0.259325D-05	-0.362748D-04	-0.398619D-02
-0.508086D-06	-0.650282D-06	-0.896836D-06	0.216518D-05	-0.571781D-05	-0.599715D-04	-0.845945D-02
-0.878237D-06	0.587860D-07	-0.502642D-06	0.102705D-05	-0.161517D-04	0.164613D-04	0.746200D-02
0.404242D-05	-0.227799D-05	0.504975D-05	-0.674563D-05	-0.228517D-03	-0.629012D-02	
-0.138041D-04	0.604712D-05	-0.310530D-04	-0.402403D-03	-0.254527D-02		
0.985300D-04	-0.240777D-03	0.533797D-03	-0.306006D-01			

W

-0.704251D-01	-0.705132D-01	-0.707713D-01	-0.712042D-01	-0.717681D-01	-0.725913D-01	-0.708835D-01
-0.704033D-01	-0.705150D-01	-0.707695D-01	-0.711990D-01	-0.718022D-01	-0.725062D-01	-0.701685D-01
-0.703486D-01	-0.705176D-01	-0.707704D-01	-0.712090D-01	-0.717766D-01	-0.726843D-01	-0.713024D-01
-0.702557D-01	-0.705100D-01	-0.707679D-01	-0.711860D-01	-0.717965D-01	-0.717550D-01	-0.106962
-0.704383D-01	-0.705157D-01	-0.707725D-01	-0.713268D-01	-0.705694D-01	-0.635293D-01	
-0.702778D-01	-0.705023D-01	-0.707177D-01	-0.673691D-01	-0.673824D-01		

-0.713347D-01	-0.673999D-01	-0.765025D-01	0.333057D-01

RHO

0.985498	0.985037	0.983587	0.980940	0.976752	0.970548	0.961272
0.985498	0.985037	0.983587	0.980940	0.976752	0.970552	0.961128
0.985498	0.985037	0.983587	0.980940	0.976749	0.970575	0.960724
0.985498	0.985037	0.983587	0.980941	0.976748	0.970531	0.964900
0.985497	0.985037	0.983587	0.980941	0.976697	0.970057	
0.985504	0.985037	0.983587	0.980860	0.976417		
0.985315	0.985027	0.983630	0.979229			

I41

0.0	0.0	0.0	0.0	0.0	0.0	0.0
0.458433D-03	-0.161458D-04	0.712666D-04	-0.156991D-03	0.942427D-03	-0.190524D-02	0.246353
0.546006D-03	-0.501029D-04	0.684147D-04	-0.407749D-03	0.957345D-03	-0.203864D-01	-0.868861
-0.121722D-02	0.832886D-04	-0.305941D-03	0.187953D-02	0.175308D-02	0.316720	-12.2044
0.105164D-02	0.938457D-03	0.623352D-04	-0.251907D-03	0.981753D-01	1.97554	
-0.750274D-02	0.134243D-02	-0.941986D-02	0.472353	0.841808		
0.804304	0.542841D-01	-0.132592	5.31792			

I42

0.0	-0.290402D-05	0.185892D-04	-0.378290D-03	-0.175973D-03	0.199953D-01	1.30318
0.0	-0.840743D-04	0.950684D-04	0.455629D-03	-0.548275D-03	0.430866D-01	1.57403
0.0	-0.557563D-04	0.887974D-04	-0.995297D-03	0.695391D-03	0.979506D-02	0.924831
0.0	-0.408153D-03	0.540892D-03	0.426734D-04	0.230208D-01	-0.508807D-01	-16.3306
0.0	-0.389304D-03	-0.680823D-04	-0.515471D-02	0.155289	2.22835	
0.0	0.358583D-02	0.645752D-02	0.356167	-0.925469D-01		
0.0	-0.266422	0.383790	4.79723			

I43

0.0	-0.516051D-06	0.208380D-04	-0.278198D-03	-0.279168D-03	-0.239512D-03	-0.923869D-01
0.0	0.421909D-05	-0.303485D-04	0.475208D-03	0.652035D-04	0.437510D-02	0.720345D-01
0.0	0.865213D-05	-0.256463D-04	-0.306738D-03	-0.233955D-02	0.912666D-02	-0.465875
0.0	0.227153D-04	-0.200342D-03	0.243120D-02	-0.327376D-02	0.940177D-01	0.306067
0.0	-0.796045D-05	0.203456D-03	-0.549697D-02	-0.410976D-01	0.858281	
0.0	0.781328D-04	-0.180376D-02	0.514509D-01	0.548752		
0.0	-0.132995D-03	0.141948D-02	-0.309001D-01			

I44

-0.169809D-04	-0.717098D-05	0.646656D-05	0.544590D-04	0.831895D-03	0.225152D-01	1.61043
0.165350D-03	0.264054D-05	0.867689D-04	-0.301432D-03	-0.639796D-03	0.374657D-01	1.71141
-0.670127D-04	-0.322530D-04	0.608523D-04	-0.162655D-03	0.337936D-02	0.101125D-01	1.12270
0.358449D-02	-0.439597D-04	0.336272D-03	-0.197170D-02	0.211771D-02	0.261737D-01	-22.0746
-0.444677D-02	0.615535D-04	-0.284278D-03	0.354429D-02	0.272076	2.02960	
-0.132671D-02	0.328754D-03	-0.399913D-03	0.532200	-0.358759D-01		
0.677712	0.328211D-01	0.835870D-01	7.26436			

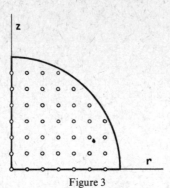

Figure 3
Distribution of 43 grid points across the integration domain.

tion. On the other hand, the numerical values of the functions $\tilde{\alpha}$, $\tilde{\beta}$, $\tilde{\sigma}$, $\tilde{u}$, $\tilde{v}$, $\tilde{w}$, and $\tilde{\rho}$ differ insignificantly from those demonstrated in Figs. 1 and 2.

The question now arises what is the cause of the fast increase of the functions I_μ^4 at the periphery of the integration domain. The roundoff errors and the truncation errors of the applied Runge–Kutta integration method may be with certainty excluded, because the integration was stopped after a few steps, and what is more important, because no discrepancies were found in the numerical results in Figs. 1 and 2 when the integration had been run also with time decreasing from $\tilde{t}_{max} = 0.8125$ to $\tilde{t} = 0.2500$ and from $\tilde{t}_{max} = 0.5625$ to $\tilde{t} = 0.0$, respectively. There exist thus only two reasons for that increase: the errors are due either to the applied Lagrange differentiation formula, or to the back-scattering of radiation (the errors of the differentiation are certainly responsible for the increases of I_μ^4 at and around the

TABLE 1a

The spatial distribution of the functions $\tilde{\alpha}$, $\tilde{\beta}$, $\tilde{\sigma}$, $\tilde{u}$, $\tilde{v}$, $\tilde{w}$, of the mass density $\tilde{\rho} \equiv$ RHO, and the functions $I_\mu^4 \equiv I4\mu$ at the initial moment $\tilde{t} = 0$ within a spherical rotating object of the radius $\tilde{R} = 7/16$. Initial values are the same as in Fig. 1. The $\tilde{r}$-coordinate increases in the horizontal direction from the left to the right, the $\tilde{z}$-coordinate increases in the vertical direction downwards. The interval between two equally spaced grid points $h = 1/16$.

TABLE 1b

The spatial distribution of the functions at $\tilde{t} = 0.1875$ within a spherical object of the Table 1a.

TABLE 2a

The spatial distribution of the functions $\tilde{\alpha}$, $\tilde{\beta}$, $\tilde{\sigma}$, $\tilde{u}$, $\tilde{v}$, $\tilde{w}$, of the mass density $\tilde{\rho} \equiv$ RHO, and of the functions $I_\mu^4 \equiv I4\mu$ at the initial moment $\tilde{t} = 0$ within a spherical rotating object of the radius $\tilde{R} = 7/16$. Initial values are the same as in Fig. 2. The $\tilde{r}$-coordinate increases in the horizontal direction from the left to the right, the $\tilde{z}$-coordinate increases in the vertical direction downwards. The interval between two equally spaced grid points $h = 1/16$.

TABLE 2b

The spatial distribution of the functions at $\tilde{t} = 0.1250$ within a spherical object of the Table 2a.

center of the integration domain). In both cases, when the integration domain was covered by 15 as well as by 43 grid points, the author computed the spatial derivatives using in the Lagrange differentiation formula all the points in each row and each column in order to take into account the fact that all the functions were even functions in r and either even or odd functions in z. However, since the Einstein field equations are of hyperbolic type, it is plausible that the Lagrange differentiation formula may give more accurate results if the derivatives are computed from a lesser number of points. From the standpoint of the numerical analysis the back-scattering of radiation means that the partial derivatives with respect to the spatial coordinates at a given point must be computed from the values of the function on both sides of that point. Both reasons, the errors of the differentiation formula and the back-scattering, are closely related to each other, but the author believes that the back-scattering is the primary source of the fast increase of I_μ^4 at the periphery of the integration domain. These errors are also responsible for the fast increase of the mass density at one point at the periphery of the integration domain which stopped the computer in Figs. 1 and 2. The question what is the source of the fast rise of I_μ^4 can be decided by increasing the number of grid points at least to 144 so that the distribution of the mass density and of the initial values of the metric can be chosen such that the field at the periphery of the integration domain is as weak as required in Secs. 1.4 and 1.5 (there is then no back-scattering at the periphery, because the background metric is here Minkowskian).

The author emphasizes once again that the extent of programming and of running the programs was limited by the research grant. A further work is needed to find out the most appropriate Lagrange differentiation formulas and the most appropriate integration method for the system of ordinary differential equations. However, once these two problems have been solved, the accuracy of the numerical results will then be limited mainly by the extent of the computer memory (it determines the highest number of grid points) and by the speed of computer operations.

One further conclusion can be drawn from Figs. 1 and 2 and Tables 1a, 1b and 2a, 2b: The only reliable indicator of the accuracy of the numerical integration of Einstein equations are the functions I_μ^4. The above mentioned very small difference in the numerical results when the integration is carried out with time increasing from $t = 0$ to t_{max} and then decreasing from t_{max} to $t = 0$ cannot be considered as a test of the reliability of the numerical integration. Therefore the author does not trust to any numerical integration of Einstein equations when its reliability has not been proved by the functions I_μ^4.

The programmer tried to elaborate also the initial data program (i) of Sec. 1.7. Unfortunately, he was not able within the research grant budget to devise a program for the numerical integration of the quasilinear partial differential equation of the first order which would give results with the accuracy required for the time evolution program.

The numerical computation of Figs. 1 and 2 and of Tables 1a, 1b and 2a, 2b was carried out on the computer IBM360–67.

4. A NEW RELATIVISTIC EFFECT

Figure 2 and Tables 2a, 2b demonstrate an interesting relativistic effect discussed already in [9] and mentioned in [10]. When the angular velocity of the rotating body is so high that the centrifugal acceleration can stop the radial contraction and revert it to a new expansion (a well-known Newtonian effect occurring when the parameter κ^2, defined by (51), surpasses a certain critical value depending in Newtonian dynamics upon the global configuration of the body [15]), then this high radial acceleration generates a gravitational wave in the perpendicular, axial, direction (a relativistic effect caused not by the non-linearity of Einstein equations, but by their hyperbolic character; remember the radiation generated in Maxwell electrodynamics by an accelerated motion of an electric charge) which stops the axial contraction and reverses it to a new expansion.

Figure 2 and Tables 2a, 2b clearly show the decrease of the mass density caused by the increase of the functions, α, β, σ, i.e., by the increase of the axial and radial dimensions of the body. When the angular velocity is not high enough to stop the radial contraction (Fig. 1 and Tables 1a, 1b), then the corresponding gravitational wave generated in the axial direction by the accelerated radial motion cannot stop the axial contraction. Of course, Fig. 2 and Tables 2a, 2b must not be considered as an unambiguous proof that the expansion effect does exist, because the I_μ^4 take inadmissibly high values at the periphery of the integration domain, but merely as an indication that its existence is very plausible.

It seems that Eddington [16] in 1935 was right when he believed that some effect would be found preventing a star to collapse into a black hole.

Notes Added in Proof

Since the results of numerical integration as described in Part 4 were unsatisfactory, the author decided to withdraw the manuscript of the paper [4] and to rewrite it after a thorough investigation of different numerical methods that would give more reliable results. With respect to the available amount of the research grant the research could be resumed in September 1974 in collaboration with Dr. D. A. Swayne, a post-doctorate fellow in computer science. The efficiency of the methods was checked by developing and using a far simpler program for a spherically symmetric collapse of a relativistic star. The results of this investigation are very encouraging and will be described elsewhere. They also resulted in the following modifications of the computer programs for the integration of axially symmetric Einstein equations.

Sec. 1.7. The unknown functions v_4 and w_4 were chosen in order to minimize the numerical errors in the initial data program, but an improper choice of other Cauchy data occurring in the program gives rise to singularities in the functions v_4 and w_4. Therefore another program was developed with α and w_4 as unknown functions.

Sec. 2.1. It is more advantageous to map the integration domain on a quadrant with a unit radius. The reduction formulas (46) are still valid, but the constant ρ_i has a slightly different meaning and $\tilde{\rho} \neq 1$ at $z = r = t = 0$.

Sec. 2.3. It has been found that the Lagrange differentiation formula is responsible for the majority of truncation and roundoff errors. Therefore the functions that have to be differentiated are expanded into finite series of Gram orthogonal polynomials, the numerical data of those functions are smoothed using the least squares technique in order to diminish the errors of numerical analysis, and then the polynomials are analytically differentiated [17].

Sec. 3. When the I_μ^4 functions surpass a certain critical value, the results of the numerical integration are no longer reliable. In order to continue the integration, four elements of the Cauchy data are to be adjusted to satisfy the conditions (9). When these functions again surpass the critical level, the procedure is repeated. In this way we obtain a sequence of solutions that do not correspond exactly to the original initial data, but which nevertheless may be considered as a set of solutions closely related to the original problem. The advantage of this procedure is that it permits a continuation of integration for a longer period of time than would be otherwise possible. The procedure was suggested in [10]; it has been successfully applied in the spherically symmetric problem, so there is a fair expectation that it may be adapted also for the axially symmetric case.

Sec. 4. The simultaneous reversion of the radial and axial contraction of a rotating body into an expansion is caused by the hyperbolic character of Einstein equations (not by their non-linearity) and by the presence of four Lichnerowicz conditions (9) restricting the free choice of Cauchy data. The term that corresponds to the centrifugal acceleration of Newtonian dynamics is coupled through these conditions with the local Cauchy data determining the axial contraction. The effect will be discussed thoroughly in the paper [18] that will replace the withdrawn manuscript [4].

REFERENCES

1. R. PENROSE, *Phys. Rev. Lett.* **14**, 57 (1965).
2. C. LANCZOS, *Phys. ZS* **31**, 112 (1925); E. LIFSHITZ, *J. Phys.* **10**, 116 (1946).
3. J. PACHNER, *J. Comput. Phys.* **15**, 385 (1974).
4. J. PACHNER, *J. Comput. Phys.* (withdrawn and replaced by [18]).
5. A. LICHNEROWICZ, *Théories relativistes de la gravitation et de l'électromagnétisme*, Masson et Cie., Paris, 1955, pp. 1–6, 31, 59–63.
6. J. PACHNER, *Bull. Astron. Inst. Czech.* **19**, 33 (1968).

7. J. PACHNER, *Gen. Rel. Grav.* **1**, 281 (1971).

8. K. GÖDEL, *Rev. Mod. Phys.* **21**, 447 (1949).

9. J. PACHNER, *Can. J. Phys.* **51**, 447 (1973).

10. J. PACHNER and R. TESHIMA, *Can. J. Phys.* **51**, 743 (1973).

11. J. L. SYNGE, *Relativity: The General Theory,* North-Holland Publ. Co., Amsterdam, 1966, pp. 39–40.

12. V. FOK, *The Theory of Space, Time and Gravitation,* 2nd revised edition, Macmillan Co., New York, 1964, p. 193.

13. R. COURANT and D. HILBERT, *Methods of Mathematical Physics,* Vol. 2, Interscience Publ., New York, 1962, pp. 208–210, 568–569, 735.

14. R. G. McLENAGHAN, *Proc. Camb. Phil. Soc.* **65**, 139 (1969).

15. P. LEDOUX, in *Encyclopedia of Physics* (S. FLÜGGE, ed.), Vol. 51, Springer, Berlin, 1958, pp. 621–624.

16. A. S. EDDINGTON, *The Observatory* **58**, 38 (1935).

17. A. RALSTON, *A First Course in Numerical Analysis,* McGraw-Hill, New York, 1965, pp. 229–232, 240–243.

18. J. PACHNER and D. A. SWAYNE (in preparation).

Tests of Theories of Gravity in the Solar System

JEAN-PAUL RICHARD

Department of Physics and Astronomy, University of Maryland
College Park, Maryland, U.S.A.

1. INTRODUCTION

This paper is essentially a review of recent or forthcoming experiments which test theories of gravity. The development of space techniques and the improvement of radar and low temperature techniques has led to an impressive increase of activity in that area in the last ten years. We have also seen in recent years many new theories of gravitation and an improved theoretical framework to help analyse and confront theories and tests. Among all the tests, the Eötvos and the gravitational red-shift experiments are given special importance as bringing support to the concept of a metric theory. Recent results on the former are familiar and will not be reviewed here again. Recent development in gravitational red-shift experiments will be reviewed first, and then, light bending, time delay, perihelion advance, gyroscopic effects and others.

2. GRAVITATIONAL RED SHIFT

The first attempt (Table 1) at measuring the time-dilation effect was made by observing shift of spectral lines originating at the surface of the Sun. The most successful measurement was the one by Brault (1962). He observed the strong sodium D_1 line emitted high in the Sun atmosphere above the highly convective zones. The shift was measured to be the General Relativity (GR) value $\pm 5\%$.

A more precise local measurement was made using the Mössbauer nuclear resonance absorption of the 15 keV line of Co^{57} by Fe^{57}. The latest of a series of measurements gave the GR value $\pm 1\%$ (Pount and Snider, 1965). The possibility

TABLE 1

Time Dilation Measurements

		Accuracy
	Performed	
Brault (1962)	Solar sodium D_1 line	5%
Pound and Snider (1965)	Mössbauer (local) Co^{57}	1%
Jenkins (1966)	GEOS-1 satellite (quasilocal)	10%
Hafele and Keating (1972)	Gesium clocks on plane (local)	14%
	Future	
Alley (1975)	Cesium and rubidium clocks on planes (local)	1%
Vessot (1975)	Maser on scout rocket (local)	0.002%
	Maser on Mercury–Venus spacecraft (non-local)	0.1%
	Clock on high-eccentricity (0.3) solar orbit (non-local):	$1/10^6$
	1st order ($\sim 10^{-8}$)	
	2nd order ($\sim 10^{-16}$)	
Pound	Tantalum isomer shift	0.02%

of using the much narrower line of zinc has been reviewed again recently by Pound and again found to be impractical. The tantalum isomer shift of 6 keV seems however promising (Pound, 1974). The observed line would be approximately four times narrower than the Co^{57} line. Preliminary experiments suggest that a light pipe could be used to allow a great vertical distance between the emitter and the absorber without the $1/r^2$ loss of intensity. If so, accuracies of the order of 0.02% could be achieved.

It has been suggested quite some time ago (Singer, 1956) that a time standard in orbit around the Earth be used to measure the time-dilation effect. The only measurement to date has been reported by Jenkins (1969). He observed the variation in frequency of a temperature controlled crystal oscillator aboard a GEOS-1 NASA satellite, as the satellite moved from perigee to apogee. The orbit semi-major axis was 1.26 Earth radii and the eccentricity 0.07. One hundred and two measurements distributed over four days in 1966 gave a 99% confidence level that the relativistic variation was there and a best fit of the residuals gave 0.95 ± 0.09 times the GR effect.

The latest measurement of the time-dilation effect was reported in 1972 (Hafele and Keating, 1972). Four cesium beam clocks were flown around the world once westward and once eastward on commercial flights. The delays predicted by GR were respectively -59 and $273 \cdot 10^{-9}$ sec, of which $\sim 150 \cdot 10^{-9}$ was gravitational. They were observed with an estimated error of $\pm 20 \cdot 10^{-9}$ sec or 14% of the gravitational effect. A similar experiment is being prepared by C. O. Alley where three cesium and three rubidium clocks will be flown at low speed at an altitude of 20 km for a few hours. The clocks will be temperature controlled. Laser contact will be made from ground during the flight to establish in a continuous fashion the marking of time by the clocks. The stability of the clocks has been determined to be five parts in 10^{14} over three hours. The accuracy of the measurement should be close to 1% of the GR effect which is two parts in 10^{12}.

A rocket probe experiment is being projected for September 1975 by R. Vessot (1974). In this experiment, a hydrogen maser will be flown aboard a Scout rocket to an altitude of approximately 3.5 Earth radii. The total flight time will be a few hours. A maser controlled 2.203 MHz signal will be transmitted to the Earth. The first order doppler will be measured by a transmitter transponder system and electronically subtracted from the maser signal on ground, leaving second order terms only to be analyzed. The stability of the maser will be a few parts in 10^{15}. The accuracy of the test is expected to be 0.002% of the GR effect assuming that corrections for atmospheric effects can be accomplished with corresponding accuracy.

More experiments have been proposed. They involve launching highly stable time standards, masers or rubidium clocks, in orbit around the Earth or around the Sun. Earth orbits of high eccentricity can provide changes in gravitational potential of the order of 10^{-9} in a few minutes and allow accurate tests with clocks of modest stability (Richard, 1965). A hydrogen maser in such an eccentric orbit of mean synchronous velocity could yield a test to 0.001% of the GR effect (Kleppner et al., 1970). A standard of similar stability around the Sun aboard a Mercury/Venus spacecraft could yield a test with an accuracy of 0.1% over a non-uniform region of the Sun gravitational field (Shapiro, 1971). On a solar orbit of eccentricity 0.3 and an apohelie of 1 A.U., the second order time-dilation effect would be about two parts in 10^{16}. Such an effect is beyond present technology capability.

3. THE PARAMETRIZED POST NEWTONIAN FORMALISM

The red-shift experiments which have been performed support the hypothesis that the "real" theory of gravity is a metric theory. A general frame to analyze the experimental predictions of such theories had been developed (Eddington, 1922; Robertson, 1962; Schiff, 1962) to cover experiments where spherical symmetry was an adequate approximation. This framework has been extended recently by Nordtvedt and Will (Nordtvedt, 1968a, b, c; Will 1971; Will and Nordtvedt, 1972). In addition to the familiar parameters β and γ which measure the non-linearity in the superposition law of gravity and the space curvature produced by unit rest mass, β_1–β_4 measure the gravity produced by kinetic energy, gravitational energy, internal energy and pressure, ζ and η measure the anisotropy in the production of gravity by velocity and stress, Δ_1 and Δ_2 measure the dragging and anisotropy in dragging of inertial frame by momentum. To each theory correspond a set of value for these parameters. For theories which conserve angular and linear momentum, $\alpha_1 = \alpha_2 = \alpha_3 = 0$ and the following combinations of the previous parameters are also zero:

$$\zeta_1 = \zeta$$

$$\zeta_2 = 2\beta + 2\beta_2 - 3\gamma - 1$$

$$\zeta_3 = \beta_3 - 1$$

$$\zeta_4 = \beta_4 - \gamma$$

The functions

$$\alpha_1 = 7\Delta_1 + \Delta_2 - 4\gamma - 4$$

$$\alpha_2 = \Delta_2 + \zeta - 1$$

$$\alpha_3 = 4\beta_1 - 2\gamma - 2 - \zeta$$

are zero in a theory with no preferred frame effects.

Preferred frame effects include an anisotropy of G which would show as a daily variation of g (Will, 1971b). Gravimeter measurements suggest an upper limit to such an effect. Also, preferred frame effects would include an additional perihelion advance to which present experimental data impose an upper limit. The experiments impose the following set of limits on presently acceptable values of α's:

$$|\alpha_1| < 0.2$$

$$|\alpha_2| < 0.03$$

$$|\alpha_3| < 2 \cdot 10^{-5}$$

A large number of theories have been shown to be in conflict with these results (Will and Nordtvedt, 1972).

4. LIGHT BENDING

Next to the perihelion advance, the deflection of light rays is the oldest test of GR. The deflection is maximum (1.75 sec of arc) for rays grazing the solar surface. For other values of the impact parameter p, it decreases as (Ward, 1970)

$$\Delta\theta = 1.749 \frac{R_\odot}{p} \frac{(1 + \cos\theta)}{2} \quad \text{sec of arc}$$

where $R_\odot$ is the Sun radius and θ is the angular distance from the Sun to the observed object as seen by the observer.

The bending of light rays has been observed in the visible and at radio frequencies.

Numerous observations of the effect have been made during solar eclipses. Figure 1 shows the results of many experiments as reexamined by Mikhailov (Mikhailov, 1959). The dispersion of the results is appreciable. No strong conclusion can be drawn from these results. (They however brought support to the GR value as opposed to half that value as calculated for a photon falling in the Newtonian field of the Sun.)

The best measurements of the light bending so far have been performed at cm wavelength using VLBI techniques as suggested by Shapiro (1967). Atmospheric and solar plasma effects limit the accuracy of these experiments.

The strong nearly point source like radio sources 3C273 and 3C279 have been observed most often. 3C279 is occulted by the Sun every October and 3C273, 11° apart, serves as a reference (Fig. 2). The angular separation of the sources is obtained from the difference in phase in their signals. In such a differential measurement, the atmospheric effects are greatly reduced but are still important. Figure 3 shows the fluctuations around the best fit curve in the 1970 NRAO observation of 3C273 and

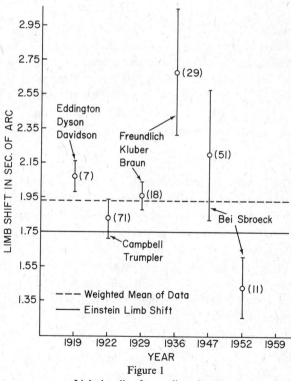

Figure 1

Light bending from eclipse data.

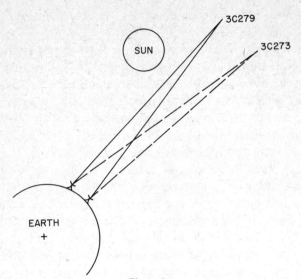

Figure 2

VLBI two-antenna configuration for light-bending experiment.

3C279 with a baseline of 2.7 km and a frequency of 8.1 GHz. The atmospheric fluctuations can be compared to the relativistic effect which is the difference between the dashed and solid line. In an effort to reduce even further atmospheric, instrumental and mechanical phases errors, the NRAO group (Sramek and Fomalon, 1974) has observed a group of three radio sources. P1306-09, passes within eight solar radii from the Sun in October. P1245-19 and P1330-02, approximately 12° from P1306-09 lie on a straight line. Figure 4 shows a significant reduction of noise for that experiment.

Atmospheric errors could also a priori be reduced by using longer baselines. In two antennas experiments, observations of each source alternate and phase have to be extrapolated between each observation. This "phase connection" has an ambiguity of 2π. In four antenna configurations (Fig. 5) closely located pair of antennas follow the sources. Phase connections are fewer, and fluctuations in the rate of the time standards are less important.

At radio frequencies, the electron population of the solar corona acts as a refraction medium. Its contribution to the relativistic bending and time delay can be important. It can be calculated from a two parameter expression for the electron density (Van de Hulst, 1950; Newkirk, 1967):

$$N_e(r) = \frac{A}{r^6} + \frac{B}{r^{2+\varepsilon}}$$

where r is the distance from the center of the Sun. The equivalent index of refraction

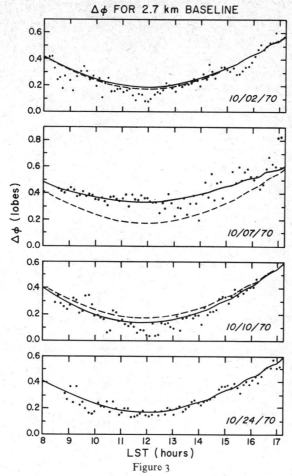

Figure 3

Phase difference between 3C279 and 3C273 on four different days. Solid line is best fit.

of the medium is then:

$$n^2 = 1 - \frac{4\pi e^2 N_e(r)}{m_e f^2}$$

where e and m_e are the charge and mass of the electron and f, the frequency of the signal observed. The values of A, ε and B have been derived at first from corona scattering observations during solar eclipses. They can also be determined simultaneously with the relativity parameters in light bending and time delay experiments.

Because of their dependence in $1/f^2$, these effects can also be made negligible by using very high frequencies (> 10 GHz). In two-frequencies experiments, the

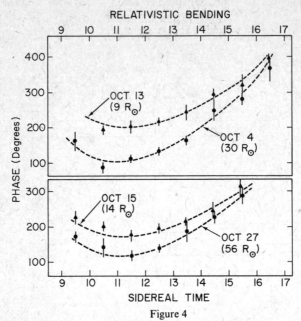

Figure 4

Phase difference in the October, 1973, P1306-09 experiment on four different days. Dotted line is best fit.

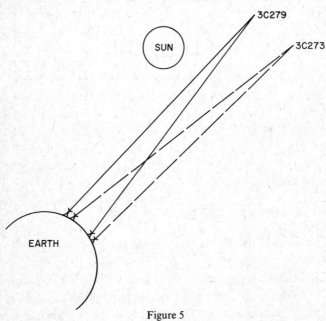

Figure 5

VLBI four-antenna configuration for light-bending experiment.

TABLE 2
Light-bending Experiments

Date	Frequency (MHz)	Baseline (km)	Antennas	$(1 + \gamma)/2$	Accuracy	Reference
1969	2400	21	Goldstone	1.04	0.15	Muhleman et al., 1970
	9602	1	Owens Valley	0.99	0.12	Seielstad, 1970
1970	5000	0.7	Mullard	1.14	0.3	Hill, 1971
	2700	1.4	R. A. Obser.	1.07	0.17	
1970	2695	2.7	NRAO	0.90	0.05	
1971	8085	1.9		0.97	0.08	Sramek, 1971, 1972
		0.8				
1972	8100	845	Haystack-NRAO	0.99	0.03	Shapiro et al., 1974
1973	8085	35	NRAO	1.01	0.08	Sramek and Fomalon, 1974
1974	8085	35	NRAO		0.03	Sramek, 1974
	2695					
1974	5000	1.4	WRST		~ 0.03	Raimond, 1974
	1400		(Netherlands)			

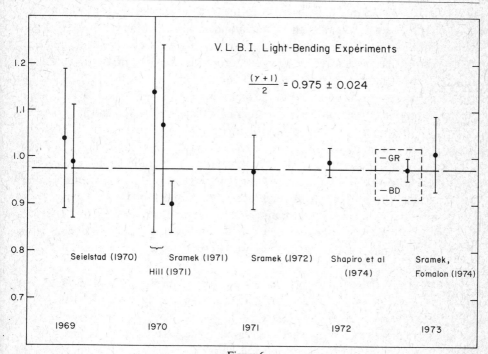

Figure 6
Statistics of VLBI light-bending experiments.

relativity and plasma effects can be directly separated. Table 2 lists observations of the relativistic bending experiments. A weighted fit of these results except for the long baseline (845 km) result gives 0.957 ± 0.04. A weighed fit of all short and long baseline results gives 0.975 ± 0.024 (Fig. 6). The errors given are the weighted average of the errors indicated by the authors and are consistent with the scattering of the data. The GR and Brans–Dicke values are also indicated.

5. TIME DELAY

The possibility of measuring a relativistic delay in the propagation of radio signals was first suggested by Shapiro (Shapiro, 1964; see also Muhleman and Reichly, 1964). The parameter tested here is again γ. The effect for rays passing near the Sun has a logarithmic dependence on the distance from the center of the Sun which helps greatly in separating it from orbital effects. For this reason, this measurement has been conducted during superior conjunctions of planets and probes. The effect is maximum for rays grazing the solar surface. The increase in the two way travel time is equivalent to an additional optical path of 72 km. The delays introduced by the Earth atmosphere are negligible.

Time-delay experiments have been performed by measuring the two way time of travel of radio signals from Earth to transponders aboard Mariner 6 and 7 space-crafts in orbit around the Sun. The Mariner transponders operate at a frequency of 2.2 GHz and can be tracked as close as 1° from the Sun. The time-delay measurements are in themselves consistent at the 3 m level (Muhleman et al., 1971). The plasma effects have to be modeled in the analysis of measurements close to the Sun. The errors on that correction are of the order of 800 m near the Sun and 30 m far from it. These errors impose a limit of approximately 1% to the accuracy with which the relativistic parameter $(1 + \gamma)/2$ can be determined. The non-gravitational forces have been more important sources of errors. These accelerations of the spacecraft are at the level of 10^{-7} m/sec^2. Fluctuations of the solar radiation pressure and gas leakage from attitude control systems account for most of it (Anderson, 1973). These random fluctuations produce an error in the predicted position of the space-craft which grows like $t^{3/2}$ and amounts to ~ 1 km after 3 months. This limits the length of the orbital arc which can be usefully processed for relativistic effect to $\leq \sim 3$ months.

The results of the experiments performed during the superior conjunctions of Mariner 6 and 7 gives $(1 + \gamma)/2 = 1.00$ and 1.01 respectively, with a formal error of 1.5% and a "realistic" error of 4% (Anderson, 1973).

The time-delay effect has also been observed in the round trip travel time of radio signals between the Earth and planets. Radar pulses have been bounced off the surfaces of Mercury, Venus and Mars from Goldstone (430 MHz) and Haystack

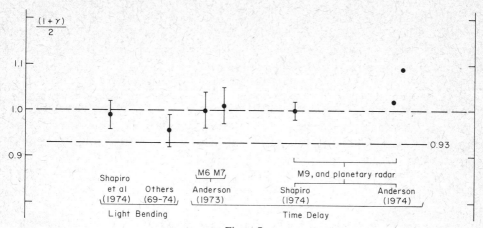

Figure 7

Statistics of VLBI and time-delay experiments.

(7.8 GHz). Also, 2 GHz measurements have been made with Mariner 9 in orbit around Mars. In all these cases the non-gravitational forces are not a problem. In the case of Mariner 9, however, the analysis of long arcs of trajectory requires the modeling of the gravitational field of Mars. The arecibo data (430 MHz) are taken far from the Sun ($> 15°$) and used for orbit determination and plasma calibration. From the planetary and Mariner 9 data, Shapiro's result is $(1 + \gamma)/2 = 1.00 \pm 0.02$ (Shapiro, 1974). Anderson's analysis however indicates a high corrolation of γ^* with the plasma effects. Values of $\gamma^* = (1 + \gamma)/2$ from 1.02 to 1.09 appear in the multiparameter solutions with little changes in rms value of the residuals (Anderson, 1974).

Figure 7 shows simultaneously the results of light-bending and time-delay measurements for $(1 + \gamma)/2$. A weighted fit of these results (except for those of Anderson for which "no realistic" errors were suggested) gives $(1 + \gamma)/2 = 0.993 \pm 0.014$ where the error is the weighted average of the errors indicated by the authors. This standard deviation is consistent with the scattering of the results. This value of $(1 + \gamma)/2$ agrees with GR but does not support a Brans–Dicke value of 0.93.

6. TIME DELAY–DOPPLER

If a time standard moving towards superior conjunction is observed from Earth, the relativistic corrections to the optical path in the field of the Sun will produce an apparent shift of the frequency of the source (Richard, 1966). This effect could possible be observed if a pulsar is used as such a time standard (Richard, 1968). The effect would be $\sim 10^{-10}$. It has not been practical yet to measure this effect.

7. CELESTIAL MECHANICS

7.1. Perihelion advance and solar oblateness

The relativistic time-delay effect is most important in radar measurements performed near superior conjunction of the targets. In the rest of the planetary and
Mariner 9 data and the previously obtained optical data the perihelion advance
is still the most important relativistic orbital effect. Its value depends on the parameters β and γ

$$\dot{\omega}_R = \frac{(2 + 2\gamma - \beta)\,nm}{a(1 - e^2)}$$

where n is the mean angular orbital velocity, m the mass of the Sun, a the semi-major
axis, and e the eccentricity of the orbit. A quadrupole moment of the Sun would
produce an advance of the perihelion ω and a precession of the node Ω given by

$$\dot{\omega}_Q = -\dot{\Omega}_Q = -\frac{3}{2}\frac{J_2 n R_\odot^2}{a^2(1 - e^2)^2}$$

where J_2 is the quadrupole moment and $R_\odot$ the radius of the Sun. Table 3 shows
the two effects for four planets.

It has not been practical to determine the quadrupole moment of the Sun from the
precession of the nodes of planetary orbits because of the low inclination of these
orbits in the plane normal to the axis of rotation of the Sun. The experiments are
in fact, more sensitive to the different radial dependences of the quadrupole and
relativistic effects.

The radar data to Mercury, Venus, Mars and Mariner 9, together with previous
optical data, have been analyzed by Shapiro (1974) and Anderson (1974). The
results are shown in Table 4. The JPL analysis gives many dissimilar results. Three
of the solutions are shown. The residuals for these three solutions do not differ

TABLE 3
Perihelion ($\dot{\omega}$) and Node ($\dot{\Omega}$) Precessions
(sec of arc/century)

	GR, $\dot{\omega}$	$J_2 \doteq 2.7 \cdot 10^{-5}$, $\dot{\omega} = -\dot{\Omega}$
Mercury	42.95	-3.4
Venus	8.6	-0.35
Earth	3.84	-0.11
Mars	1.35	-0.03

TABLE 4

Perihelion Advance and Solar Quadruple Moment

Radar data to Mercury	(assumed: $\beta = 1, J_2 = 0$)	$\gamma = 1.02 \pm 0.002$
Venus and Mars	$\beta = 2.14$	$J_2 = 10.8 \cdot 10^{-5}$, $\gamma = 1.09$
Mariner 9 ranging (Anderson, 1974)	$\beta = 2$	$J_2 = 12 \cdot 10^{-5}$ $(\gamma = 1)$
Radar data alone (Shapiro, 1974)	$\lambda_p = 0.98 \pm 0.04$	$J_2 = (0.5 \pm 1.3) \cdot 10^{-5}$,
Dicke–Goldenberg	Quadrupole moment	$J_2 = (2.7 \pm 0.5) \cdot 10^{-5}$
	Optical oblateness	$(5 \pm 0.7) \cdot 10^{-5}$
Hill, 1974	Optical oblateness	$(0.5 \pm 0.66) \cdot 10^{-5}$
	Uniformly rotating Sun, optical oblateness	$0.88 \cdot 10^{-5}$

significantly although the discrepancies between the various solutions are much larger than the formal errors associated with them. From the radar data alone, the MIT analysis gave a value for $\lambda_p = (2 + 2\gamma - \beta)/3$ which is consistent with General Relativity.

A measurement of the solar oblateness has also been reported by Hill et al. (1974) and Stebbins (1974). The instrument used is a vertical telescope with a single lense objective of 12 m focal length and 12.5 cm aperture. Mirrors provide altitude and azimuth coverage. An image of the Sun of coronographic quality is produced at the focal plane. Stars close to the Sun can also be observed. Four experiments are projected: the solar oblateness, the bending of light, the time variation of G and the Earth perihelion advance. The two last measurements are expected to be performed over a period of ten years.

Figure 8 shows the arrangement for the solar oblateness measurement. Two narrow slits (one second of arc radial opening) are driven at a frequency ω. The position of each slit is adjusted so that the second harmonic of the signal passing through it is zero. The distance between the two slits is measured as the "diameter" of the Sun. Figure 9 shows variations of the measured diameter of the Sun as the amplitude of the motion of the slits is changed. This is a response to the structure of the edge of the Sun. If the Sun intensity would vary as a step function at the edge, the measured diameter of the Sun would be independent of the amplitude of oscillation of the slits.

The oblateness has been measured for two amplitudes of oscillation of the slits, 6 and 24 arc seconds. The difference in the two measurements reflect the presence

Telescope Objective

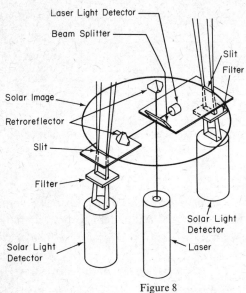

Laser Light Detector
Beam Splitter
Slit
Filter
Solar Image
Retroreflector
Slit
Filter
Solar Light
Detector
Solar Light
Detector
Laser

Figure 8

Schematic of slit and interferometer arrangement along with the solar-light detection apparatus.

of an excess equatorial brightness:

$$\Delta I = I_{\text{equator}}(r) - I_{\text{pole}}(r)$$

where r is the distance from the center of the Sun. Figure 9 shows how the difference of the two oblateness measurements varies with time. According to Hill, the high level of excess equatorial brightness seen in November 1973 could explain a solar oblateness twice as large as the one observed by Dicke and Goldenberg for the procedure they used. Figure 10 shows the oblateness measured during the days where the excess equatorial brightness was low (3% of its average value). The x-axis is essentially the time of the day. The sinusoidal term is instrumental. The bias contains an instrumental bias and the measured oblateness. The result is a visual oblateness of 10.4 ± 12.4 millisecond of arc. A detailed analysis of the various corrections involved has been given by Hill and Stebbing (1974). This result can be compared (Table 4) with Shapiro's result, with the oblateness calculated for a uniformly rotating Sun, and with the result reported by Dicke and Goldenberg (1967).

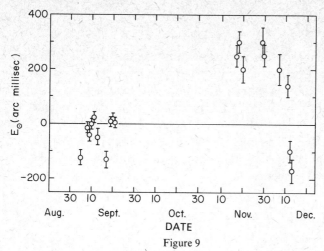

Figure 9

Difference of two oblateness measurements (slit amplitude of oscillation, 6 and 24 seconds of arc).

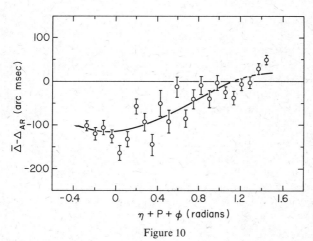

Figure 10

Solar oblateness measurements corrected for atmospheric contribution and best fitted.

7.2. Remark

The separation of orbital relativistic effects due to β and classical effects due to a solar oblateness would a priori be easier in time delay experiments performed from Earth to spacecrafts on "quasiradial" orbits such as sections of a grand tour to outer planets due to the different radial dependence of the two corrections. The relativistic correction due to β in such experiments could be of the order of 1 km (Richard, 1972). Observation of such an effect would require "drag-free" probes. Even then, interplanetary gravitational noise could limit the accuracy of the experiment.

7.3. Time variations of G

An experimental limit on the time variation of G can also test theories of gravity. Two measurements of that quantity have been reported recently (Table 5). The first one (Shapiro, 1974) is extracted from the radar and optical observations of the planets. It basically compares the mean angular motions of Mercury with atomic time. The topography fluctuations limit the accuracy of the result. The second has been reported by Van Flandern (1974) from observations of the secular acceleration of the moon. Both results overlap. Continuous collection of radar data should rapidly improve the former result.

TABLE 5

Time Variations of G

From the secular acceleration of the Moon (Van Flandern, 1974)	
Total acceleration deducted from 1955 occultations of stars	-83 ± 10 sec of arc/cent^2
Part due to tidal friction	
—from ancient eclipses (Newton)	
—newly discovered data on ancient eclipses (Moller and Stephenson)	-41 ± 4 sec of arc/cent^2
—meridian observations of the Sun, Moon and planets since 1913 (Oisterwinter, Cohen)	
Result:	$\dot{G}/G = (-1.2 \pm 0.3) \cdot 10^{-10}$ year^{-1}
Radar observations of Mercury	
Effect observed	$\dot{G}/G = -4 \pm 8 \; 10^{-11}$/year
Corresponding displacement	~ 200 m
Topography fluctuations	~ 1 km

7.4. Orbital equivalence principle

Dicke (1964) and Nordtvedt (1968a) have discussed the possibility that gravitational binding energy contributes to inertial and passive gravitational mass in a different way. This effect would be important for a body of planetary size. Nordtvedt (1971) has considered the implications of such a breakdown of the equivalence principle in the solar system. Shapiro (1974) reports equivalence at the level of 1 to 0.1 % from past observations of the Sun–Jupiter–Mars–Earth system.

The effect would also lead to a polarization of the orbit of the Moon. The accuracy of the laser ranging experiment seems adequate, although the modeling of the lunar orbit is complex. At present, a fit of 4.5 years of data has been obtained with rms residuals of 0.7 m (Williams, 1974). If a Nordtvedt term is introduced, the amplitude determined in the fit is of 1 m and improves the residuals by 6 to 7%. This preliminary result suggests that the value of 67 m predicted by Ni's theory is incorrect. With improved modeling, residuals are expected to be reduced at the level of 15 cm and the and the accuracy in the amplitude of the Nordtvedt term correspondingly increased.

8. ANGULAR MOMENTUM PRECESSION

The experiments previously discussed do not involve diagonal components of the metric at their present level of accuracy. The gyroscope experiment suggested by Schiff (1960) and developed at Stanford is designed to test for these components which reflect the effect of the motion of matter on the gravitational field it creates. In this respect, the gyroscope experiment will explore a new phenomenological aspect of gravity. Within the context of the PPN formalism, the instantaneous precession of a gyroscope in orbit around the Earth involves the parameter γ for the geodetic part and Δ_1 and Δ_2 for the Lense–Thirring part (O'Connell, 1972):

$$\mathbf{\Omega} = \left(\frac{\gamma + 1}{2} \right) \frac{m}{r^3} (\mathbf{r} \times \mathbf{v}) + \frac{1}{8} (7\Delta_1 + \Delta_2) \frac{I}{r^3} \left[\frac{3\mathbf{r}}{r^2} (\boldsymbol{\omega} \cdot \mathbf{r}) - \boldsymbol{\omega} \right]$$

where $\mathbf{r}$ and $\mathbf{v}$ are the position and the velocity of the gyroscope, m, ω and I are the mass, the rotation and the moment of inertia of the Earth. The second term is the mass-current effect and it is maximized for a polar orbit (Fig. 11). In GR $\Delta_1 = \Delta_2 = \gamma = 1$. The geodetic effect is ~ 7 sec of arc/year and the mass-current effect is ~ 0.05 sec of arc/year. The design goal is an accuracy of 0.001 sec of arc/year.

A schematic of the present design of the experiment (Everitt, 1971) is shown in Fig. 12. A telescope is used to refer the gyro axis to the stars system. The telescope and the gyro assembly are kept at superfluid helium temperature and are made of quartz to insure mechanical stability (at 1 m, 0.001 sec of arc corresponds to 50 A). Two gyroscopes spinning in opposite directions are used for the study of each effect. In addition a zero G reference mass is included. The London moment associated with the rotation of the superconducting surface of the gyro is exactly along the rotation axis. This moment contributes to the quantized flux through a superconducting loop attached to the telescope. Any precession and change in the orientation of the London moment gives rise to a current which depends on the inductance of the rest of the closed circuit. Modulation of this inductance will result in a modulation of this current which is then more easily amplified and detected.

A preliminary experiment is now planned for 1978–1979. It would last for three months. The orbit would be inclined at 37° and the altitude would be 500 km. The accuracy goal is 0.1 sec of arc for a 2% observation of the GR geodetic effect. The

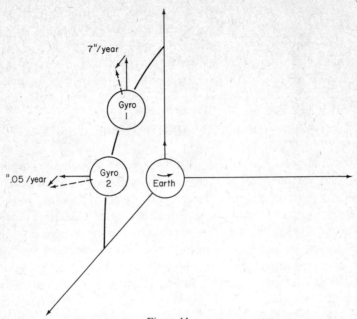

Figure 11

Spin orientation and precession predicted by GR for the Stanford gyroscope experiment in a 800 km polar orbit.

Stanford Gyroscope Experiment.

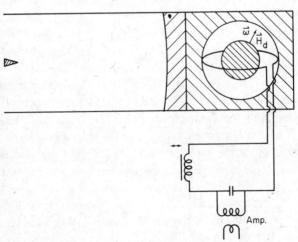

Figure 12

Detection system for gyroscope experiment.

final one year flight in a 800 km polar orbit would follow after final improvements of the system.

The design goal would allow a measurement of the GR geodetic effect to better than 0.1 % and a measurement of the GR mass current effect to a few per cent.

Note Added in Proof

E. B. Fomalont and R. A. Sramek have recently reported the result of their 1974 light-bending experiment using the 35 km N.R.A.O. baseline at 8085 and 2695 Mhz to observe three colinear sources (*Astrophys. J.*, in press). The observed bending is 1.015 ± 0.011 times the general relativity effect and "virtually excludes" a value of $\gamma < 0.96$ or a scalar coupling constant $\omega < 23$. This result is consistent with the results of previous light-bending and time-delay experiments. The current status of the analysis of the lunar ranging data collected over the last five years, incorporating an improved modeling of the lunar librations and the general relativistic accelerations, implies a negligible amplitude for the Nordtvedt term with a systematic error (at the 70 % confidence level) of 30 cm and with the formal statistical error of 5 cm. Such a result is to be compared with a predicted value of $970/(\omega + 2)$ cm for a scalar-tensor theory and would also exclude small values of ω.

REFERENCES

ANDERSON, J. D., Washington APS meeting, 1973.

ANDERSON, J. D., Fifth Cambridge Conference on Relativity, 1974.

BRAULT, J. W., "The Gravitational Redshift in the Solar Spectrum," Thesis, Princeton University, 1962.

DAVIS, R. W., ed., Proceedings of the Conference on Experimental Tests of Gravitation Theories, Nov. 11–13, 1970, California Institute of Technology J.P.L. technical memorandum, pp. 33–499, 1971.

DICKE, R. H., in *Gravitaion and Relativity* (H. Y. CHIU and W. F. HOFFMANN, eds.), W. A. Benjamin, New York, 1964.

DICKE, R. H. and GOLDENBERG, H. M., *Phys. Rev. Letters* **18**, 813 (1967).

EDDINGTON, A. S., *The Mathematical Theory of Relativity*, Cambridge University Press, Cambridge, England, 1922.

EVERITT, C. W. F., in DAVIS, *op. cit.*, p. 68.

HAFELE, J. C. and KEATING, R. E., *Science* **177**, 166 and 168 (1972).

HILL, H. A., CLAYTON, P. D., PATZ, D. L., HEALY, A. W., STEBBINS, R. T., OLESON, J. R. and ZANON, C. A., June 8 SCLERA report, Department of Physics, University of Arizona, 1974.

HILL, H. A. and STEBBINS, submitted to *Ap. J.*, 1974.

HILL, K. M., *Mon. Not. R. Astr. Soc.* **153**, 7 (1971).

JENKINS, R. E., *Astron. J.* **74**, 960 (1969).

KLEPPNER, D., VESSOT, R. F. C. and RAMSEY, N. F., *Astrophys. Space Science* **6**, 13 (1970).

MIKHAILOV, A. A., *Mon. Not. R. Astr. Soc.* **119**, 593 (1959).

MUHLEMAN, D. O., EKERS, R. D. and FOMALON, E. B., *Phys. Rev.Lett.* **24**, 1377 (1970).

MUHLEMAN, D. O., ANDERSON, J. D., ESPOSITO, P. B. and MARTIN, W. L., in DAVIS, *op. cit.*

NEWKIRK, G., *Am. Rev. Astron. Astrophys.* **5**, 213 (1967).

NORDTVEDT, JR. K., *Phys. Rev.* **169**, 1014 (1968a).

NORDTVEDT, JR. K., *Phys. Rev.* **169**, 1017 (1968b).

NORDTVEDT, JR. K., *Phys. Rev.* **170**, 1186 (1968c).

NORDTVEDT. JR. K., in DAVIS, *op. cit.*, p. 32.

O'CONNELL, R. F., in *Proceedings of the Sixth International Conference on Gravitation and Relativity,* General Relativity and Gravitation, **3**, 19 (1972).

OLESON, J. R., ZANONI, C. A., HILL, H. A., HEALY, A. W., CLAYTON, P. D. and PATZ, D. L., *Applied Optics* **13**, 206 (1974).

POUND, R. V. and SNIDER, J. L. *Phys. Rev.* **140**, B788 (1965).

POUND, R. V., fifth Cambridge Conference on Relativity, June 10, 1974.

RAIMOND, E., Fifth Cambridge Conference on Relativity, June 10, 1974.

RICHARD, J.-P., *C. R. Acad. Sci. Paris* **260**, 4151 (1965).

RICHARD, J.-P., *Nature* **212**, 601 (1966).

RICHARD, J.-P., *Phys. Rev. Lett.* **21**, 1483 (1968).

RICHARD, J.-P., *Phys. Rev. Lett.* **6**, 961 (1972).

ROBERTSON, H. P., in *Space Age Astronomy* (A. J. DEUTSCH and W. B. KLEMPERER, eds.), Academic Press, New York, 1962.

SCHIFF, L. I., *Proc. Nat. Acad. Sci.* **46**, 871 (1960).

SCHIFF, L. I., *J. Indust. and Appl. Math.* **10**, 795 (1962).

SEIELSTAT, G. A., SRAMEK, R. A. and WEILER, K. W., *Phys. Rev. Lett.* **24**, 1973 (1970).

SHAPIRO, I. I., *Phys. Rev. Lett.* **13**, 789 (1964).

SHAPIRO, I. I., *Science* **157**, 806 (1967).

SHAPIRO, I. I., in DAVIS, *op. cit.*, p. 136.

SHAPIRO, I. I., Fifth Cambridge Conference on Relativity, June 10, 1974.

SHAPIRO, I. I., COUNSELMAN III, C. C., KENT, S. M., KNIGHT, C. A., CLARCK, T. A., HINTEREGGER, H. F., ROGERS, A. E. and WHITNEY, A. R., *Phys. Rev. Lett.* **33**, 1621 (1974).

SINGER, S. F., *Phys. Rev.* **104**, 11 (1956).

SRAMEK, R. A., *Ap. J. Lett.* **167**, 55 (1971).

SRAMEK, R. A., Fourth Cambridge Conference on Relativity, June 1972.

SRAMEK, R. A., Fifth Cambridge Conference on Relativity, June 10, 1974.

SRAMEK, R. A. and FOMALON, E. B., VLBI Symposium, Caltech., Feb. 4–6, 1974.

VAN DE HULST, H. C., *Bull. Astron. Inst. Netherlands* **11**, 135 (1950).

VAN FLANDERN, T. C., submitted to *Mon. Not. Roy. Soc.*, 1974.

VESSOT, R. F. C., Fifth Cambridge Conference on Relativity, June 10, 1974.

WARD, W., *Astrophys. J.* **162**, 345 (1970).

WILL, C. M., *Astrophys. J.* **163**, 611 (1971a).

WILL, C. M., *Astrophys. J.* **169**, 141 (1971b).

WILLIAMS, J. B., Fifth Cambridge Conference on Relativity, June 10, 1974.

The Texas Mauritanian Eclipse Expedition

BRYCE DeWITT

University of Texas, Austin, Texas, U.S.A.

The deflection of light by large masses is one of the most important effects predicted by currently viable theories of gravity, and an accurate observation of the effect is of fundamental importance in the determination of which theory is correct. Of the various alternative methods that have been devised for measuring the deflection none has as simple a concept or as little dependence on secondary parameters as the classical method which makes use of photography at optical wavelengths during a solar eclipse (the deflecting mass being, of course, the sun). Because of its difficulty, however, this method has often been criticized as being unreliable and insufficiently accurate for modern needs. Inaccuracies in the older observations stemmed from a number of problems including an inadequate number of eclipse-field stars, incorrect plate processing, failure to use identical optics for eclipse and reference exposures, failure to use identical optics for eclipse and reference exposures, failure to obtain night plates with exactly the same instrumental set-up as used for day plates, asymmetry of star fields, and the unavailability of modern microdensitometric techniques. Most and probably all of these defects have been overcome in the successful photography executed at the 30 June 1973 eclipse.

Advance members of the Texas eclipse team arrived at the oasis of Chinguetti, Mauritania (lat. 20°N) on 18 May 1973. A semipermanent well-insulated building was first assembled to house the telescope and to provide a dust-free and thermally controlled environment in the desert. The building was of plywood and styrofoam construction braced with bolted 2 × 4's. A gasketed movable roof section provided access to about one hour of sky and 30° of declination. Otherwise the building was completely caulked and sealed. It was provided with a two-door entrance "air lock" (to permit entry and egress during sandstorms), a workbench, tool racks, a darkroom, an evaporative cooler (to provide filtered air and to maintain the telescope typically at night temperatures), and an air-conditioner (to provide darkroom temperature control at 68°F).

The telescope itself was assembled as soon as the building was completed. Alignment was performed near zenith through the roof and at lower elevation through removable ports, until there was no detectable stellar drift during a half hour of running time. The drive was activated by an electronically controlled synchronous motor run by a 12 V battery source. The motor was mounted on a 4 foot tangent arm equipped with a high precision screw that allows a 30 minute free run. No manual guiding was attempted at any time, and examination of the eclipse plates reveals no trace of tracking error. A preload on the drive was used during the actual eclipse run.

The anticipated outdoor temperature at mid-eclipse was 85° F. (It turned out, as a consequence of a morning sandstorm, to be 96°.) Photographic plate focus runs and adjustments were made with the lens at precisely this temperature. Thermal invariance was achieved by holding the building at 85° for several hours until equilibrium was reached and by exposing the hooded lens to the night sky only for brief periods. At all other times the open ends of the lens hood was covered by several layers of aluminum foil. A similar aluminum blanket covered the plate end of the telescope except during exposure runs. The entire telescope tube itself, including the hood, was wrapped in fiberglass insulation, and the uniformity of its temperature was monitored by means of thermistors mounted along its length, and on the lens elements as well.

The lens was a 4-element blue corrected astromet, made by REOSC for Danjon and never previously used. It had a focal length of 2.1 meters and an 8 inch diameter, stopped down to $6\frac{1}{2}$ inches for image improvement and as insurance against vignetting. After exhaustive (3 months) tuning and adjustment in Austin, knife edge and other tests showed excellent quality. The lens was hand-carried to Chinguetti and treated with gentle loving care en route. Image quality on field plates indicates that the excellent optics were maintained intact.

At eclipse time rigid temperature control was maintained for 24 hours prior to second contact. During the night, after reaching thermal equilibrium, the eclipse plates were prepared for use. Each was provided with a rectangular grid of artificial star images, or fiducial marks, contact printed from a master plate (prepared with laser technology by Texas Instruments, Inc.) that was also in thermal equilibrium with the ambient air. The lattice spacing of the grid was 1 cm and the artificial images were remarkably uniform and circular, each with a diameter of 50 μ. The chief purpose of the grid was to control emulsion creep errors, but it is not difficult to imagine other uses for the grid in the plate reduction process.

In addition to the grid, each plate was provided with step-tablet wedge patterns on two opposite corners and in the center, and with small-spot sensitometric scales on the other two corners. Each pattern was placed about an inch from the nearest plate edge, to avoid emulsion and development anomalies. The patterns are being used as sensitometric standards in the plate reduction process, in order to control image displacements arising from nonlinear emulsion response and nonuniform background. The Texas expedition was the first ever to standardize its plates in this fashion, which should now make it feasible to do meaningful sub-micron densitometry of every important star image.

Atmospheric conditions at eclipse time were, unfortunately, far from ideal. A sandstorm was raging up until about ten minutes before second contact, at which time the meteorological effect of the 150-mile-wide shadow took over and halted the wind in its tracks. During totality, extinction due to dust in the sky was over 85%, leading to a loss in the number of star images on the plates by a factor of perhaps eight or nine, with consequent reduction in the hoped-for statistical precision by a factor of perhaps as much as three. Nevertheless each plate bears between seventy and eighty good star images, symmetrically distributed around the sun, a circumstance favored by the sun's location in the Milky Way for this particular eclipse. A rotating sector was used to reduce saturation near the sun so that images of a few of the brighter stars might be secured in the outer coronal region. The corona was not large (quiet sun year) but there are several prominent streamer images on the plates. It should be noted that the use of a sector rather than a graded filter eliminated an unnecessary additional optical element.

Atmospheric dust produced a bright sky, but this problem was admirably met with the use of fine-grained slow plates (Kodak III-0). The plates are $\frac{1}{4}$ inch, 12×12, micro-flat, each yielding a $7.5° \times 7.5°$ field with the Danjon optics. Each was exposed to the eclipse field for 60 seconds and to a comparison field 10° away in declination (nearly identical in altitude and likewise in the Milky Way) for 30 seconds. Three plates in all were taken during the 6 darkest minutes of totality, the exposure sequence being as follows:

Plate 1: Comparison field, eclipse field.
Plate 2: Eclipse field, comparison field.
Plate 3: Eclipse field, comparison field.

The distribution of comparison field exposures over the period of totality allows some control over systematic errors arising from temporal changes in temperature, atmospheric refraction, etc. Moreover, choice of a comparison field at identical altitude and only 10° away from the eclipse field reduces to a minimum the systematic errors (that have to be unearthed by the least-squares analysis) arising from changes in tube and lens flexure and atmospheric refraction. Finally, the star images in a comparison field 10° away can be adequately corrected for their trace of gravitational deflection.

The most important and, indeed, indispensable role of the comparison fields is to fix the plate scale, a quantity that is very difficult to disentangle from the gravitational shift in a least squares analysis. The presence of a comparison field on each plate allows one to determine, with much greater precision, the gravitational deflection by direct measurement of the night-time plates against the eclipse plates. The telescope was constructed to register all plates identically with the lens. During exposure each micro-flat plate rested against steel buttons at the back of the telescope, *independent of the plateholder*. Plate-to-plate differences should reflect only secular changes.

After eclipse day the building was sealed and placed under guard, with the telescope left in exactly the final eclipse position (except that the tangent drive was returned to the position it had at the beginning of the eclipse run). The building was reopened by a second team headed by David S. Evans, which included, for continuity, Drs. Jones and Sy from June. They found everything in perfect order, and the guard on duty, upon arrival 6 November. Generators, refrigerators, air-conditioners and telescope tracking system all worked and first observations were attempted through clouds on 7 November. Last observations were taken the night of 16 November. Several nights in this interval, including that when the moon was centered in the eclipse field were lost due to overcast sky and even rain. The first observations of importance were made after the housing interior was maintained for 48 hours at its temperature of mid-eclipse, 85°. For the last two nights the temperature was that of current outside ambient, 75°. The key series of observations at each temperature precisely duplicated the hour angle and time sequence of the eclipse exposures, with the addition of three extra eclipse-reference exposures (three extra plates): one before and two after the eclipse matching series. Contrary to preliminary fears, seeing appears to have been excellent, even on the plates taken when the air inside the building was considerably warmer than that outside. As in June, the change of temperature by 10° required a change of focal setting, in order to obtain well-focussed images, by 3 millimeters. In all, 33 plates were exposed, including many exposures of starfields, such as Pleiades, for calibration purposes. Processing of the plates was carried out effectively as in June, and the disassembly of equipment was not undertaken until the last plate was seen to be good. New plates, coated in October, were used almost exclusively, but at least 18 plates were available from June, and tested on-site. They showed observably higher fog levels, and almost a stellar magnitude more sensitivity. As in June, the new plates were almost literally hand-carried to the site.

The lens was hand-carried to Austin, so as to be available for further tests. The balance of the equipment was seen onto a truck to Dakar.

The June plates have been copied and the originals are consigned to Burton Jones at the Royal Greenwich Observatory, along with four of the comparison plates. He has measured these as intensively as possible with Herstmonceux machines, a Zeiss-Abbé comparator and the GALAXY automatic-centered machine. These preliminary reduction efforts should provide preliminary values of the gravitational shift. It is assumed that a detailed microdensitometric study of every star image will eventually be necessary for the final reduction. This, with detailed photoelectric studies of all the stars—required for understanding the variation of refraction effects over the large field—will be carried out at Texas. The large amounts of digital data of these several studies will be handled straightforwardly, though expensively, by available computers. If necessary the lens can be set up for further photography.

The members of the Texas eclipse expedition were:

R. ALLEN BRUNE, JR.	Engineering and procurement
CHARLES L. COBB	1st field team
BRYCE S. DEWITT	Principal investigator; leader of 1st field team
CÉCILE DEWITT-MORETTE	1st field team; liaison
DAVID S. EVANS	Leader of 2nd field team
JOHNNIE E. FLOYD	Engineering; telescope design
BURTON F. JONES	1st and 2nd field teams; astrometry and data reduction
RAYMOND V. LAZENBY	Photography
MAURICE MARIN	Optics; engineering
RICHARD A. MATZNER	1st field team; photography and réseau research
ALFRED H. MIKESELL	Principal coordinator; 2nd field team
MARJORIE R. MIKESELL	2nd field team
RICHARD I. MITCHELL	1st field team; optics; thermometry
MICHAEL P. RYAN	Photography and réseau research
HARLAN J. SMITH	Principal invistigator
ALASSANE SY	1st and 2nd field teams; liaison
CHARLES D. THOMPSON	Engineering; housing design

The team also included the secretarial, fiscal, engineering and maintenance staffs of the Department of Astronomy and Center for Relativity of the University of Texas. Important help and advice came from associates at Princeton University: David T. Wilkinson and Phillipe Crane who devised a plate rocker and experimented with artificial star images and Robert Dicke who provided a wide range of advice and encouragement. Dr. Robert S. Harrington of the U.S. Naval Observatory checked the team's astrometric concepts and preliminary test plates. Drs. Sy and Jones were able to participate because of the kindness of the University of Orléans and Royal Greenwich Observatory, respectively.

Direct financial support, additional to that assessed against various departments of the University of Texas, came from several sources. Of key importance was a grant from the Research Corporation, New York, of sufficient size to appoint staff specifically for the project and maintain it for the preliminary chores so essential to such an endeavor. The largest part of the direct cost was met with a grant from the National Science Foundation. The very important help of the European team-members, concerning the peculiar needs of locating in Africa on two occasions, was possible only because of sizable grants from NATO, Brussels, and the National Geographic Society, Washington. Much of the special research in photographic technique was assisted by a grant from the Dean's Office, the University of Texas.

Of primary significance to the undertaking was the logistic support of the National Science Foundation, under the direction of Ronald LaCount. In particular the project feels its debt to the NCAR staff under George W. Curtis and Nelder Medrud, Jr., who saw that its fifty packages weighing close to six tons all arrived on time and in perfect order.

The expedition is also extremely grateful to the Government of the Islamic Republic of Mauritania, which made arrangements for the observatory guard and did its utmost in many other ways to assist the project.

Special acknowledgement is due Professor R. Michard, director of the Paris Observatory, and Professor Jean Texereau, for making available the astrometric telescope. Many manufacturers supplied special consideration to the needs of a hastily assembled expedition whose goal was the ultimate in astrometric precision. Of these particular mention must be made of the Eastman Kodak Company whose Special Photographic Products and Research Divisions stood by at all stages of the endeavor. Mr. Edward J. Hahn, on behalf of Kodak, reflected the Company's deep concern in pure science, regardless of immediate financial return. Scientists of six American observatories have provided freely out of their experience, and sometimes of their material assets, as called upon. The whole venture represented an amazing expression of confidence and cooperation on the part of so many people, that those of us who manipulated the telescope in Chinguetti can only feel most humbly grateful to them.

Note Added in Proof

Since this talk was given the plate analysis and data reduction have been completed. The analysis reveals that there was much more "noise" on the plates than we had hoped, resulting directly from the bad atmospheric conditions at eclipse time. Both the GALAXY measurements and the microdensitometric studies confirm this. In consequence many of our special precautions and innovations (e.g., artificial grid and sensitometric standards) have proved irrelevant. The final result of our measurements is $\gamma = .95 \pm .11$ where $\gamma = 1.0$ corresponds to Einstein's prediction.

"Exotic" Black-Hole Processes[*]

WILLIAM H. PRESS

California Institute of Technology, Pasadena, California, U.S.A.[†]

1. INTRODUCTION

This report is a telegraphically brief summary of a subject that will be discussed rather more thoroughly in a forthcoming review (Eardley and Press, 1975). The first word of my title means simply: processes whose immediate observational significance is not so compelling as to lead to their inclusion in the excellent review by M. J. Rees elsewhere in this volume. Accordingly, some rather speculative results can be included here. Another useful recent review is that of Bardeen (1974).

2. UNIQUENESS OF THE KERR SOLUTION

Half of the story lies in the *Cosmic Censorship Hypothesis* of Penrose (1969, 1973). Does realistic collapse always produce a black hole (with an event horizon) rather than a naked singularity (with regions of infinite curvature visible from infinity)? The past three years seem to have brought little progress in answering this question. It is not known just how "realistically" the bulk properties of matter must be modeled for the hypothesis to have a chance of being true; for example, matter whose sound velocity goes to zero at high densities can exhibit "shell-crossing" type singularities whose origin is essentially Newtonian, but which technically violate the letter of the hypothesis (Yodiz et al., 1973).

Almost as useful as a rigorous proof of the hypothesis would be a collection of numerical examples of realistic collapse events, showing that horizons do (or do

[*] Supported in part by the National Science Foundation grants GP-36687X, GP-40682.
[†] Present address: Princeton University, Princeton, New Jersey, U.S.A.

not!) in fact form. One such effort in progress is the collision of two black holes by Smarr (1974). The problem of collapse of a spheroidal gas cloud may also be studied by Smarr and Ostriker (unpublished).

The other half of the story is as follows: *assuming* that a horizon forms, and *assuming* that at late times the exterior region becomes asymptotically stationary, what black hole states are available? Certainly, as a number of people have stressed, the Kerr solution is not unique, because the generic collapse may be left with a disc of orbiting matter whose gravitation effect on the central black hole might be important.

Even restricting attention to the vacuum case, the Kerr solution has not been known to be unique. Carter's (1973) famous result shows that such solutions must come in discrete families, each family allowing an infinite range of masses and angular momenta. Is the Kerr family in fact the only one? Some recent work by Demianski and independently by Arkuszewski (unpublished) sought to prove that the final state must be of Petrov type D (two pairs of degenerate principal null directions). Kerr uniqueness would then follow from Kinnersley's (1969) catalog of *all* type D solutions.

Unfortunately, the proofs contain technical errors. Even more recently, Robinson (1975) has offered a direct proof of Kerr uniqueness, within the Carter formalism, which may be the last word on this interesting mathematical problem.

3. STABILITY OF THE KERR SOLUTION

The astrophysical significance of the Kerr stability problem has been discussed elsewhere (Press and Teukolsky, 1973). The method for solving the problem has been numerical integration of the remarkable Teukolsky (1973) equation for perturbations of the Kerr background. Here let me first mention some recent technical advances: the decoupled variables for which the Teukolsky equation completely separates are related to the Newman–Penrose (1962) curvature components ψ_0 or ψ_4. The equation in ψ_0 allows the physical flux of ingoing radiation to be read off by inspection, and the equation in ψ_4 treats outgoing radiation likewise, but it had been a bit of a sticky problem to get both fluxes out of *one* equation. Starobinsky and Churilov (1973) first discovered how to do this at radial infinity. Their method was extended to arbitrary radius by Teukolsky (see Teukolsky and Press (1974)). In another line of attack, Wald (1973) has shown that not only the radiation fluxes, but also the entire perturbed metric (except for two constants of integration corresponding to mass and angular momentum) can be reconstructed from the single Teukolsky variable. And very recently Chrzanowski (1974) has taken steps towards giving a *constructive* procedure for computing the metric coefficients thus determined.

The question of stability is answered by searching for a pole in the "superradiance function" of wave frequency. Such a pole in this function, the ratio of scattered wave energy to incoming energy, in the upper half complex frequency plane (growing modes) corresponds to an outgoing wave which can support itself with *no* ingoing wave, hence an instability. Numerical work (Press and Teukolsky, 1973; Teukolsky and Press, 1974) shows no such pole: the maximum superradiant amplification for real-frequencies is 138% for gravitational waves, and is 4.4% for electromagnetic waves. The sufficiency of looking for poles only for *real* frequencies (marginal instabilities) but for all values of the hole's specific angular momentum *a* has been shown by Wilkins and Hartle (1974). By purely analytic techniques Friedman and Schutz (1974), and independently Teukolsky, have proved that there are no axisymmetric instabilities. A general analytic proof of stability, complementing the numerical results, is still lacking.

Although charged black holes are probably not astrophysically plausible, some theoretical work bears mentioning here. In the zero angular momentum (Reissner–Nordstrom) case, the spherical symmetry guarantees separability of the coupled gravitational and electromagnetic perturbations into spherical harmonics. But until recently it has been thought impossible to *decouple* the electromagnetic and gravitational wave modes. Indeed, coupled equations—which by their nature allow parametric conversion of electromagnetic waves to gravitational waves near the hole—have been studied by Zerilli (1974) and others. Now Moncrief (1974a, 1974b) has found certain perturbation combinations for which the equations *do* decouple; in a sense these are the natural eigenmodes of the parametric wave conversion. Moncrief has gone on to prove that the Reissner–Nordstrom hole is stable. This result has one peripheral but novel interpretation: a Reissner–Nordstrom black hole initially unperturbed and at rest is *not* subject to the self-acceleration catastrophe which has burdened the point-charge solutions of classical electromagnetism. Seen in perturbation order, a self-accelerating black hole would have unstable outgoing wave solutions carrying off linear momentum; these do not exist.

Proving that a black hole is stable *in vacuo* does *not* prove that it is stable in an astrophysical environment. An example is the "black-hole bomb" (Press and Teukolsky, 1972): surround a rotating black hole by a distant spherical mirror with a small porthole in it. An electromagnetic wave inside which is alternately scattered superradiantly from the hole and reflected back from the mirror will grow in amplitude exponentially in time. The radiation escaping through the porthole to infinity also grows exponentially. In this example the reflecting (i.e., conducting) mirror has changed the nature of the problem's boundary conditions and has introduced a growing mode where none previously was present. The mirror configuration here is artificial; but it is not impossible that more natural configurations also lead to instabilities. D. M. Eardley (unpublished) has suggested that a hole surrounded by a thin, conducting-plasma, orbiting disc of accreting matter (cf. Pringle and

Rees, 1972; Shakura and Sunyaev, 1973; Novikov and Thorne, 1973) would be an interesting configuration to study. Can the disc, acting as a sort of ground plane (like the Earth does for radio waves $\lesssim 1$ MHz) "impedance match" an unstable growing wave out to radial infinity?

4. QUANTUM PROCESSES NEAR A HOLE

Zel'dovich (1971, 1972) and Starobinsky (1973) first noted that the existence of classical wave amplification by a rotating black hole implies that the hole must also be a spontaneous emitter of quanta. Furthermore, the lowest modes, with wavelengths of order the size of the hole GM/c^2, are classically amplified by an amount of order unity and this means that the spontaneous emission proceeds about as rapidly as the phase space allows:

Luminosity $\sim$ (energy per quantum) (probability per phase space) (phase space)

$\sim (\hbar\omega)\,(1)\,(\omega)$

Putting $\omega \sim c^3/GM$,

$$\text{Decay time} \sim \frac{Mc^2}{\text{Luminosity}} \sim M^3 \frac{G^2}{\hbar c^4} \sim \left(\frac{M}{10^{15}\,\text{g}}\right)^3 10^{10}\,\text{years} \qquad (1)$$

Notice that Eq. (1) also says that a hole of one Planck mass $((\hbar c/G)^{1/2} \sim 10^{-5}$ g) decays in one Planck time $((\hbar G/c^5)^{1/2} \sim 10^{-43}$ sec) by the emission of about one quantum, and the results scale up from there.

Complementary to this "dimensional" work have been the attempts of Unruh (1974), Parker and Fulling (1974) and others to arrive at a consistent formalism for the detailed computation of quantum fields in a curved space background. Recently, important contributions have been made by Hawking (1974, 1975), who proposes that not only rotating holes, but also nonrotating holes (which have no classically amplified wave modes) "evaporate" by quantum emission at a rate comparable to Eq. (1).

Hawking's calculation considers not a hole which has existed to the infinite past, but one which was formed at some finite time by a spherical collapse. At very late times there are outgoing null geodesics just "peeling off" from outside the horizon. Following one of these backwards, we find in general that it goes through the dynamical region of collapse and out to past null infinity (before the collapse). Now go forwards along this ray, this time keeping track of its quantum occupation number: zero (vacuum state) when it came into the collapse; but the strong field region of collapse mixes up the definition of creation and annihilation operators and the occupation number becomes nonzero. At late times, when the ray peels off to infinity, it carries a spectrum of quanta which, according to Hawking, is independent of the details of the collapse and depends only on the "surface gravity" κ (derived from the mass and angular momentum) of the hole. The spectrum is essentially that of a blackbody, with temperature $kT = \hbar\kappa/2\pi \sim 10^{-6}(M_\odot/M)$ °K. These striking conclusions will bear considerable further investigation.

5. WHITE HOLES

White holes are naked singularities which expell matter into the universe. In the context of a big bang cosmology, they are "delayed cores," pieces of the initial singularity whose expansion begins much later than the average universe. They have occasionally been put forward as quasar models.

The trend of recent thought is towards disbelieving that white holes are possible in a realistic cosmology. Zel'dovich (1974) has argued that the delay of a white hole is strongly limited by a quantum process: pair creation in the strong gravitational field of the singularity is supposed to result in a large mass-loss rate and correspondingly short lifetime. However, some parts of this argument are not yet well substantiated (cf. Novikov, 1973).

Eardley (1974) has recently found a classical mechanism which probably rules out white holes. Roughly, an infalling test electromagnetic field (e.g., a single photon) becomes strongly blueshifted as it comes too near the as-yet-unexpanded white hole singularity. Its gravitational influence is then not negligible, and a detailed calculation shows that it has the effect of turning the white hole into a black hole: a horizon forms and the unexpanded core never does expand into the external universe.

6. HIGHLY NONSPHERICAL OR CHAOTIC COLLAPSE

Here lie the most interesting problems of the future (at present very little is known). Chaotic collapse is probably the *generic* case, because it is well known that departures from spherical symmetry grow unstably in a spherical collapse. Bekenstein (1973) has argued that since gravitational radiation from many different multipoles will all be important in a chaotic collapse, the radiation will carry off some net linear momentum in a random direction. Hence the final black hole may end up moving at some substantial velocity. Making some assumptions about the relative strengths of quadrupole and octupole radiation, Bekenstein estimates that a velocity of 1/300 the speed of light might be typical.

M. J. Rees (unpublished) has suggested that the strong waves produced by chaotic collapse could have another effect: matter immediately outside the collapse might be accelerated outward by its interaction with the wave. In fact, a simple dimensional analysis gives this result: if the characteristic velocities of the inner regions of collapse are of order c, and if some fraction of order unity of the total mass goes to gravitational waves, then the net changes in velocity imparted to nearby matter by the interaction of the wave with its chaotic turbulent dissipation is *also* of order c; this suggests that matter ejection is not beyond possibility; Teukolsky and I are presently looking at some details of the process, with some speculative applications to violent galactic nuclei. This matter ejection mechanism complements the well-known tidal-force tube-of-toothpaste effect which has previously been discussed by Wheeler (1971) and others.

The question has now arisen, just what fractional conversion of matter to gravitational waves *does* occur in a chaotic collapse? In hopes of arousing some substantive work on this question, let me wildly conjecture an answer.

Start with the known result of a small particle of mass m falling into a larger black hole of mass M (Davis et al., 1971),

$$E_{\text{radiated}} = 0.0104 \, mc^2 \, \frac{m}{M} \tag{2}$$

Scaling this up to equal mass objects, and guessing that the precisely radial infall of Eq. (2) is not optimal, one gets, say,

$$E_{\text{radiated}} \sim (0.01 \text{ to } 0.1) \, Mc^2 \tag{3}$$

But this is allowing only "one degree of freedom." A *chaotic* collapse ought to have some larger number of degrees of freedom, all radiating simultaneously. How many? A guess is that it is determined by the equation of state of the matter; for example, "clumps" smaller than the Oppenheimer–Volkoff mass limit $M_{\text{O-V}}$ for a neutron star $(m_P(\hbar c/Gm_P^2)^{3/2} \sim 2M_\odot)$ can support themselves against local collapse. An estimate of the number N of "independent collapse centers" is then

$$N \sim M/M_{\text{O-V}}$$

A little thought now shows that these extra degrees of freedom ought to increase (3) by a factor $\sim \ln N$. Method (i): Collide the collapse centers in pairs, giving (3); then collide the results in pairs, giving (3) again; this can be done $\log_2 N$ times in all. Method (ii): Imagine each center falling simultaneously into its nearest neighbor, into the system of its two nearest neighbors, then three and so on; Eq. (2) gives then $Mc^2(1 + \frac{1}{2} + \ldots + 1/N) \sim Mc^2 \ln N$.

The final conjecture for the gravitational energy radiated by a collapse of mass M is therefore

$$E_{\text{radiated}} \sim (0.01 \text{ to } 0.1) \, Mc^2 \ln (M/M_{\text{O-V}}).$$

This would correspond to efficiencies of order unity for $M \gtrsim 10^6 M_\odot$.

REFERENCES

BARDEEN, J. M., in *Gravitational Radiation and Gravitational Collapse* (C. DEWITT-MORETTE, ed.), Reidel, Dordrecht, 1974.

BEKENSTEIN, J. D., *Astrophys. J.* **183**, 657 (1973).

CARTER, B., in *Black Holes* (C. DEWITT and B. S. DEWITT, eds.), Gordon and Breach, New York, 1973, p. 205.

CHRZANOWSKI, P., contributed paper at GR7 (1974).

DAVIS, M., RUFFINI, R., PRESS, W. H. and PRICE, R. H., *Phys. Rev. Lett.* **27**, 1466 (1971).

EARDLEY, D. M., *Phys. Rev. Lett.* **33**, 442 (1974).

EARDLEY, D. M. and PRESS, W. H., *Ann. Rev. Astron. and Astrophys.* **13**, in press (1975).

FRIEDMAN, J. L. and SCHUTZ, B. F., *Phys. Rev. Lett.* **32**, 243 (1974).

HAWKING, S. W., *Nature* **248**, 30 (1974).

HAWKING, S. W., *Comm. Math. Phys.*, in press (1975).

KINNERSLEY, W., *J. Math. Phys.* **10**, 1195 (1969).

MONCRIEF, V., *Phys. Rev. D* **9**, 2707 (1974*a*).

MONCRIEF, V., *Phys. Rev. D.* **10**, 1057 (1974*b*).

NEWMAN, E. T. and PENROSE, R., *J. Math. Phys.* **3**, 566 (1962).

NOVIKOV, I. D., preprint *I.P.M. No. 93*, Moscow, 1973.

NOVIKOV, I. D. and THORNE, K. S., in *Black Holes* (C. DEWITT and B. S. DEWITT, eds.), Gordon and Breach, New York, 1973, p. 343.

PARKER, L. and FULLING, S. A., *Phys. Rev. D* **9**, 341 (1974).

PENROSE, R., *Revista del Nuovo Cimento*, **1**, 252 (1969).

PENROSE, R., *Ann. N.Y. Acad. Sci.* **224**, 125 (1973).

PRESS, W. H. and TEUKOLSKY, S. A., *Nature* **238**, 211 (1972).

PRESS, W. H. and TEUKOLSKY, S. A., *Astrophys. J.* **185**, 649 (1973).

PRINGLE, J. E. and REES, M. J., *Astron. and Astrophys.* **21**, 1 (1972).

ROBINSON, D. C., *Phys. Rev. Lett.*, in press (1975).

SHAKURA, N. I. and SUNYAEV, R. A., *Astron. and Astrophys.* **24**, 337 (1973).

SMARR, L. L., contributed paper at GR7 (1974).

STAROBINSKY, A. A., *Zh. ETF* **64**, 48 (1973).

STAROBINSKY, A. A. and CHURIKOV, S. M., *Zh. ETF* **65**, 3 (1973).

TEUKOLSKY, S. A., *Astrophys. J.* **185**, 635 (1973).

TEUKOLSKY, S. A. and PRESS, W. H., *Astrophys. J.* **193**, 443 (1974).

UNRUH, W. G., *Phys. Rev. D.* **10**, 3194 (1974).

WALD, R., *J. Math. Phys.* **14**, 1453 (1973).

WHEELER, J. A., in *Nuclei of Galaxies* (D. J. K. O'CONNELL, ed.), North Holland Publishing Co., Amsterdam, 1971.

WILKINS, D. and HARTLE, J. B., to be published (1974).

YODIZ, P., SEIFERT, H. J. and MULLER ZUM HAGEN, H., *Comm. Math. Phys.* **34**, 135 (1973).

ZEL'DOVICH, YA. B., *Pisma v Zh. ETF* **14**, 270 (1971).

ZEL'DOVICH, YA. B., *Zh. ETF* **62**, 2076 (1972).

ZEL'DOVICH, YA. B., in *Gravitational Radiation and Gravitational Collapse* (C. DEWITT-MORETTE, ed.), Reidel, Dordrecht, 1974.

ZERILLI, F. J., *Phys. Rev. D.* **9**, 860 (1974).

Observational Effects of Black Holes

M. J. REES

Institute of Astronomy, Madingley Road
Cambridge, England

1. INTRODUCTION

I shall attempt to summarize the evidence—due primarily to X-ray astronomers—which has led many people during the last year towards the belief that a black hole has probably been discovered in the system Cygnus X-1, and try to indicate how future observations of this and similar objects may aid studies of gravitation. Then I shall make some more speculative comments about the possible existence of black holes much more massive than stars, or (alternatively) very much *less* massive than a star, concentrating on developments which postdate the earlier reviews by Peebles (1972a) and Bardeen (1974).

2. STELLAR-MASS BLACK HOLES

Peebles (1972a) has summarized the astrophysical arguments which suggest that up to 10 % of the mass of the galactic disc may reside in collapsed stellar remnants of $\sim 10 M_\odot$. Recently Shapiro (1973a,b, 1974) has refined and extended Schwartzman's (1971) earlier calculations of accretion by these black holes. The expected angular momentum of infalling interstellar gas is likely to be so low that the flow is more or less spherically symmetrical. The radiation would be bremsstrahlung, synchrotron emission, and γ-rays from decay of π_0 resulting from particle collisions. The overall spectrum of the resulting radiation is hard to predict, mainly because it is uncertain how the magnetic field behaves, but the typical expected luminosities (which are more or less the same for Schwarzschild and Kerr black holes) are unpromisingly low.

203

However, Schwartzman (private communication) is carrying out a search for such objects, which he believes should have colours similar to white dwarfs, but be distinguished by rapid (millisecond) variability.

Zeldovich and Guseynov (1965) suggested that accretion onto a black hole in a *binary system* might cause strong X-ray emission. There are two main reasons for this expectation: (i) A much greater supply of material is available from the companion star's atmosphere than from the tenuous interstellar gas surrounding an isolated stellar-mass black hole. (ii) The accreted material would have too much angular momentum to be swallowed by the hole, and would instead form an accretion disc, guaranteeing efficient conversion of gravitational energy into radiation via viscous dissipation. The main development since Peebles' (1972a) review at GR6 is the discovery of an object—Cygnus X-1—which seems to answer this description.

The arguments that Cygnus X-1 may involve a black hole have been reviewed elsewhere (see, for example, Giacconi (1974), Rees (1974)). The spectrum and rapid irregular variability of the X-rays are consistent with an interpretation involving an accretion disc around a black hole. Recently Rothschild et al. (1974) have claimed evidence for variability on time scales as short as a millisecond, but the statistical significance of their data (Press and Schechter, 1974) is hard to quantify. The X-ray emission is associated with the companion of HDE 226868. This companion is inferred to have a mass $\gtrsim 6M_\odot$, which exceeds the limiting mass of a non-rotating white dwarf or neutron star (assuming general relativity or almost all viable gravitation theories whose post-Newtonian predictions agree with experiment (Malone and Wagoner, 1974)). Thus, *if* a single compact object is involved, a black hole seems the only option (unless one invokes a differentially rotating white dwarf or neutron star, supported largely by centrifugal forces—but viscosity or dynamical instabilities may in any case cause such an object to evolve rapidly into a black hole).

But the argument is certainly by no means watertight. For instance, the rapid variability does not necessarily imply that the whole source is compact—it could be due to independent flares on different parts of the surface of a larger body. Another possibility (Fabian, Pringle and Whelan, 1974; Bahcall et al., 1974) is that the companion of HDE 226868 is itself a binary, consisting of an ordinary $5-10M_\odot$ main sequence star orbited by a neutron star, the latter being the actual X-ray source.

We shall be able to draw firmer conclusions about whether black holes exist when a larger sample of X-ray binaries has been discovered, but as long as Cygnus X-1 remains unique one cannot exclude "ad hoc" models which do not involve a black hole. On the other hand, unless one believes that black holes are intrinsically absurd, the black hole interpretation seems the least contrived and most plausible. The formation of systems such as Cygnus X-1 can then be quite naturally interpreted in the context of binary star evolution (see, for example, van den Heuvel and de Loore (1973)), whereas the origin and stability of 3-body systems pose serious problems; and they certainly raise fewer difficulties than the origin of systems like Her X-1 which are believed to contain neutron stars.

The discovery of a firm black hole candidate would signify that strong gravitational fields actually existed. It would also permit—at least in principle—several tests of rival gravitation theories. If the X-rays from Cygnus X-1 are emitted by an "accretion disc", it is therefore interesting to consider whether, by detailed observations, one can diagnose some properties of the metric around the black hole. The theory of accretion discs has been discussed by many people (e.g., Pringle and Rees, 1972; Shakura and Sunyaev, 1973; Novikov and Thorne, 1973; Lightman, 1974). If the disc is in a steady state and its outer radius is very large compared to GM/c^2, the luminosity is assumed to be $\sim \varepsilon F c^2$, where F is the accretion rate, and ε is the binding energy of the innermost stable circular orbit (radius $r_{\min}$). The value of ε varies from 6% for a Schwarzschild metric ($r_{\min} = 6GM/c^2$) to 42% ($r_{\min} = GM/c^2$) for a "maximal Kerr" metric whose angular momentum is aligned with that of the disc.

Bardeen (1970) pointed out that, if the material accreted from the disc carries with it the specific angular momentum corresponding to the innermost stable orbit, the angular momentum/mass ratio of the hole secularly increases towards "maximal Kerr" ($a/m = 1$). The time scale for this to occur, $\sim M/\dot{M}$, would typically be $\gtrsim 10^8$ yrs, which probably greatly exceeds the duration of the mass transfer. Thus the metric, and consequently the efficiency ε, would be determined by how the black hole formed rather than by what happened to it later. Thorne (1974) has recently modified Bardeen's work to take account of the net angular momentum of the radiation from the disc which falls into the hole, and finds that the limiting value of a/m becomes ~ 0.998 rather than unity. Although this may seem a minor refinement, the location of the innermost stable orbit, and therefore ε, is very sensitive to a when a/m is close to unity, and Thorne's correction would imply that the limiting efficiency was ~ 0.3 rather than 0.42. Thorne argues that a black hole is unlikely to form with $a/m > 0.998$; but if it did, a/m would initially *decrease* on a time scale $\ll M/\dot{M}$.

These efficiencies refer to *uncharged* black holes. A black hole surrounded by unmagnetised plasma could not acquire a charge Q exceeding $\sim GMm_p/e$ (the value for which the electric and gravitational forces on a test proton are comparable). The fractional electrostatic contribution to the mass of the hole is then only $\sim 10^{-40}$. A black hole accreting dense rotating *magnetised* plasma may, however, acquire a larger charge, because a large electric field then arises in frames that do not move with the fluid.

The *spectrum* of the radiation from accretion discs is hard to calculate, even when F, ε and M are given, because it depends on the density and vertical structure of the disc, and thus on the very uncertain viscosity. For luminosities $\sim 10^{37}$ erg sec^{-1} and masses of 5–$10M_{\odot}$, the radiation is predominantly in the X-ray band, the harder emission coming predominantly from the inner regions (see the discussions by Shakura and Sunyaev (1973) and Lightman (1974)). Moreover, we have no independent estimate of F—even if the efficiency has the minimum value of 6%, the required accretion rate in Cygnus X-1 is far below the mass transfer rates inferred in more normal binary systems.

Minimum time scale of variability

The period of the innermost stable orbit is $2\pi(6)^{3/2} GM/c^3$ ($\sim 0.5M/\dot{M}_\odot$ ms) for a Schwarzschild metric, but only $4\pi GM/c^3$ ($\sim 0.06\ M/M_\odot$ ms) for "maximal Kerr". The minimum epicyclic periods are also close to these values. Since the value of M can already be pinned down at least to within a factor ~ 2, this period (if it could be determined) would tell us something about the metric. It has been argued (Sunyaev, 1973; Bardeen, 1974) that if the disc were irregular this is the minimum time scale that would be expected in any observed quasi-periodic fluctuations. Although the existing data show no evidence for any "pulse trains" (and it is easy to think of many instabilities which could cause rapid but irregular X-ray flickering) improved future experiments might conceivably reveal periodicities. On the other hand, it is questionable whether the "innermost stable orbit" retains any precise significance in a realistic model: the X-ray luminosity of Cygnus X-1 is high enough that radiation pressure forces are comparable with gravity in the inner parts of the disc; there are also instabilities in the inner part of the disc if the viscosity has certain forms (Lightman and Eardley, 1974); also, it is by no means clear that the radiation emitted from *within* r_{min} is necessarily negligible, and this could vary on even shorter time scales.

Deviations from axisymmetry

If the angular momentum of a (Kerr) black hole is oriented obliquely with respect to that of the accreted material, the disc structure is more complicated (Bardeen and Petterson, 1974). The orbital planes would precess owing to the Lense–Thirring effect, with a period $\sim (a/M)^{-1} (rc^2/GM)^{3/2}$ times longer than the Keplerian period. This precession will have no effect at large radii when its time scale is longer than the infall time scale (v_θ/v_r times the Keplerian period). At smaller radii, precession would cause the inner part of the disc to align with the hole's equatorial plane.

How could this effect be discerned observationally? The X-ray luminosity from a disc emerges anisotropically, the intensity being greatest in directions perpendicular to the plane. This effect cannot be detected directly, however, because—even when Lense–Thirring effect occurs—the disc orientation remains almost fixed in space. But there are two indirect possibilities: (i) The amount of heating of the companion star at different orbital phases gives us, in principle, some estimate of the X-ray emission in directions other than our own line of sight. (ii) If (as is likely) electron scattering is a dominant cause of X-ray opacity in the disc, the X-rays will be linearly polarized, with the electric vector being (for an optically thick disc) preferentially in a direction perpendicular to the plane of the disc. Thus it is in principle possible, by measuring polarization, to determine the orientation of the disc. The precession effect mentioned above might cause the hard X-rays (which come from the inner regions) to have a different polarization direction from the softer X-rays.

Consequences of alternative theories of gravitation

In some theories of gravity the equivalence principle is violated for self-gravitating bodies. The resulting effect (Nordtvedt effect) is being searched for in the orbits of bodies in the solar system. For a neutron star or black hole, this effect could amount to several per cent, the Keplerian velocity of a compact object around a binary companion being several per cent different from that of a test particle in a similar orbit. Possible observable consequences of this in X-ray binaries would include the following: (i) Gas streaming from the companion star would have an amount of angular momentum relative to the compact object which depended on the gravitation theory adopted. (ii) The orbits of fluid elements in the accretion disc would be ellipsoidal and not circular. (iii) Masses obtained by dynamical arguments and by stellar atmosphere arguments may be incompatible. Although these all look somewhat unpromising, I mention them because they may stimulate people to think of some "cleaner" signature of equivalence principle violation which I have overlooked. (See note (1) added in proof.)

In the next few years we can expect much better X-ray data on known X-ray binaries, and maybe the discovery of many more — HEAO B, planned to be launched before 1980, should surpass the sensitivity of the UHURU satellite by $\sim 10^4$. Detailed observations of X-ray variability (in particular, correlations between the behavior in different energy bands), and polarization, should yield some understanding of the structure and stability of accretion discs.

If the fluid dynamical and radiative aspects of discs were better understood, one would then have some chance of using the X-ray observations to learn something about strong gravitational fields. But my personal (rather pessimistic) guess is that it will be a long time before we can determine the magnitude and orientation of the angular momentum of the putative black hole in Cygnus X-1 or any similar system. To use such systems to discriminate between different gravitation theories is perhaps an even more remote goal.

3. MASSIVE BLACK HOLES

Numerous models have been proposed to explain quasars and other manifestations of "violent activity" in galactic nuclei. A feature common to many of them is that the estimated duration of the phenomenon is $\ll 10^{10}$ years, and that the formation of a massive black hole seems a likely end result. Massive black holes ("dead" quasars) might therefore perhaps lurk in the centres of some galaxies. Another independent motivation for searching for such objects relates to the proposal (Hoyle and Narlikar, 1965; Ryan, 1972) that massive *primordial* black holes might have constituted the "seeds" around which elliptical galaxies condensed.

If accretion of interstellar gas occurred, these massive holes would develop scaled-up version of the accretion discs in X-ray binaries, except that the radiation would emerge not as X-rays but at optical, ultraviolet, or infrared wavelengths (Lynden-Bell and Rees, 1971). Even if there were no accretion, the gravitational field of a massive black hole might affect the motion and spatial distribution of the surrounding stars. The possible existence of massive black holes in elliptical galaxies has been discussed by Wolfe and Burbidge (1970). The velocity dispersion of stars in the inner regions sets a fairly firm upper limit of $\sim 10^{10} M_\odot$ to a hypothetical central dark mass. The effects on the stellar density distribution are less straightforward because the relaxation time in elliptical galaxies is typically $\geqslant 10^{10}$ years—it is not obvious that the introduction of a black hole need necessarily cause *any* extra concentration of stars in its vicinity. (Peebles (1972a, b) has discussed the analogous situation in globular clusters—the stellar relaxation time is here $\lesssim 10^{10}$ yrs, and so the absence of a "cusp" in the light distribution does set limits on a central black hole.)

Black holes *outside* galaxies might be detectable if their masses were very large. Since the accretion rate depends on M^2, a very massive ($\gtrsim 10^{12} M_\odot$) hole would be highly luminous even if it were surrounded by very tenuous intergalactic gas (Pringle, Rees, and Pacholczyk, 1973). If such objects were numerous enough to make a major contribution to the mass of the universe, some should have been detected via the gravitational lens effect (Refsdal, 1970; Press and Gunn, 1973).

4. MINI-HOLES

One cannot envisage any plausible astrophysical process, occurring at the present epoch, which could produce collapsed objects of $\ll 1 M_\odot$. But such objects could conceivably have formed in the very early stages of the "big bang" when the primordial material greatly exceeded nuclear densities. The mass encompassed within the particle horizon of a Friedmann universe is $M_H \simeq 10^5 t_{\text{sec}} M_\odot$; and this is itself below a solar mass throughout the "hadron era" of the standard "hot big bang" cosmology. There is no reason to believe that the universe was even approximately "Friedmannian" in the hadron era—the amplitude of small-scale inhomogeneities, or curvature fluctuations, could have been so great that some regions promptly recollapsed, forming black holes. (Zeldovich and Novikov conjectured that a primordial black hole forming in this fashion would accrete so rapaciously from its surroundings that its mass remained $\sim M_H \propto t$ throughout the radiation era, eventually reaching $\gtrsim 10^{15} M_\odot$. Recent work by Carr and Hawking (1974) shows that this conjecture is incorrect, except under very special initial conditions, and that primordial holes do not accrete fast enough to grow significantly.) Furthermore, some physicists favour a "soft" equation of state (i.e., $P/\rho \ll \frac{1}{3}c^2$) at supernuclear densities. If this were correct, and the Jeans mass were many orders of magnitude smaller than M_H, then initial inhomogeneities might recollapse before being stabilized by pressure gradients even if their amplitude were $\ll 1$ when they first came within the horizon. Carr (1975) has attempted to estimate the mass spectrum of these black holes, for various assumptions about the equation of state and the nature of the primordial fluctuations.

If these mini-holes exist, they could have played a role in the process of galaxy formation (Meszaros, 1975) because, being almost unaffected by radiation drag and pressure gradients, they could have started aggregating into clusters as a result of gravitational instability before the end of the radiation era.

Black holes of *very* low masses (Schwarzschild radii comparable to elementary particle dimensions) are of special interest because they are the only ones for which the particle creation effects recently discussed by Hawking (1974a, b) could be observationally significant. (See the accompanying article by Press for a summary of this work.) According to Hawking, a black hole of mass m g radiates like a black body of temperature $T \simeq 10^{27} m^{-1}\,^{\circ}\mathrm{K}$. The radiation rate is then $\propto m^{-2} f(m)$, where $f(m)$ is a function that takes account of the number of species contributing to the radiation. The predicted temperatures for stellar-mass black holes are so low ($\sim 10^{-6}\,^{\circ}\mathrm{K}$) that the radiative effects (for which the time scale is 10^{60} Hubble times) are completely swamped by accretion. When $T \ll 10^{10}\,^{\circ}\mathrm{K}$, only zero mass particles contribute to the radiation. However for higher temperatures (i.e., lower masses), an increasing number of species can contribute: the radiation rate thus depends on m *more steeply* than $\propto m^{-2}$. The lifetime is thus a steeper function of m than m^3. The only black holes whose "evaporation time" is predicted to be $\lesssim 10^{10}$ yrs are those with $m \lesssim 10^{14}$ g; and for these the temperature is unfortunately so high that $f(m)$ is very uncertain. If Hagedorn's hypotheses were correct, $f(m)$ would tend to infinity as m decreased toward a few times 10^{13} g; on most other assumptions $f(m)$ would increase, but remain finite, as m decreases. The duration and intensity (and hence the potential detectability) of the final "explosion" in which a mini-hole effectively annihilates itself is plainly sensitive to the form of $f(m)$. This is, of course, a question primarily in the domain of particle physics; and it is also unclear how Hawking's calculations would apply to a situation where the radiation is mainly being emitted by many different species whose mass-energy is $\gg kT$, and whose thermal velocities are very low. Even if this difficult question could be settled, it is still uncertain in what form the energy actually escapes (though it should be feasible to resolve this latter uncertainty by doing some calculations). The "exotic" particles would presumably decay, and the hole may be surrounded by a "photosphere" with effective temperature ~ 0.5 MeV within which electron-positron pairs provide a large optical depth. The mean energy of the escaping γ-rays might however be $\gg 0.5$ MeV if the particles in this "photosphere" were themselves moving relativistically outwards. (See note (2) added in proof.)

The total energy density in the cosmic X-ray and γ-ray background is only 10^{-8} times the total mass-energy density of an Einstein–de Sitter universe. This sets a stringent upper limit to the number of black holes exploding at cosmic times $\gtrsim 10^9$ yrs. One cannot place such straightforward limits on the number of black hole explosions at earlier cosmic epochs, because the universe would not then have been transparent to X-rays and γ-rays. In fact any radiation arising from black holes that decayed in the first $\sim 10^3$ yrs of the expansion would have been completely thermalized, and this perhaps suggests a possible origin for the cosmic black body radiation. Whether this hypothesis is compatible with the stringent upper limits on the amount of unthermalized radiation produced at recent epochs depends on two uncertain factors: the initial mass spectrum of the primordial black holes, and the dependence of their lifetime on their mass.

Notes Added in Proof

1. D. Eardley has recently emphasized that, in (e.g.) the Brans–Dicke theory, a binary system containing a compact component will emit *dipole* gravitational waves; and that in consequence, the size of the orbit can shrink much more rapidly than general relativity predicts. Although unimportant for Hercules X-1 and Cygnus X-1, this effect could be significant for the recently discovered binary pulser (Hulse and Taylor, *Astrophys. J. Lett.* **195**, L. 51 (1975) , and for the very short-period binaries that may be associated with some transient X-ray sources.

2. D. Page has completed a more detailed calculation of the temperature-versus-mass relation for the radiation by "mini-holes". He finds that the temperature corresponding to a lifetime of $\sim 10^{10}$ years is ~ 20 Mev. This is well below the temperature at which uncertainties about the equation of state become crucial, and implies that the limits on the gamma-ray background impose firm constraints on the permitted number and mass spectrum of "mini-holes". B. Carter and his associates have attempted to estimate the spectrum of the particles emitted by a "mini-hole" during its final evaporation. If particle creation effects are neglected, one can calculate the accretion flow onto a "mini-hole" embedded in equilibrium radiation at the temperature appropriate to its mass. Carter and his colleagues assume that an isolated "mini-hole" generates a wind which is conjugate to this accretion solution. They find that for a Hagedorn-type equation of state, there is an effective photosphere which prevents the energy of escaping particles from becoming much more than ~ 100 Mev; whereas for a "hard" equation of state, the Lorentz-factor of the wind becomes very large as the hole loses mass, resulting in the production of very energetic particles.

REFERENCES

BAHCALL, J. N., DYSON, F. J., KATZ, J. I. and PACZYNSKI, B., *Astrophys. J. Lett.* **189**, L. 1 (1974).

BARDEEN, J. M., *Nature* **226**, 64 (1970).

BARDEEN, J. M., in *Gravitational Radiation and Gravitational Collapse* (C. DEWITT, ed.) Reidel, Holland, 1974.

BARDEEN, J. M. and PETTERSON, J., *Astrophys. J. Lett.* **195**, L. 65 (1974).

CARR. B., preprint (1975).

CARR, B. and HAWKING, S. W., *MNRAS* **168**, 399 (1974).

FABIAN, A. C., PRINGLE, J. E. and WHELAN, J. A. J., *Nature* **247**, 351 (1974).

GIACCONI, R., in *Gravitational Radiation and Gravitational Collapse* (C. DeWitt, ed.), Reidel, Holland, 1974.

HAWKING, S. W., *Nature* **248**, 30 (1974*a*).

HAWKING, S. W., *Proc. Roy. Soc.* (in press) (1974*b*).

HOYLE, F. and NARLIKAR, J. V., *Proc. Roy. Soc.* **290A**, 177 (1966).

LIGHTMAN, A. P., *Astrophys. J.* **194**, 429 (1974).

LIGHTMAN, A. P. and EARDLEY, D. M., *Astrophys. J. Lett.* **187**, L. 1 (1974).

LYNDEN-BELL, D. and REES, M. J., *MNRAS* **152**, 461 (1971).

MALONE, R. C. and WAGONER, R. V., *Astrophys. J. Lett.* **189**, L. 75 (1974).

MESZAROS, P., *Astron. Astrophysics* (in press) (1975).

NOVIKOV, I. D. and THORNE, K. S., in *Black Holes* (C. DeWitt, ed.), Gordon and Breach, New York, 1973, p. 343.

PEEBLES, P. J. E., *Gen. Rel. and Grav.* **3**, 63 (1972*a*).

PEEBLES, P. J. E., *Astrophys. J.* **178**, 371 (1972*b*).

PRESS, W. H. and GUNN, J. E., *Astrophys, J.* **185**, 397 (1973).

PRESS, W. H. and SCHECHTER, P., *Astrophys. J.* **193**, 437 (1974).

PRINGLE, J. E. and REES, M. J., *Astron. Astrophysics* **21**, 1 (1972).

PRINGLE, J. E., REES, M. J. and PACHOLCZYK, A. J., *Astron. Astrophysics* **29**, 179 (1973).

REFSDAL, S., *Astrophys. J.* **159**, 357 (1970),

REES, M. J., *Highlights of Astronomy* (G. CONTOPOULOS, ed.), Reidel, Holland, 1974, p. 89.

ROTHSCHILD, R. E., BOLDT, E. A., HOLT, S. S. and SERLEMITSOS, P. J., *Astrophys. J. Lett.* **189**, L. 13 (1974).

RYAN, M. P., *Astrophys. J. Lett.* **177**, L. 79 (1972).

SCHWARTZMAN, V. F., *Soviet Astr.– A. J.* **15**, 377 (1971).

SHAKURA, N. I. and SUNYAEV, R. A., *Astron. Astrophysics* **24**, 337 (1973).

SHAPIRO, S. L., *Astrophys. J.* **180**, 531 (1973*a*).

SHAPIRO, S. L., *Astrophys. J.* **185**, 69 (1973*b*).

SHAPIRO, S. L., *Astrophys. J.* **189**, 343 (1974).

SUNYAEV, R. A., *Soviet Astr.–A. J.* **16**, 941 (1973).

THORNE, K. S., *Astrophys. J.* **191**, 507 (1974).

VAN DEN HEUVEL, E. P. J. and DE LOORE, C., *Astron Astrophysics* **25**, 387 (1973).

WOLFE, A. M. and BURBIDGE, G. R., *Astrophys. J.* **161**, 419 (1970).

ZELDOVICH, Y. B. and GUSEYNOV, O. R., *Astrophys. J.* **144**, 840 (1965).

Progress in Relativistic Statistical Mechanics, Thermodynamics, and Continuum Mechanics*

JÜRGEN EHLERS

Max-Planck-Institut für Physik und Astrophysik
Munich, Fed. Rep. of Germany

1. INTRODUCTION

Statistical mechanics is concerned with the relations between fine-grained and coarse-grained descriptions of physical systems, and in particular with the derivation of laws governing the gross, macroscopic behaviour of matter and radiation from microscopic laws which are considered as fundamental. Since the laws governing the behaviour of particles and fields depend in an essential way on the assumed space-time structure, the replacement of non-relativistic, Galilean space-time by the flat or curved pseudo-Riemannian space-times of relativity demands the creation of appropriately generalized forms of statistical mechanics. Apart from the question of principle thus raised, there are now also concrete reasons for interest in special- and general-relativistic statistical mechanics. In connection with astrophysics one has to deal, e.g., with relativistic plasmas, with superdense, reacting mixtures of energetic particles in a rapidly expanding space, and possibly with relativistic star clusters. The long-standing tradition in general relativity theory to restrict the description of matter to writing down the stress-energy-momentum tensor of a perfect fluid has, therefore, to be broken. (This change is clearly visible in the recent books by Weinberg, Zeldovich and Novikov, and Misner, Thorne and Wheeler.)

---

* The following text is a considerably extended version of the talk delivered at GR7. Remarks by A. Peres and H. P. Künzle which I gratefully acknowledge have influenced the written version of the review of predictive, Poincaré-invariant particle mechanics in Sec. 2.

Unfortunately, there is not yet a (special- or general-) relativistic statistical mechanics, if by that term one means a consistent theory based on physically reasonable and mathematically meaningful assumptions and containing some well-established, significant results. There are, however, various attempts to create such a theory. Moreover, there is a less ambitious, but now well developed relativistic kinetic theory, and also a relativistic continuum mechanics, and some progress in these areas has been made within recent years.

The purpose of this talk is to survey some of these developments and to mention some problems. I shall not try to push a particular point of view nor give many formal details, but rather wish to outline the general situation. I have to mention that my own degree of understanding of some of these matters is anything but perfect, and consequently the following presentation does not claim to be a complete, balanced review. Also, the references are not meant to be exhaustive; they are only meant as representative and are so chosen that they can be used as a guide to further literature. Therefore, I quote recent reviews rather than original papers if such are available.

2. BRIEF REVIEW OF NON-RELATIVISTIC STATISTICAL MECHANICS

In non-relativistic classical mechanics most important dynamical systems are *Hamiltonian* ones, i.e., the set P of all possible initial states (Cauchy data), called *phase space*, is a (finite dimensional) manifold with a symplectic form $\Sigma \, dp_a \wedge dq^a$, and the evolution is given by a Hamiltonian phase flow,

$$\dot{q}^a = \frac{\partial h}{\partial p_a}, \qquad \dot{p}_a = -\frac{\partial h}{\partial q^a} \tag{1}$$

Gibbs ensembles (macrostates) are then characterized by a *distribution function* $f(p_a, q^a, t)$, interpreted as a probability density (with respect to the Lebesgue measure $\Pi \, dp_a \wedge dq^a$ on P) which evolves according to *Liouville's equation*

$$\partial_t f = [f, h] \tag{2}$$

and *expectation values* (macroscopic quantities) are obtained as

$$\langle g \rangle_f = \int g f \, dp \, dq \tag{3}$$

In quantum mechanics one has a formally similar structure. Pure states are represented by rays in a complex Hilbert space, their evolution is given by a Schrödinger equation. Gibbs ensembles are characterized by *statistical operators* (density matrices) F, observables by Hermitean operators G, and Eqs. (2) and (3) are replaced by the *von Neumann's equations*

$$\partial_t F = i[F, H] \tag{2'}$$

and

$$\langle G \rangle_F = \text{Tr}\,(FG) \tag{3'}$$

respectively. (For recent accounts of statistical mechanics see, e.g., [1], [2], [3].)

The quantum-statistical formalism can be cast into a pseudoclassical phase-space language by means of the *Weyl–Wigner correspondence* (see, e.g., [4], ch. VI, §3 or [5], appendix). Then $(3') \to (3)$, whereas $(2')$ goes over into an equation which has the form of (2), but with the Poisson bracket replaced by the Moyal bracket which passes over into the P.b. in the classical limit $h \to 0$.

On the basis of (2) or $(2')$ and assumptions about the Hamiltonian (in most cases, a decomposition $h = h_0 + h_{\text{int}}$, or at least the existence, for a given h, of a simpler comparison Hamiltonian h_0 [6]), methods have been developed to deduce (generalized) *master equations* and *hierarchies* for reduced distribution functions (BBGKY, Klimontovich), and with additional assumptions about the initial distributions f or F and/or the smallness of correlations, the weakness and range of interactions, etc., various *kinetic equations* and techniques for their (approximate) solution have been worked out. If these are combined with (3) or $(3')$ for suitably chosen macro-observables g or G, one obtains *macroscopic laws* (hydrodynamics, e.g., [6], [7]; in particular [6] contains an important new way ("subdynamics") to relate and contrast the microscopic dynamical description to the macroscopic description, and a generalized second law).

The scheme just outlined indicates how, in principle, non-relativistic statistical mechanics is capable of deriving contracted descriptions from an underlying microscopic theory. It is to be emphasized that the "philosophy" of the theory [1], [2], [3]— initial information provides f or F at $t = 0$, (2) or $(2')$ determines f or F at $t > 0$, and (3) or $(3')$ gives statistical predictions for later observations—rests on the fact that the *initial value problem* of the microscopic theories is, again in principle, well understood.

The phase space P plays two roles: It is the set of all initial states and, therefore, the *sample space* carrying all possible probability measures, and it is the *container of all possible histories* (integral curves of the vector field

$$\frac{\partial h}{\partial p^a}\frac{\partial}{\partial q^a} - \frac{\partial h}{\partial q^a}\frac{\partial}{\partial p^a}$$

on P) of the dynamical system.

Non-Hamiltonian dynamical systems can, of course, also be subjected to statistical treatment (see, e.g., [8]), but since many more techniques have been elaborated within the canonical framework than outside of it (and because it exhibits a close correspondence between classical and quantum theories), it is reasonable to use the canonical formalism whenever it is available.

For a broad survey of the development of statistical mechanics and thermodynamics see, e.g., Landsberg [46].

3. TOWARDS SPECIAL-RELATIVISTIC STATISTICAL MECHANICS

While the general framework outlined in Sec. 2 does not presuppose a particular space-time structure, its physical importance for relativistic theories depends on whether realistic models of physical systems in these latter theories can be cast into Hamiltonian form. Since the meaning of "Hamiltonian" in this connection is far from obvious (see [9], [10], [11]), I shall first consider the meaning of this term.

Let a particular space-time structure with an associated symmetry group G be assumed. Also, let a dynamical system S consisting of particles and (or) fields be given so that a dynamically possible history of S consists of a number of world lines and (tensor or spinor) fields on space-time X. Let Z be the set of all these histories. The dynamics of S admits G as an invariance group if and only if G acts, in an obvious sense, on Z. Assuming the dynamics of S to be deterministic and choosing a frame of reference Σ in X we may set up a one-to-one correspondence between Z and the set Z_Σ of all initial data belonging to the instant $t = 0$ relative to Σ (see Fig. 1 for illustration).

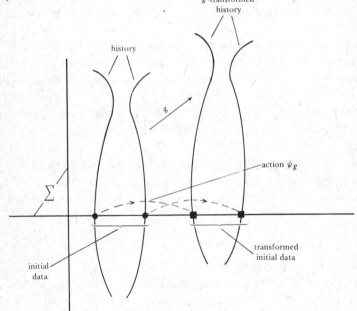

Figure 1. Action of G on Z and Z_Σ.

Clearly the action of G on Z induces an action of G on Z_Σ (which is, in fact, independent of the particular frame Σ if only G-equivalent frames are permitted). We call this action $\psi : G \times Z_\Sigma \to Z_\Sigma$ the *dynamical action** of G belonging to S. ψ, of course, depends on the dynamics of S; conversely ψ determines the dynamics as well as the transformation properties of the variables specifying the initial data or "states" under G. Thus, a G-invariant dynamical system can be fully specified by ψ, i.e., by the structure of the set of initial data (= state space, domain of ψ) and G's action on it.

If Z_Σ can be given the structure of a (possibly infinite-dimensional) symplectic manifold such that the action ψ is in terms of canonical transformations, we call S *fully Hamiltonian.* This property is *much stronger* than the requirement that the evolution can be described, in terms of suitable variables (p_a, q^a), by Hamiltonian equations (1). In the fully Hamiltonian case, ψ can essentially be characterized by a canonical representation of the Lie algebra of G in terms of a Poisson algebra of functions (generators) on the *phase space* $P \equiv Z_\Sigma$. (In the quantum case, P would be the ray-space of all pure states, and ψ would be a unitary or antiunitary action.) This consideration establishes the relation between Hamiltonian dynamical systems and canonical realizations (or semiunitary representations) of the space-time group G.

In the *non-relativistic* case (Galilean space-time) it is easy to construct fully Hamiltonian systems of N interacting particles with a $6N$-dimensional phase space P. The reason is that the Galilean group G is a *semidirect product* of the 9-dimensional normal subgroup F of those transformations which leave the absolute time fixed, and the 1-dimensional group R of time translations (see Appendix). In particular, in G boosts and spatial translations commute, so that in a canonical (or unitary) representation of the Lie algebra of G the generators K_a and P_b of boosts and translations have the Poisson brackets (or commutators)

$$[K_a, P_b] = M\delta_{ab} \tag{4}$$

(M is a constant). Therefore, one can pass from a representation of a non-interacting system to one for an interacting system by leaving the generators of F—P_a, J_a, K_a— unchanged, and adding to the time-translation generator H an interaction term which commutes with the generators of F. That is easily accomplished.

In the case of *special relativity*, i.e., the Poincaré group P, the simple procedure of passing from a representation belonging to non-interacting particles to one involving interactions just described does not work since, in P, the subgroup of those transformations which map a hypersurface of constant time into itself is only the six-dimensional Euclidean group of rotations and (spatial) translations, whereas boosts

---

* It is important to distinguish this "dynamical" action of G from the "kinematical" action which tells one how the components of local objects like tensors relative to Σ change if these objects are carried along by the active transformation from an event x to $g(x)$.

are "mixed up" with time translations. Instead of (4) one has

$$[K_a, P_b] = \delta_{ab} H \tag{4'}$$

To change the evolution, H, in order to incorporate interactions requires, therefore, to change also the K's or the P's, or both. (Usually one changes the K_a and leaves the P_a unaffected by the interaction*.) That is the group-theoretic origin of the difficulty in constructing interacting special-relativistic systems. (Compare similar discussions in [9] and [12], pp. 105–107. The fundamental paper is [13].)

One drastic consequence of (4') is that the only fully Hamiltonian (as defined above!), P-invariant dynamical systems of finitely many particles obeying "Newtonian" equations of motion

$$\ddot{x}_r^{(i)} = f_r^{(i)}(x_s^{(j)}, \dot{x}_s^{(j)}) \tag{5}$$

($i = 1, \ldots, N; r = 1, 2, 3$) for which the position coordinates (as coordinates on Z_Σ) have vanishing Poisson brackets, are systems of free particles. This famous "no interaction theorem", due in this generality to Leutwyler and Hill, has been carefully analyzed in [10]. This result has sometimes been interpreted to mean that, in (special) relativity, interactions *have* to be transmitted via *fields*. This conclusion, however, is not warranted. For one thing, one can give up the requirement of Hamiltonian equations of motion, or at least relax the definition of fully Hamiltonian systems given above in various ways, and then *there are* P-invariant, deterministic particle systems [10], [11]. But one may even keep the canonical structure and also maintain vanishing Poisson brackets between the position coordinates if one parametrizes the space Z of histories of an N-particle system not in terms of initial data referring to one instant of time t in an inertial frame, but instead in terms of positions and velocities taken at relatively lightlike events on the particle worldlines. At least for $N = 2$, Künzle has shown [11] that in this way one can construct P-invariant systems with a 12-dimensional phase space (which, now, is defined as the set of non-instantaneous "initial" data just mentioned) which carries a symplectic structure invariant under the action of P. Unfortunately, in Künzle's examples the two particles appear to play intrinsically different roles; moreover, it is still an open question whether such systems exist for more than two particles, whether the particles can be treated as indistinguishable, and whether the interaction can be chosen to be separable (in the sense that interactions vanish for large distances). The mathematical problems posed by predictive P-invariant (Hamiltonian) mechanics have been very clearly exposed in [10], [11]. For a related approach which is also concerned with relativistic quantum 2-particle systems see [14]. It emerges that whereas at present no P-invariant Hamiltonian theory of many interacting particles exists, the task of

---

* Then H, K_a are "dynamical", and P_a, J_a are "kinematical" generators, in contrast to the Galilean case.

creating such a theory is neither finished nor hopeless, and future work may well influence statistical mechanics as well as quantum theory.

It is well known that some important Poincaré-invariant (classical) fields or systems of *particles interacting via fields* can be described, at least formally, in a fully Hamiltonian way (in the above sense). This is the case, e.g., for the free electromagnetic field and for point charges interacting with such a field. The generators of the corresponding canonical representation have been given by Dirac [13]. The corresponding statistical mechanical formalism, in particular for a relativistic plasma, has been elaborated by Balescu and Kotera [15], where earlier work is cited. Just as (2) describes the effect of time translations on the distribution function f, so similar equations describe the behaviour of f under boosts, e.g.,

$$\partial_s f = [f, K_1] \tag{6}$$

where s is the rapidity of the boost $\exp(-sK_1)$ in the (x^1, x^4)-plane. With this formalism one can derive, e.g., the values of macroscopic quantities as ensemble averages, study their evolution and transformation properties, and obtain macroscopic balance equations [15]. The Liouville operator and its analogues for boosts have been studied in detail, as is necessary for perturbation theory. Moreover, Balescu has studied the generalized master and kinetic equations of this theory, and verified formally their appropriate invariances.

However, this formalism, being formulated in terms of bare (mass-unrenormalized) particles and a single "total" field, is plagued by infinite self-energies. A related, but perhaps more serious objection to this approach is this: The (as far as one can tell) rigorous formulation of the classical theory of electromagnetically interacting point charges proposed by Rohrlich [16] strongly suggests that one fundamental assumption made by Balescu and Kotera, that initial values of the canonical variables (of the particles and the field) can be specified arbitrarily at a finite time (in some inertial frame), is fictitious. If this were indeed so, then the whole theory (which conforms to the "philosophy" outlined at the end of Sec. 2) would break down. The point is that the essential non-locality of the complete set of basic laws of (renormalized) classical electrodynamics with point sources (asymptotic condition! See [16], in particular chapter 6 B and sections 7-1, 9-1 and 9-3) seems to imply that no exact formulation of it can be strictly canonical. The crucial unsolved problem is the *initial value problem* or, more generally, the problem how to describe, in some explicit way, the set Z of dynamically possible (particle + field) histories of the theory. In connection with regularization rules and perturbation theory the Balescu–Kotera formalism may be very useful, but in view of the difficulties of principle just discussed other approaches should also be considered.

I only mention that Balescu has also initiated the study of a relativistic quantum statistical plasma theory by means of the Weyl–Wigner correspondence [17].

A completely different, non-Hamiltonian, manifestly Lorentz-covariant framework for classical relativistic statistical mechanics has been elaborated by Hakim [18]. In this formulation the intrinsic space-time geometrical structures are put into the foreground, and the use of quantities related to extraneous things like spacelike hyperplanes, inertial frames, etc., is avoided. This formulation, therefore, can be generalized (as far as the general structure is concerned) even to general relativity, as long as the gravitational field (the metric g_{ab}) is treated as a given external field or as a self-consistent collective field. (In the latter case the problem of justifying this mean field approximation remains, of course, outside of the theory.)

Let a system consist of N particles interacting via a field ϕ (and possibly be embedded in an external field, e.g., a metric field g_{ab}), with some specified equations of motion. If the equations of motion for the particles contain the space-time coordinates x^a, the 4-velocities u^a, and the 4-accelerations $\dot{u}^a = b^a$, but no higher derivatives of the functions $x^a(\tau)$ describing the world lines, the tangent bundle $T(X \times X \times \ldots \times X)$ (N factors X, where X = space-time) is called the N-particle phase space (dimension $8N$). If also the (absolute) derivatives $\ddot{u}^a$ enter (as they do if radiation reaction is taken into account in the electromagnetic case), one takes instead the product of N factors $\{x^a\} \times \{u^a\} \times \{b^a\}$ as phase space P. The possible values of the field ϕ will be cross sections of some tensor (or spinor) bundle over X; the space of these cross sections is then called the *field phase space* $\hat{P}$. Then $P \times \hat{P}$ is the *total phase space*. A solution of the equations of motion will be a collection of N world lines, parametrized in terms of N (independent) proper times $\tau_1, \ldots, \tau_N$, and a field $x^a \to \phi(x^a)$. In other words, it will consist of an N-dimensional submanifold ("N-worldline") of P and a "point" of $\hat{P}$. Thus $P \times \hat{P}$ "contains" all dynamically possible histories of the system (recall the remarks at the end of Sec. 2). (If there is no field and one assumes Fokker-type actions-at-a-distance, the phase space is just P.) For a theory where one knows which Cauchy data (e.g., on a spacelike hypersurface of X) determine uniquely a history of the system one could also define a *space of initial data*, which would be different from $P \times \hat{P}$. Since for the theories of interest here (e.g., renormalized classical electrodynamics or Wheeler–Feynman electrodynamics) this is not so, one cannot introduce such a space. Nevertheless, there is the (huge) set of all histories, Z, and that is naturally taken as the *sample space* for the statistics of the system. (If a space of initial data were known, it would of course be bijectively related to Z, and could itself be taken as sample space, as in ordinary statistical mechanics.) A *Gibbs ensemble* is a probability measure on Z.*

To simplify the writing, let $X_i = (x_i^a, u_i^a, \ldots)$ be the phase coordinates of the ith particle, so that $X = (X_1, \ldots, X_N) \in P$. An element of Z can then be characterized either by an "N-worldline" (a term introduced by Künzle [11])

$$(\tau_1, \ldots, \tau_N) \to (X_1(\tau_1), \ldots, X_N(\tau_N))$$

---

* Such measures may not exist, but a *mean* would be sufficient, see [47], Appendix 7.

and the accompanying field ϕ, or by a *proper-time dependent microscopic density*

$$R_N(X_1,\ldots,X_N;\tau_1,\ldots,\tau_N) = \sum_\pi \prod_{j=1}^N \delta(X_j - X_{i_j}(\tau_j)) \tag{7}$$

where the summation extends over all permutations $\pi:(1,\ldots,N)\to(i_1,\ldots,i_N)$ and we assume indistinguishable particles, again together with ϕ. By means of formulae analogous to (7) one can introduce *reduced densities* $R_1, R_2,\ldots$; e.g.,

$$R_1(X,\tau) = N^{-1}\sum_{i=1}^N \delta(X - X_i(\tau)) \tag{8}$$

or

$$W_3^2(X_1,X_2,X_3;\tau_1,\tau_2,\tau_3) = \{N(N-1)\}^{-1}\times$$

$$\times \sum_{\substack{1\le i,j\le N\\ i\ne j}} \delta(X_1 - X_i(\tau_1))\delta(X_2 - X_i(\tau_2))\delta(X_3 - X_j(\tau_3))$$

If a Gibbs ensemble is given, one has not one history (R_N, ϕ) but a collection $\{(R_N(\omega), \phi(\omega))\}$ of such indexed by ω, and one may consider the *stochastic microscopic density* $R_N(X_1,\ldots;\tau_1,\ldots;\omega)$ together with the *stochastic field* $\phi(x^a, \omega)$ as characterizing the ensemble. In addition to these basic random fields, there will be others like R_1 and W_3^2. It is clear that ensemble averages $\langle R_N\rangle$, $\langle\phi\rangle$ can be defined. In particular

$$D_1 = \langle R_1\rangle \tag{8'}$$

is a τ-dependent average one-particle density, and

$$f_1(x^a, u^a) = \int_{-\infty}^{+\infty} D_1\,d\tau \tag{8''}$$

is the usual *single-particle distribution* function on the single-particle phase space $T(X)$. Clearly one can also define *current densities* of physical quantities within this framework, like a kinetic energy-momentum tensor.

The microscopic equations of motion (for examples, see [18]) imply coupled equations of motion for the τ-dependent microscopic densities $R., W.$; and hence for their averages, the *reduced distributions*, either the τ-dependent or the τ-independent ones. (The τ-dependent quantities are nothing but tools, convenient for the adaptation of non-relativistic methods to the relativistic case; at the end one gets rid of the τ's by integration.) In this way, Hakim has been able to establish a *relativistic analogue of the Klimontovich hierarchy*, an *infinite* system of linear partial differential-integral equations coupling a suitably chosen set of generalized reduced distribution functions ([18], see also [19]). These can be used as a basis for deriving kinetic equations. So far, only some first steps have been made to carry out the enormous programme thus indicated.

The theory outlined manages to provide, at least in principle, a conceptually clear, manifestly covariant statistical description of a relativistic system which does not presuppose a knowledge of the nature of the Cauchy initial data of the system. The price paid for this remarkable achievement is that the description is non-Hamiltonian, and that the equations for the reduced distributions are non-local.

No corresponding quantum-statistical theory (which one might envisage as an extension of the pseudoclassical Weyl–Wigner–Moyal description mentioned above) appears to be known. In this important respect the Hakim theory is inferior to the "Balescu-type" theory which does have a close (formal) quantum analogue.

Before closing this section I wish to mention that the *relation between microscopic and macroscopic electrodynamics*, classically as well as quantum mechanically, at the non-relativistic as well as at the special-relativistic level, has been extensively investigated, in particular with regard to the old and frequently discussed question of the macroscopic material and field energy-momentum tensors, by de Groot and Suttorp [4]. It turns out, not surprisingly, that T^{ab} is much more complicated than any of the previously proposed (guessed) candidates (by Minkowski, Abraham, Einstein–Laub,…), if long range correlations between particles are taken into account. De Groot and Suttorp treat the electromagnetic field as a functional of the particle world lines and use a covariant averaging over systems of world lines (à la Hakim) to obtain the macroscopic quantities and relations. The method leads to macroscopic laws—ref. [4] contains the most detailed treatment of macroscopic electrodynamics on a relativistic microscopic-statistical basis of which I am aware—but the macroscopic equations of motion do not form a closed system and are therefore supplemented by phenomenological equations by the authors.

4. STATISTICS AND GENERAL RELATIVITY

Whereas the theory of Hakim outlined in the preceding section can be adapted, in a rather obvious way, to systems of particles and non-gravitational fields in a given (or self-consistent, mean) gravitational field, a theory which includes g_{ab} in the set of "microscopic" variables to be subjected to statistics does not seem to have been attempted. In principle it should be possible to take the Hamiltonian form of general relativity—say, the ADM formulation—as a basis for a *statistics of gravitational fields*. One could try to introduce a distribution functional $f[g_{ij}(\mathbf{x}), \pi^{rs}(\mathbf{x}); t]$ obeying an "Einstein–Liouville equation"

$$\frac{\partial f}{\partial t} = [f, \mathscr{H}] N + [f, \mathscr{H}_i] N^i \tag{9}$$

where t labels the members of a 1-parameter family of spacelike hypersurfaces (formal time), $\mathscr{H}$, $\mathscr{H}_i$ are the super-Hamiltonian and supermomentum, respectively,

N, N^i are the lapse and shift functions, and the r.h.s. of (9) stands for an integral. The mathematical problems of such a theory would be similar to those of a superspace quantum gravity theory [20], [21], [47], though probably somewhat simpler. Superficially such a theory would be analogous to the Balescu version of statistical electrodynamics considered in Sec. 3. The roles of H and K_a correspond to that of $\mathscr{H}$, whereas P_a, J_a correspond to $\mathscr{H}_a$ (dynamical and geometrical generators); the Poincaré group corresponds (formally) to the group of deformations of a spacelike hypersurface in space-time [22]. It might be interesting to develop a statistical theory of gravitational fields of a restricted (e.g., cosmological) type, i.e., in a "mini-super-space".

In connection with perturbation theory, statistical arguments have been used in general relativity. An important example is Isaacson's definition of an effective energy-momentum tensor for short-wave gravitational waves superimposed on a smooth background by means of the Brill–Hartle averaging method [23].

In this connection the following remarks may be useful. Averaging over metric tensors $\langle g_{ab} \rangle$ may be a questionable procedure since linear combinations of Lorentzian metrics are, in general, not metrics at all. (As pointed out by P. G. Bergmann, one can generate Lorentzian metrics from spin-matrices; the latter do form a closed set with respect to averaging.) However, averaging over linear connections $\langle \Gamma^a_{bc} \rangle$ is meaningful, since the set of linear connections on a manifold is closed under convex combinations. Also, $\Gamma^a_{bc} - \langle \Gamma^a_{bc} \rangle$ is a tensor which plays an important role in perturbation theory. If the realizations of a random connection $\Gamma(\omega)$ are metric, the average connection $\langle \Gamma(\omega) \rangle$ will in general not be metric, however. Therefore it may be useful in considering an ensemble of metrics, $\{g(\omega)\}$, to define a splitting $g(\omega) = g(\text{background}) + h(\omega)$, *not* by requiring $\langle h \rangle = 0$, but by postulating $\langle \Gamma(\omega) \rangle = \Gamma$, where $\Gamma(\omega)$, Γ are the Riemannian connections of $g(\omega)$ and $g(\text{background})$, respectively.

A problem of interest is to create a gravitational analogue of Lorentz's derivation of macroscopic electrodynamics from microscopic electrodynamics, particularly in view of the justification and refinement (correlation energy?) of the Vlasov-type approximation used so far to describe stellar systems.* This seems not to have been attempted. A difficulty arises from the fact that point particles are not (strictly) compatible with Einstein's field equation, as is clear from the Kruskal extension of the Schwarzschild space-time.

A truly "first principle derivation" of Einstein's macroscopic theory of gravitation from an underlying microscopic theory can be hoped for only once the relation of general relativity to quantum theory is better understood, of course.

---

* "Microscopic" in the gravitational case should not be taken literally; stars might play the roles of "molecules", or even galaxies in a cosmological context.

5. GENERAL-RELATIVISTIC KINETIC THEORY OF GASES

It is well known how one can set up, in a given space-time (X, g_{ab}) which may also contain an external electromagnetic field F_{ab}, a kinetic theory of gases, using plausibility arguments like Boltzmann to obtain a *Boltzmann equation* for the single-particle distribution function $f_1(x^a, u^b)$. For systematic presentations see, e.g., [24]–[26].

The single-particle distribution function f_1 of kinetic theory may be identified with the function defined in Eq. (8'') within the covariant statistical mechanics of Hakim. The definition given there implies, as is easily verified, that f_1 measures the average density of world lines in a Gibbs ensemble representing a macrostate of the gas considered. If all particles have the same proper mass, f_1 contains a factor $\delta(p_a p^a + m^2)$ ($p^a = mu^a$), and one may replace f_1 by an ordinary function $f(x^a, p^a)$ defined on the mass-shell bundle $\{(x^a, p^a) | p_a p^a = -m^2\}$ over space-time. This f can be scaled so that it coincides numerically with the ordinary $(\mathbf{x}, \mathbf{p})$-phase space density of *any* local Lorentzian observer, and is the *basic quantity* of kinetic theory (which has also the corresponding interpretation in the limiting case $m = 0$).

For a collisionless gas, the *collisionless Boltzmann equation* (single-particle Liouville equation)

$$Lf = \left(p^a \frac{\nabla}{\partial x^a} + eF^a_b p^b \frac{\partial}{\partial p^b} \right) f = 0 \tag{10}$$

($\nabla/\partial x^a$ denotes the horizontal lift of $\partial/\partial x^a$ into TX restricted to the mass-shell) follows immediately from the definitions (8), (8'), (8'') and the Lorentz–Einstein equation of motion

$$\frac{Dp^a}{ds} = eF^a_b p^b \tag{11}$$

Similar derivations of the relativistic Vlasov and Landau equations from statistical mechanics have been given, even with the inclusion of radiation reaction terms related to the collective emission of radiation [18]₄.

For the relativistic Boltzmann equation (with collisions due to short range interactions), a "quasi-derivation" similar to the non-relativistic ones does not seem to exist. An outline of a derivation on the basis of the Fock-space formalism of quantum theory has been given in [27], somewhat improved in [26]₁.

The possibility to include straightforwardly the gross effects of *inelastic collisions* (nuclear reactions) in mixtures and of *quantum statistics* (induced processes, e.g.) into kinetic theory make it a particularly useful tool for astrophysics. It would be desirable to have a more satisfactory microscopic justification of the relativistic Boltzmann equation, however.

If one accepts, presently on the basis of plausibility only, the gravitational-Vlasov approximation (approximation to what? one does not know), then the *basic equations for a gravitating gas* are

$$G^{ab} = 8\pi \int p^a p^b f \pi$$

$$Lf = \text{Boltzmann collision integral} \tag{12}$$

The microscopic data specifying such a system are cross sections entering the r.h.s. of $(12)_2$. (I have here disregarded a collective electromagnetic field which could easily be included in (12).)

These laws imply conservation laws and an H-theorem, as is well known. As in the non-relativistic case, one can determine local and global equilibrium distributions from the requirement of vanishing entropy production, and thus reestablish the old results of Jüttner. If the distributions of a mixture differ only slightly from local thermal equilibrium,

$$f_j = f_j^0 (1 + g_j), \qquad f_j^0 = \exp\{-\alpha_j + \beta u_a p^a \pm 1\} \tag{13}$$

$(\beta^{-1} = T = \text{abs. temperature}, \alpha_j \beta^{-1} = \text{chemical potentials})$, the relations

$$s = F(\rho, n_1, \ldots, n_R)$$

$$ds = T^{-1} d\rho - \sum \alpha_j \, dn_j \tag{14}$$

between entropy density s, mass-energy density ρ, particle number densities n_j, etc., which hold exactly at local thermal equilibrium, remain valid to within first order in the perturbations g_j; this establishes part of irreversible thermohydrodynamics of gases. Eqs. (14) can be re-expressed thus: near equilibrium, the entropy current 4-vector S^a is given by

$$S^a = su^a + \frac{\partial F}{\partial \rho} q^a + \frac{\partial F}{\partial n_j} i_j^a \tag{14'}$$

where $s = F(\rho, n_j)$ is a thermostatic potential of the system and u^a, q^a, i_j^a are, respectively, the mean velocity (see (13)), heat flow, and diffusion currents.

To complete the non-equilibrium theory, one needs approximation methods to determine the g_j and the transport and reaction laws and coefficients (see [24]–[26] and the references given there).

The following recent improvements of the theory are of importance:

1) D. Bancel and Y. Choquet [28] have established the local existence, uniqueness and stability (i.e., stable dependence on initial data) of solutions to the *Cauchy initial value problem* for the coupled Boltzmann–Maxwell–Einstein system (Eqs. (12) + electromagnetic terms).

2) J. M. Stewart has largely clarified the question of how fast *deviations from*

equilibrium can *propagate* in a relativistic gas, using the relativistic Grad-method of 14 moments [24]. This work solves an old causality puzzle, at least for dilute gases.

3) S. R. de Groot and W. A. van Leeuwen have treated non-reacting mixtures of isobaric Maxwellian particles, and have deduced *Onsager relations* for such systems [29].

4) W. Israel and J. N. Vardalas have computed transport coefficients of relativistic quantum gases [30].

5) Relativistic kinetic theory has been extended from structureless particles (characterized by m and scalar charges only) to a) particles with a magnetic moment [31] and spin [32], and b) photons with spin (which usually has been left out of account) [33].

In both cases one has to deal with an enlarged phase space to accommodate the internal degrees of freedom. As Israel [32] points out, "an anisotropic universe containing material with a large store of internal spins (elementary particles, turbulent eddies, primeval black holes) will tend to dissipate its anisotropy by rotational viscosity. The importance of this effect remains to be estimated."

In case b the polarization states for fixed x^a, p^a form a 2-dimensional complex Hilbert space $\{t^a | p_a t^a = 0; \langle t_1, t_2 \rangle = \bar{t}_1^a t_{2a}\}$, hence one has a *distribution matrix* (2×2 Hermitean), $F(x^a, p^b)$, the trace of which is the ordinary distribution function $f(x^a, p^b)$ considered above. Aquista and Anderson have derived a transfer equation for partially polarized light passing through a stationary or moving scattering medium; in particular, they have considered Compton scattering. (As an aside I wish to point out that partially polarized radiation is an example of a system with non-commuting, macroscopic observables—the matrices corresponding to the Stokes parameters.)

6) J. L. Anderson and H. R. Witting [34] have elaborated a new relaxation time model for the collision term of the Boltzmann equation for a single-component gas, and have used it to determine relativistic quantum transport coefficients.

Further, technical improvements of the mathematical tools used to compute transport coefficients have been achieved by J. L. Anderson; however I shall not go into that.

7) J. Guichelaar, W. A. van Leeuwen and S. R. de Groot have investigated the propagation of sound in a dissipative relativistic gas [35].

8) A. Mangeney has given a general-relativistic extension of the Chew–Goldberg–Low equations governing a strongly magnetized collisionless plasma [45]. This theory may be of use in studying the behaviour of the plasma in pulsar magnetospheres close to the velocity-of-light cylinder.

Kinetic theory has been applied to various cosmic processes such as dissipative processes in Galaxy formation and the behaviour of neutrinos in the early universe. For an introduction to these problems and references see, e.g., Stewart [36] and the books quoted in the introduction.

6. CONTINUUM MECHANICS AND THERMODYNAMICS

Basic for the *kinematics* of a deformable medium in a general space-time (X, g_{ab}) is a congruence K of timelike curves describing the *mean motion*. The latter may be characterized by the 4-velocity field u^a. Its gradient determines derived kinematical variables like rate of expansion, vorticity, etc.; for a complete review see [37]. The members of K may be identified with the "particles" (in a macroscopic sense) of the medium. The distances between these particles are measured by $h_{ab} = g_{ab} + u_a u_b$. An *instantaneous state* of the medium corresponds to data associated with a spacelike cross section Σ of K. In particular, the restriction of h_{ab} to Σ is the relativistic analogue of a "state of strain" of the medium. In order to judge (Lagrangean) changes of local quantities associated with the medium relative to the medium (in contrast to changes relative to a local inertial frame), one uses a suitable modification of the Lie derivative $\mathscr{L}_{u^a}$, the convective derivative (see [38] or [26]$_2$ for careful discussions). By means of these concepts one can adapt to the relativistic case in an elegant and satisfactory way those local notions of non-relativistic kinematics (like finite or infinitesimal deformation, irrotationality, etc.) which are prerequisites for a *dynamics* of deformable media.

Relativistic continuum mechanics and thermodynamics, like its non-relativistic predecessor, is based on a few general principles and various constitutive laws characterizing particular models of materials. As *general principles* one can take the *existence, symmetry* and *balance of an energy-momentum tensor*,

$$T^{[ab]} = 0, \qquad T^{ab}_{;b} = 0 \tag{15}$$

and the existence of an *entropy 4-current* S^a satisfying the *Clausius–Duhem inequality*

$$S^a_{;a} \geq 0 \tag{16}$$

Whereas Eq. (15) is generally accepted as a basic law of all metric theories of gravity, the proposal to regard the inequality (16) as a *general expression* of the local law of the non-decrease of entropy in relativistic, phenomenological thermodynamics of deformable media is perhaps still controversial. To me, (16) appears well motivated and almost unavoidable, for the following reason. With any macroscopic state of a medium, i.e., a set of Cauchy data on a spacelike hypersurface Σ for the macroscopic equations of motion, there should be associated a numerical information-entropy (which one could compute on a microscopic basis if such a basis were known) $S[\Sigma,\ldots]$, a functional of that state which should be additive with respect to Σ. If this is so, and if D is a space-time region bounded by two spacelike hypersurfaces, then $S[\partial D,\ldots]$ measures the entropy production, or loss of state-information, associated with irreversible processes in D. Hence $S[\partial D,\ldots]$ should be of the order of the space-time volume of D. These assumptions which seem hardly avoidable in any macroscopic description which is a coarse-grained version of an underlying (unknown) microscopic description, combined with mild smoothness assumptions on the functional $S[...]$, *imply* the existence of a 4-vector field S^a, a local functional of the state variables, such that $S[\Sigma,...] = \int_\Sigma S^a \sigma_a$, so that (16) follows. (The proof is a straightforward adaptation to space-time of Cauchy's classical argument establishing the existence of a stress tensor from assumptions about the distribution of surface forces in a 3-dimensional body.) This general reasoning is illustrated by kinetic theory, but I wish to emphasize that (16) is motivated not only by analogy to that special microscopic model; on the contrary, the general argument explains why one has (16) particularly for dilute gases.

No general law can be expected to link the 14 variables T^{ab}, S^a. The general laws (15), (16) and possibly further general conservation laws like baryon conservation provide a framework only into which all specific models of matter (perfect fluids, viscous fluids, elastic solids,...) must fit.

In order to characterize particular materials, one postulates the existence of *material currents* and *balance equations* for them, e.g., conservation laws like

$$N^a_{j;a} = 0 \tag{17}$$

and possibly additional fields like polarization densities, etc. Moreover, one imposes restrictions on the (so far independent) quantities T^{ab}, S^a, N^a_i,....

A *multi-component fluid*, e.g., can be characterized as follows: There exists a *mean 4-velocity* u^a and a *thermostatic potential* $F(\rho, n_j)$ such that (14′) holds (for all processes of the material). (The quantities s, ρ, n_j, q^a, i^a_j are defined in a standard manner in terms of the basic fields T^{ab}, S^a, N^a_j, and u^a: $s = -S_a u^a$, $\rho = T_{ab} u^a u^b$, $n_j = -N^a_j u_a$, $q^a = -u_b T^{bc}(\delta^a_c + u^a u_c)$, $i^a_j = N^b_j(\delta^a_b + u^a u_b)$.)

This assumption (of local thermal equilibrium) introduces simultaneously the mean velocity u^a and a caloric equation of state $(14)_1$. The *Gibbs equation* $(14)_2$ then serves to *define* the temperature T and the α_i's. (Whereas in kinetic theory (14), (14′) *follow* as approximations from the kinetic equation, with completely determined F, they are *postulated* in the phenomenological continuum theory, with a disposable F restricted by some inequalities, e.g., $\partial F/\partial \rho > 0$.)

Elastic solids can be defined by means of a constitutive assumption quite analogous to the one given above for fluids, except that the configuration variables n_j in F have to be replaced by the strain tensor h_{ab} (defined in the first paragraph of this section). For details, see [38], [39], [26]$_2$, [40].

If (14′) (or its analogue for solids), (15) and (17) are inserted into (16) one obtains an inequality for pairs of variables connected with irreversible processes. In the particular case where the last (diffusion) term in (14′) vanishes, (16) can be rewritten in the form

$$(su^a)_{;a} \geq -(q^a/T)_{;a}$$

which is a precise version of the somewhat vague, but traditional formula $T\delta S \geq \delta Q$.

The constitutive laws introduced so far still do not specify a material completely. As a last step needed to obtain a complete set of macroscopic equations of motion one has to postulate *transport* and *reaction equations*, etc. These are suggested and restricted by the inequality (16). A complete model of matter should permit a *Cauchy problem* compatible with relativistic causality (see [41]), and the subset of its solutions satisfying $S^a_{;a} = 0$ should give a physically reasonable set of *equilibrium states*. Moreover it should have an acceptable non-relativistic limit.

The adaptation of the highly developed non-relativistic, non-linear theory of materials (due to W. Noll, C. Truesdell, C. Coleman and others) to relativity has been initiated by various authors (Bragg, Bressan, Carter, Eringen, Grot, Leaf, Maugin and others). For an application of relativistic elasticity theory to the theory of gravitational wave detection, and many references to modern work on non-relativistic and relativistic continuum mechanics, see [40].

It appears that the conceptual and mathematical apparatus now available for describing relativistically matter in bulk is sufficiently well developed and flexible to accommodate most of what has been done non-relativistically. However, whether such largely formal adaptations of classical theories are physically correct in the relativistic regime must be considered an open question until such theories can either be confronted with observations, or be linked to relativistic, quantum-mechanical many-body theories.

In the presentation sketched in this survey *thermodynamic quantities* $S^a, T, \ldots$ have been introduced, both at the kinetic and the continuum level, by means of *local laws* assumed to be valid at each point of space-time. In this approach, entropy appears either as a 4-vector S^a or as a density $s = -S_a u^a$ or as specific entropy $\sigma = s/n$ (n = particle number density). The temperature T appears as a scalar, just like s, n and σ. These quantities are often referred to as "measured by a comoving local observer". This procedure seems to be simpler, less ambiguous and, in view of astrophysical applications, more realistic than the traditional way of considering laws for *integral quantities* like U, V, S for a fluid enclosed in a box. There does not appear to be any need to introduce in addition relative temperatures T' for non-comoving observers. How should they be measured? In which way do such quantities help to describe physical states or processes? The physical content of the laws of thermodynamics seems to be adequately expressed by the local laws (15), (16) in conjunction with constitutive equations as exemplified above. Classical relations like $dU + pdV = TdS$ could be obtained by integrating local laws over appropriate space-time domains representing gases in boxes. (A simple example is given in [42]. The point of view indicated here is well illustrated by the work of Horwitz and Katz [49] on the thermodynamics of rotating fluids in relativity.)

For the reasons given I do not wish to comment on the "Planck–Ott controversy", but refer to Møller [43] and Balescu [44] for detailed discussions.

The validity of all such relativistic continuum theories can be questioned on the grounds that one cannot, at present, derive them in sufficient generality from a microscopic theory or check them experimentally. The mere formal resemblance to classical laws and the covariance do not, of course, guarantee their physical validity. However, classical continuum theories were and still are developed and successfully applied prior to and independently of their microscopic justification, and particularly in view of the difficulties of relativistic microscopic theories of interacting systems, I do not see any reason for rejecting such relativistic theories of matter.

Acknowledgment: The preparation of this paper was supported in part by NATO research grant No. 613.

APPENDIX

Let (τ, a, w, D) denote the Galilean transformation

$$\begin{cases} t' = \tau + t \\ x' = a + wt + D \cdot x \end{cases}$$

where τ is a real number, a and w are column-matrices with three elements, and D is a 3×3 rotation matrix. The multiplication law

$$(\tau', a', w', D') \cdot (\tau, a, w, D) = (\tau' + \tau, a' + w'\tau + D'a, w' + D'w, D'D)$$

implies, as is easily verified, that the set $F = \{(0, a, w, D)\}$ of those transformations not affecting t is a normal subgroup, the set $R = \{(\tau, 0, 0, I)\}$ of time translations is an abelian subgroup, $G = F \cdot R$ (since $(\tau, a, w, D) = (\tau, 0, 0, I) \cdot (0, a, w, D)$), and $F \cap R = (0, 0, 0, I) = id$. Hence $G = F^{40} \times {}^{40}_h R$ is a *semi-direct product*, where the homomorphism h from R into the group of automorphisms of F is given by $h(\tau) : (0, a, w, D) \to (0, a - w\tau, w, D)$. It is easily checked that F is the *commutator group* of G.

Notes Added in Proof

1. The mathematical and conceptual problems one has to face when setting up a relativistic—particularly a general-relativistic—statistical theory of particles or fields are closely related to those encountered in some of the various attempts to *"quantize"* general relativity. It may therefore be useful to read this review in conjunction with the excellent report on "Quantum Theory of Gravity" by Ashtekar and Geroch [47]. Progress in either field will probably affect the other one.

2. An important field of research related to several parts of this review is that of *relativistic turbulence theory*. Such a (not yet existing) theory may, for example, be

relevant, in connection with cosmology, for the problem of galaxy formation. A start in this direction has been made in Newtonian cosmology by D. W. Olson and R. K. Sachs, see [48].

REFERENCES

1. O. Penrose, *Foundations of Statistical Mechanics*, Pergamon Press, Oxford, 1970.
2. A. Katz, *Principles of Statistical Mechanics*, Freeman and Co., San Francisco, 1967.
3. I. Prigogine, *Non-Equilibrium Statistical Mechanics*, Interscience, New York, 1962.
4. S. R. de Groot and L. G. Suttorp, *Foundations of Electrodynamics*, North-Holland, Amsterdam, 1972.
5. W. Kundt, *Springer Tracts in Modern Physics* **40**, 107 (1966).
6. I. Prigogine, C. George, F. Henin and L. Rosenfeld, *Chemica Scripta* **4**, 5 (1973); C. George, I. Prigogine and L. Rosenfeld, *Mat. Fys. Medd. Dan. Vid. Selsk. (Kopenhagen)* **38**, 12 (1972).
7. R. Balescu, J. L. Lebowitz, I. Prigogine, P. Resibois and Z. W. Salsburg, *Lectures in Statistical Physics*, Vol. 7 in series *Lecture Notes in Physics*, Springer, Heidelberg, 1971; W. C. Schieve and J. S. Turner, *Lectures in Statistical Physics*, Vol. 28 of same series; A. Lenard, *Statistical Mechanics and Mathematical Problems*, Vol. 20 of same series.
8. R. Kurth, *Axiomatics of Classical Statistical Mechanics*, Pergamon Press, Oxford, 1960.
9. P. Havas, in *Statistical Mechanics of Equilibrium and Non-Equilibrium* (J. Meixner, ed.), North-Holland, Amsterdam, 1965.
10. L. Bel, *Ann. Inst. Henri Poincaré* **A XIV**, 189 (1971); **XVIII**, 57 (1973).
11. H. P. Künzle, "Galilei and Lorentz Invariance of Classical Particle Interactions," to appear in *Symposia Mathematica,* Acad. Press; "Hamiltonian Description of Relativistically Interacting Two-Particle Systems," *J. Math. Phys.* **15**, 1033 (1974).
12. F. R. Halpern, *Special Relativity and Quantum Mechanics*, Prentice-Hall, Inc., Englewood Cliffs, N.J., 1968.
13. P. A. M. Dirac, *Rev. Mod. Phys.* **21**, 392 (1949).
14. L. P. Horwitz and C. Piron, *Helv. Phys. Acta* **46**, 316 (1973).
15. R. Balescu and T. Kotera, *Physica* **33**, 558 (1967); R. Balescu, T. Kotera and E. Pina, *Physica* **33**, 581 (1967); A. Pytte and R. Balescu, *Nuov. Cim.* **55B**, 51 (1968); R. Balescu, *Physica* **38**, 119 (1968); R. Balescu, M. Baus and A. Pytte, *Bull. C. Sc., Acad. Roy. Belg.* (1968).
16. F. Rohrlich, *Classical Charged Particles,* Addison-Wesley, Reading, Mass., 1965.
17. R. Balescu, A covariant formulation of quantum relativistic statistical mechanics, preprint Brussels, 1968.
18. R. Hakim, *Ann. Inst. Henri Poincaré* **6**, 225 (1967); *J. Math. Phys.* **8**, 1315 and 1399 (1967); R. Hakim and A. Mangeney, *J. Math. Phys.* **9**, 116 (1968).
19. P. Müller, Thesis, Universität Hamburg (unpublished), 1970.
20. A. E. Fischer, in *Relativity* (M. Carmeli, St. J. Fickler and L. Witten, eds.), Plenum Press, New York, 1970, p. 303.
21. B. S. DeWitt, in *Relativity* (M. Carmeli, St. J. Fickler and L. Witten, eds.), Plenum Press, New York, 1970, p. 359.

22. C. TEITELBOIM, preprint Princeton University, 1973.

23. R. A. ISAACSON, *Phys. Rev.* **166**, 1263 and 1272 (1968).

24. J. M. STEWART, "Non-equilibrium Relativistic Kinetic Theory," *Lecture Notes in Physics*, Vol. 10, Springer, Berlin, 1971.

25. W. ISRAEL, in *General Relativity* (L. O'RAIFEARTAIGH, ed.), Clarendon Press, Oxford, 1972.

26. J. EHLERS, "General Relativity and Kinetic Theory," in *General Relativity and Cosmology* (R. K. SACHS, ed.), Academic Press, New York, 1971, p. 1; J. EHLERS, "Survey of General Relativity," in *Relativity, Astrophysics and Cosmology* (W. ISRAEL, ed.), D. Reidel Publ. Co., Dordrecht, Holland, 1973, p. 1.

27. K. BICHTELER, Doctoral Dissertation, Hamburg, 1965.

28. D. BANCEL and Y. CHOQUET-BRUHAT, *Commun. Math. Phys.* **33**, 83 (1973).

29. S. R. DE GROOT, CH. G. VAN WEERT, W. TH. HERMENS and W. A. VAN LEEUWEN, *Physica* **40**, 581 (1969); **60**, 472 (1972).

30. W. ISRAEL and J. N. VARDALAS, *Nuov. Cim. Lett.* ser. I, **4**, 887 (1970).

31. CH. G. VAN WEERT, Doctoral Dissertation, University of Amsterdam, 1970.

32. W. ISRAEL, preprint, University of Alberta, 1974.

33. C. AQUISTA and J. L. ANDERSON, *Ap. J.* **191**, 567 (1974).

34. J. L. ANDERSON and H. R. WITTING, *Physica* **74**, 466 (1974).

35. J. GUICHELAAR, W. A. VAN LEEUWEN and S. R. DE GROOT, *Physica* **68**, 342 (1973).

36. J. M. STEWART, in *Proc. of the Cargese Summer School in Theoretical Physics*, Gordon & Breach, 1973, p. 175.

37. G. F. R. ELLIS, in *General Relativity and Cosmology* (R. K. SACHS, ed.), Academic Press, New York, 1971, p. 104.

38. B. CARTER and H. QUINTANA, *Proc. Roy. Soc.* **A331**, 57 (1972).

39. E. N. GLASS and J. WINICOUR, *J. Math. Phys.* **13**, 1934 (1972); **14**, 1285 (1973).

40. G. A. MAUGIN, *G.R.G. Journal* **4**, 241 (1973); **5**, 13 (1974).

41. S. W. HAWKING and G. F. R. ELLIS, *The Large Scale Structure of Spacetime*, Cambridge University Press, Cambridge, 1973.

42. A. STARUSZKIEWICZ, *Nuov. Cim.* **45**, 684 (1966).

43. C. MØLLER, Nordita publications No. 525, 1973.

44. R. BALESCU, *Physica* **40**, 309 (1968).

45. A. MANGENEY, preprint Meudon Observatory, April 1974.

46. P. T. LANDSBERG, *Physics Bulletin* **19**, 268 (1968).

47. A. ASHTEKAR and R. GEROCH, *Rep. Prog. Phys.* **37**, 1211 (1974).

48. D. W. OLSON and R. K. SACHS, *Ap. J.* **185**, 91 (1973).

49. G. HORWITZ and J. KATZ, *Ann. Phys.* **76**, 301 (1973).

Einstein and Zionism

BANESH HOFFMANN*

*Queens College of the City University of New York
Flushing, N.Y. 11367, U.S.A.*

An experienced orator once gave some excellent advice to after-dinner speakers. He said, "Think of a good start, and think of a good ending—and then keep them as close together as possible."

I am tempted to stop right here. But I don't think he meant the ending to follow the beginning quite so quickly. Besides, when it comes to giving speeches, I do not believe in the law of the excluded middle. So let me try to help you fill in the time while you wait for the end.

In a sense, what I want to tell you tonight is a story of a man finding his way to a cause. But more profoundly, it is a story of a man finding his way to part of his own self. The cause is Zionism, and the man, I need hardly say, is Einstein.

Because he made the universe seem a bit more comprehensible to us as scientists, and a bit more beautiful, we find ourselves gathered here in amity from far parts of the globe. Indeed, because of him, and because his theory is so beautiful, we can truly say that what draws us together is the attraction of gravitation. It unifies us despite our singularities.

But Einstein's spirit is present here also in a second sense. For this is the land of his forefathers. I want to tell of his feelings towards it. But in the telling I face an insuperable problem—the problem of trying to recapture for you feelings that belong to a different era. Let me try to explain the problem by means of examples.

On the island of Principe, Eddington hastened to measure his eclipse photographs and was thrilled to find a deflection of light that seemed to agree with the prediction

* I am grateful to Dr. Otto Nathan for permission to quote material belonging to the Estate of Albert Einstein; to Helen Dukas for invaluable help in providing copies of documents from the Einstein Archives and sharing with me her unrivalled knowledge of Einsteiniana; and to Professor Konrad Gries for unstinting aid in the recognition and rendering into English of subtle nuances of German.

of the general theory of relativity. Telling about it later, he said that it was the greatest moment of his life. Tell this to the man in the street, and he will probably be utterly incredulous. Recalling joys of a quite different sort, he might well exclaim, "What a drab life poor Eddington must have had!" As scientists, we can understand what Eddington meant. But even so we cannot recreate for ourselves the thrill that came to him at that time. Our best efforts yield only a pale imitation.

Again, in 1916 Einstein wrote to his friend Paul Ehrenfest, "Imagine my joy at the feasibility of the general covariance and at the result that the equations yield the correct perihelion motion of Mercury. I was beside myself with ecstacy for days." As relativists we well understand what Einstein meant. But can we really recapture the thrill? Much as we marvel at Einstein's theory, were any of us, on contemplating it, beside ourselves with ecstacy for several days? And yet we are specialists.

But my topic is one on which few of us are specialists. Let me, therefore, offer two examples that do not pertain to the physical sciences.

When we read about the Battle of Waterloo, we can recapture only faintly the emotions of Wellington, who lived through the long day not knowing how the battle would end. For, in more familiar jargon, once the wave function collapses, things are never the same.

Again, suppose we see a prisoner of war returning to his own country. He falls to his knees and kisses the ground. We are deeply moved. There may even be tears in our eyes. But we cannot feel what he feels. We cannot know in our guts what he felt in his at that precious moment. And in fact, after a while neither can he.

Please do not misunderstand me. I shall not be harrowing. Just the opposite. I fear that many of the Einstein quotations to be presented, though they tell of deep-felt things, may seem a little banal. Their emotion may seem remote, and little related to our own experience. I chose them to play a dual role: to tell us of Einstein's motives and feelings even though we cannot really recapture them; and through them, to give at least an indication of the situation and the conflicting aspirations of the Jews at the time, as if by reflected light.

If they do not do this, if I seem to have underplayed the story, I ask your forgiveness and your understanding. For, as Einstein said in 1929, "How many non-Jews [and he could easily have included Jews as well] have any insight into the spiritual suffering and distortion, the degradation and the moral disintegration engendered by the mere fact of homelessness of a gifted and sensitive people?"

To be a Zionist, it sometimes helps if one is a Jew. Einstein's parents were Jews. But they were assimilated Jews for whom the orthodox rituals were little more than a memory. In his *Autobiographical Notes* Einstein describes them as "entirely ir-religious" but I do not think we should take that too literally.

His parents sent him and his sister Maja to the local Catholic elementary school: it was cheaper than sending them to the distant private Jewish school, and much more convenient. Even so, the children learned about Judaism at home from a private instructor. And at the Gymnasium religious instruction in one's own faith was compulsory, as was attendance at services.

The children learned other things connected with being Jewish, as we see from the following excerpt from a draft of a letter written by Einstein in 1920. He was some forty years old when he wrote it, and at the height of his fame. Note how indelibly certain childhood happenings had impressed themselves on him:

"The teaching staff of the elementary school was liberal and made no denominational distinctions. Among the Gymnasium teachers there were a few anti-Semites, one in particular who never let us forget that he was a reserve officer. Anti-Semitism was evident among the children, particularly in the elementary school. . . . Physical assaults and insults were frequent on the way to school, though for the most part not really malicious. Even so, however, they were enough to confirm, even in a child of my age, a vivid feeling of not belonging."

The young Einstein was so impressed by his Jewish religious instruction that he quickly became intensely religious in a formal sense. For example, he refused to eat pork, and he was disturbed that his parents did not observe the Jewish rituals. But this phase did not last. As he wrote at age sixty-seven in his *Autobiographical Notes*:

"This deep religiosity came to an abrupt end at age twelve. Through the reading of popular scientific books I soon reached the conviction that much in the stories of the Bible could not be true. The consequence was a positively fanatical orgy of freethinking coupled with the impression that youth is intentionally deceived by the state through lies. It was a crushing impression. Suspicion of every kind of authority grew out of this experience . . . an attitude that has never left me. . . ."

It is perhaps worth mentioning here that some forty years after this crushing experience Einstein had the grace to say, "To punish me for my contempt for authority, Fate made me an authority myself."

For many years after his break with the Bible, Einstein seems to have been little concerned with Judaism. An incident in 1910 illustrates his casual attitude. He was offered the chair of theoretical physics at the German University in Prague, and, as was his custom, he listed his religious status as "unaffiliated." But it turned out that one of the requirements for any such appointment was a pledge of allegiance to the Emperor Franz Josef; and the Emperor, not without justification, felt that an oath of allegiance did not amount to much if it came from someone who had no God by whom to swear the allegiance. Einstein was thus in grave danger of being refused the professorship. How did he solve the problem? By a maneuver that reveals both his contemptuous understanding of petty authority and his casualness about formal religion. He went to the registrar and simply asked him to change the word "unaffiliated." This the registrar absolutely refused to do without proper authorization. Einstein asked by what authority the registrar had used the word in the first place, upon which the registrar said indignantly that Einstein himself had told him to use it. To the registrar that must have seemed the end of the matter, as indeed it probably would have to you and me. But Einstein, having maneuvered the registrar into citing him as an authority on this particular matter, now played his trump card. He said to the registrar that he, Einstein, now formally declared himself to be of the Hebrew faith. Since the logic was unanswerable, the registrar was persuaded to change "unaffiliated" to "Mosaic," which was the appropriate technical term. And that is how Einstein, with tongue in cheek, saved his almost lost professorship.

While in Prague, Einstein did come in contact with Zionists, but apparently he remained quite without interest in Zionism. From Prague he was invited back to Zurich, and from there, as we all know, to a prestigious position in Berlin in 1914. But still Jewish matters seem not to have occupied his mind to any significant extent.

An indication of what happened next can be gleaned from the following two quotations, even though they slightly conflict with previously quoted material. Both were written in 1929. In an article, Einstein said, "When I came to Germany fifteen years ago I discovered for the first time that I was a Jew, and I owe this discovery more to Gentiles than to Jews." And in a letter, he wrote in similar vein, "I first came to Zionism after my emigration to Berlin in 1914 at the age of 35, after I had lived in a completely neutral environment."

Let us not jump to conclusions. Einstein did not become a Zionist in 1914. Listen as I re-read his words: "I first came to Zionism *after* my emigration to Berlin at the age of 35."

How long after? Quite a while. As Kurt Blumenfeld said, "Till 1919 Einstein had no connection with Zionism or Zionistic ideas."

Who, then, is this Kurt Blumenfeld who speaks of Einstein with such assurance? He was the director of propaganda for the Zionist Union of Germany [Zionistisches Vereinigung für Deutschland]. The German Zionists had made a list of Jewish intellectuals whom they wanted to attract to the Zionist cause. Einstein's name was on the list. And because of it Blumenfeld presented himself to Einstein in the February of 1919 to talk with him about Zionism.

Note the circumstances. Two years before, the famous Balfour Declaration had promised a national homeland for the Jews. The Zionists rejoiced, of course. They could see their dream becoming a reality. But many successful assimilated Jews were bitterly outspoken in their opposition. They were afraid that a Jewish national homeland might tend to give them the status of aliens in the lands where they had achieved their success. As for Einstein, he had propounded his general theory of relativity in 1915, yet in early 1919 he was still relatively unknown to the general public. The results of the British eclipse expedition under Eddington were not officially announced till 6 November of that year, and only then did the people of the world suddenly realize that a mighty genius was in their midst. When Blumenfeld first came to Einstein in the February of 1919, Einstein's spectacular worldwide fame was still some nine months in the future, and if Einstein could not foresee it, surely neither could Blumenfeld. Obviously, the Zionists did not approach Einstein because of his world fame. But, naturally, when it came they were not at all averse to capitalizing on it.

Blumenfeld's task did not turn out to be easy. Einstein had many doubts that had to be resolved. For instance, as an internationalist he worried about the nationalistic aspects of Zionism. The months of discussion dragged on into years. Sometimes Einstein asked questions with what Blumenfeld described as "Godlike naivete." For example, when Blumenfeld started out by telling about the Jewish question and thus, presumably, of the historical persecutions of the Jews, Einstein asked what that had to do with Zionism. He asked, too—and I paraphrase his words as reported by Blumenfeld—whether it was really a good thing to separate the Jews from the intellectual pursuits for which they were born, and whether it was not a backward step to put manual skills, and above all agriculture at the center of things as Zionism did.

A month after Blumenfeld's first visit, Einstein did make this passing reference in a letter to his friend Paul Ehrenfest: "What pleases me the most is the realizing of the Jewish State in Palestine." But as Blumenfeld put it: "Einstein [warmed] up to the Zionist idea only gradually and after long deliberation. He joined the movement when he felt that it was actually a matter of a struggle for spiritual freedom, for human rejuvenation, and when he became convinced that the conquest of *Eretz Israel* for the Jewish people was a conquest through labor and that the movement was free from tendencies of profiteering and exploitation."

As we have said, Blumenfeld began his discussions with Einstein early in 1919. By 1920 Einstein was beginning to speak out in Zionistic tones, saying for example, "Only when we [Jews] have the courage to regard ourselves as a nation, only when we have respect for ourselves, can we win the respect of others."

This change of outlook had not been brought about solely by Blumenfeld. The end of World War I brought a sharp rise in anti-Semitism, and Einstein saw for himself the shattering effect it had had on the refugees who fled to Germany from Eastern Europe.

On 10 March 1921, Chaim Weizmann, in England, sent a detailed telegram to Blumenfeld. Weizmann was the leader of world Zionism—he was later to become the first President of the State of Israel. In his telegram he said that Blumenfeld should persuade Einstein to accompany Weizmann on a visit to the United States to help raise funds for Zionist causes. In particular, Blumenfeld was to stress to Einstein the need to raise money for the creation of a Hebrew University in Jerusalem.

Blumenfeld used all his arts of persuasion, but Einstein would have none of it. Right away he said, "I am no orator. I can add nothing convincing. You will only be using my name." And he defeated Blumenfeld's arguments so easily that Blumenfeld gave up the battle. But as he was about to take his leave, a last desperate thought came to him. He turned to Einstein and said, "I do not believe that we can weigh arguments against each other in this case. Our work can succeed when all of us are moved by a new spirit of national discipline. . . . I do not know what [Dr. Weizmann] would say to you in my place. But I know that he has been entrusted by the Jewish people with the responsibility of realizing the Zionist program. Dr. Weizmann, not as an individual, but as president of the Zionist Organization has ordered me to persuade you to go to America, and I have the right to expect that you subordinate your considerations to Dr. Weizmann's decision." To Blumenfeld's surprise, Einstein agreed. And that is how it came about that Einstein and Weizmann travelled together to America in the cause of Zionism.

(It is interesting to see Einstein, the instinctive rebel, here bowing to authority. I have a feeling of kinship with him. The topic of the present talk was not of my choosing. It was decided upon by the organizers of the Conference and, like Einstein, I am simply obeying orders.)

Weizmann was a distinguished scientist. Telling about the boat trip across the Atlantic, he said, "Einstein explained his theory to me every day, and on my arrival I was fully convinced that he understood it."

Einstein was received in America with extraordinary enthusiasm. Many honors were bestowed on him, and his presence on the platform turned out to be a major asset. For his unguarded account of what happened, let us look at a letter he sent from New York to his long-time friend Michele Besso, whom he had thanked in the celebrated 1905 paper that set forth the special theory of relativity. Einstein wrote:

"Two frightfully exhausting months now lie behind me, but I have the great satisfaction of having been very useful to the cause of Zionism and of having assured the foundation of the University. We found special generosity among the Jewish doctors of America (ca. 6000) who provided the funds to create the medical school. . . . I had to let myself be exhibited like a prize ox, to speak an innumerable number of times at small and large gatherings, and to give innumerable scientific lectures. It is a wonder I was able to hold out. But now it is over, and there remains the beautiful feeling of having done something truly good, and of having intervened courageously on behalf of the Jewish cause, ignoring the protests of Jews and non-Jews alike."

And in 1944, for Blumenfeld's sixtieth birthday, Einstein wrote to him saying:

"Now almost twenty-five years have passed since your first visit to me, when you persuaded me to make the journey to the United States, a journey that was good and necessary—and also pleasant once it was over."

(Actually, it had not been on Blumenfeld's *first* visit. On this point Einstein's memory had played him false. But he still vividly remembered how nice the whole thing was once it was over.)

If we look behind the banter, it is easy to see that the journey to America meant a great deal to Einstein. His commitment to Zionistic ideas was enormously strengthened and his sense of being a Jew took on a new profundity. From now on, guided by Blumenfeld, he spoke and wrote frequently in support of Zionism. No longer did he object that the Zionists were using his name. He realized that his fame was a unique asset entrusted to him by Fate, and his conscience told him that he must let it be used in the cause of Zionism, as in other causes he saw as worthy. Unlike too many other Jews in Germany, he felt a strong bond of kinship with the bedraggled, poverty-stricken Jewish refugees from Eastern Europe. He could see beyond their outward appearance, their air of vagabondage, and their lingering fear, and he recognized them not just as fellow human beings but also as fellow Jews who, therefore, had a special claim on his sympathy. He did not run from them as if fearing contamination from their contact. He saw that, unlike some of the more assimilated Jews in Germany, they retained a vivid sense of belonging—"a healthy national feeling . . . not yet destroyed by the process of atomization and dispersion." But let him speak for himself. Here are excerpts from a speech he gave in England in 1921 on his way home from the trip to America:

"When I moved to Berlin . . . I realized the difficulties with which many young Jews were confronted. I saw how, amid anti-Semitic surroundings, systematic study, and with it the road to a safe existence, was made impossible for them. This refers specially to the Eastern-born Jews in Germany. . . . These Eastern-born Jews are made the scapegoat of all the ills of present-day German political life and all the after-effects of the war. Incitement against these unfortunate fugitives, who have only just saved themselves from the hell which Eastern Europe means for them today, has become an effective political weapon, employed with success by every demagogue. When the Government contemplated the expulsion of these Jews, I stood up for them, and pointed out in the *Berliner Tageblatt* the inhumanity and the folly of such a measure. Together with some colleagues, Jews and non-Jews, I started University courses for these Eastern-born Jews, and I must add that in this matter we enjoyed official recognition and considerable assistance from the Ministry of Education.

"These and similar happenings have awakened in me the Jewish national sentiment. I am a national Jew in the sense that I demand the preservation of the Jewish nationality as of every other. I look upon Jewish nationality as a fact. . . . I regard the growth of Jewish self-assertion as being in the interests of non-Jews as well as Jews. That was the main motive of my joining the Zionist movement."

And later in the speech he says: "The main point is that Zionism must tend to enhance the dignity and self-respect of the Jews of the Diaspora. I have always been annoyed by the undignified assimilationist cravings and strivings which I have observed in so many of my friends."

As Einstein now saw it, Zionism was a unique revitalizing and unifying force for all Jews. Though steeped in tradition, it looked to the future. A few months after his return from America he said, "For the last two thousand years the common property of the Jewish people has consisted entirely of its past. . . . Now all that is changed. History has set us a great and noble task in the shape of active cooperation in the building up of Palestine . . . [which must] become a seat of modern intellectual life, a spiritual center for the Jews of the whole world. . . . A Jewish University in Jerusalem constitutes one of the most important aims of the Zionist organization."

The fact that in 1922 Einstein visited Japan is not in itself relevant to our topic. But, early in 1923, on his way home from Japan, he stopped off in Palestine, and this was an event indeed.

It is well known that Einstein was in the habit of doing calculations on any handy scraps of paper. It seems that just before his trip to Japan a friend gave him a going-away present of a travel diary. It consisted of blank pages bound together into a volume. We shall not be surprised to learn that Einstein did not waste it. What *is* surprising is that instead of using it for calculations, he actually used it as a travel diary.

Some of the entries are amusing. For example, one day in Japan, because of a mishap, Einstein had to borrow a top hat. It turned out to be so small that he had to carry it around in his hand all day. (It is not easy to imagine Einstein of the flowing mane willingly wearing a top hat in the first place. Perhaps he was secretly glad it did not fit.)

Here is a striking entry—striking because it was utterly routine. Einstein is on his way to Palestine and he writes in his diary, "1 Feb. Arrival in Port Said." Good heavens, we say to ourselves, Port Said is in Egypt! A Jew cannot travel these days to the Jewish homeland via Egypt—if his name is not Kissinger. But things were different in those days. Indeed, the diary unwittingly points up the difference further. It goes on to say that a young Jew came to the customs house in Port Said to meet the Einsteins and, with another young Jew, to accompany them on the ferry boat to Palestine. Times have changed since then.

As in Japan, so in Palestine Einstein was greeted with extraordinary enthusiasm. But in Palestine the visit had an emotional impact that could not possibly be duplicated in any other land. The diary tells of staying as the guests of the very British Sir Herbert Samuel, who was the Governor of Palestine under the British Mandate. It tells, too, of an incredible number of engagements and visits packed into a mere two weeks. There are frequent references to "rain" and "torrential rain." Here are some other excerpts:

"8 Feb. Drive to Tel Aviv. . . . The accomplishment of the Jews in just a few years in this city arouses the highest admiration. . . . An incredibly active people, our Jews."

"10 Feb. . . . Visit to Weizmann's mother, surrounded by x sons, daughters, etc." (Note how deftly Einstein here portrays Weizmann's mother. I know nothing about her beyond this sentence, yet I feel that she is no longer a stranger.)

"13 Feb. In evening German lecture in Jerusalem in packed hall with unavoidable speeches and Jewish doctoral diploma, during presentation of which the speaker grew nervous and came to a stop. Thank goodness some of us Jews are not so self-assured." In this entry Einstein goes on to say that he is pressed from all sides to settle in Jerusalem, and he adds, "The heart says yes, but the mind says no."

But the highlight was the visit to Mount Scopus, the site on which the Hebrew University was destined to be built. There, on 6 February 1923, Einstein gave a lecture in French, and remarked in his diary, "I had to begin with a greeting in Hebrew, which I read off with great difficulty." What Einstein did not mention in his diary was the joy and sense of fulfillment that his presence on Mount Scopus brought to the Jews of Palestine, and the reverence that was manifest in their invitation to him to speak from "the lectern that has waited for you for two thousand years." In these few words are packed the tragedy and triumph of Jewry.

With the rise of Nazism in Germany, the position of the German Jews became desperate. When Hitler seized power, Einstein was in California. He never returned to Germany. Instead, he severed all official German ties and spoke out against the Nazi tyranny with the fervor and fearlessness of the ancient Hebrew prophets. In Princeton, where he settled, he found ways to rescue friends and strangers from death at the hands of the Nazis.

After World War II, with the Nazis defeated, he was invited to rejoin various German orgainizations from which he had resigned. But he refused. And his words of refusal reveal how profound had become his sense of Jewish identity. For example, to one organization he said, "The Germans slaughtered my Jewish brethren. I will have nothing further to do with the Germans." And to another he said, "Because of the mass murder that the Germans inflicted on the Jewish people, it is evident that any self-respecting Jew could not possibly wish to be associated in any way with any official German institution." He never relented.

When Weizmann died, in 1952, Einstein was invited to succeed him as president of the State of Israel. It was a natural thing. For, while Einstein looked on Zionism as the only ideal powerful enough to give to the scattered Jews of the world a sense of unity, he seems not to have realized how potent a unifying symbol and source of self-confidence he had himself become for the Jews. He declined the invitation gently, citing his lack of aptitude and experience for such a post, and adding, "I am the more distressed over these circumstances because my relationship to the Jewish people has become my strongest human bond, ever since I became aware of our precarious situation among the nations of the world."

And in March of 1955, less than a month before he died, he wrote to Blumenfeld saying, "I thank you, as I should have done much earlier, for having helped me become aware of the Jewish soul."

Has Einstein a message for Israel today? I do not know. But let me cite two passages written in another time. In Los Angeles, in 1932, he said:

"The Zionist goal gives us an actual opportunity to put into practice, through a viable solution of the Jewish-Arab problem, those principles of tolerance and justice that we owe primarily to our prophets. I am convinced that the living transmission of those principles is the most important thing in Judaism."

And in a speech in 1939, given shortly after acts of violence in Palestine by Arabs against Jews, he said:

"There could be no greater calamity than a permanent discord between us and the Arab people. Despite the great wrong that has been done us, we must strive for a just and lasting compromise with the Arab people. . . . Let us recall that in former times no people lived in greater friendship with us than the ancestors of these Arabs."

Every speaker faces a delicate problem of simultaneity. His task is just to speak. The task of his auditors is harder; it is to listen. The aim of the speaker is to have their listening and his speaking come to an end simultaneously.

There is a story of a professor who said to his class: "I don't so much mind your looking at your watches during my lectures. What hurts is that some of you look at your watches, stare at them in utter disbelief, and then hold them to your ears to see if they are still running."

Well, I have said my say. I think you'll all agree with me in echoing Einstein's words and saying it's pleasant—now that it's over.

Gravitational Waves: Panel Discussion

D. Sciama (Moderator): I would like to welcome you all here to this panel discussion on gravitational waves. This, alas, is the final session of this conference where we feed the discussion into John Wheeler's computer and then, at the end of this session, will come an output from him to remind you of all you have heard during the last few days. I would like to introduce the panelists to you and then explain the format of this session. On my left, Joe Weber of Maryland. On his left, Tony Tyson from the Bell Laboratories. On my right Ron Drever from Glasgow, and on his right, Peter Kafka from Munich. The format of this session will be as follows: we will begin with formal statements from each of the participants. This will be followed by a discussion between them of what they have just said. At that point, I will invite the audience to ask questions or make comments and then finally we will move over from discussing the present situation to having a look at the future to see what possibilities there may be for designing detectors far more sensitive than the current generation of detectors. Now, we have a very tight schedule and I have promised the organizers we would finish by 11:00 o'clock absolutely. Therefore, I am forced to be a ruthless chairman. I warn everybody here about that. Now it is my great pleasure to ask Joe Weber to open the session.

Joe Weber: Thank you very much. Perhaps President Nixon should not have promised nuclear technology to the Egyptians. It might have been better if he had offered gravitational radiation detector technology. Therefore, I propose giving Tony Tyson's equipment to Egypt, Ron Drever's equipment to Syria, the Munich–Frascati equipment to Arabia, and the Douglass bar to Jordan; Israel does not need one since she has Sadeh. It is very important for these experiments to have isolation. To meet this need we can take the International Business Machines' detector together with Richard Garwin and transport them to the Moon to send back data until the sidereal anisotropy is confirmed by Egypt and Syria. Needless to say, as soon as these installations are transferred and start operating, the level of controversy in the Middle East will rise exponentially. The amount of national energy involved will be so great as to leave no chance of any more shooting wars.

243

Perhaps then I could exchange salaries with Dr. Kissinger in order to facilitate the data exchange.

The gravitational radiation antenna (detector) was developed at the University of Maryland during the period of 1958–present [1]. It is an elastic solid with a normal mode instrumented to observe strains excited by the Riemann curvature tensor. In 1968, we began a series of experiments to search for two antenna coincidences. The experiments are done by looking for coincident changes in some function of the detector oscillation amplitudes for two widely separated detectors. We count the number of coincidences and compare it with the number of chance coincidences, which are measured by inserting a time delay in one channel or the other. The desired output is a histogram similar to the ones in Particle Physics, and a positive result means that you see a larger number of coincidences without a time delay than with a time delay.

Statistically significant numbers of coincidences were observed between antennas at the University of Maryland and the Argonne National Laboratory early in 1969 [2]. Later, sidereal anisotropy was observed, suggesting the galactic center as the source [3]. Observations with a disc gave evidence for the tensor character of the interaction causing the coincidences. These data were taken by human observers studying pen and ink records.

From the very beginning all features of these experiments were intensely criticized. Nonetheless, it has been clear for some time that the early instrumentation did achieve the claimed sensitivity [4], and that the thermal fluctuations of the aluminum cylinder detector were being observed. The number of coincidences observed for extended periods exceeded the number of chance coincidences by statistically significant margins.

Now we come to a problem which has been encountered in high energy physics. The problem relates to computing and there are important moral questions here. It is a full-time job for me to develop the apparatus and maintain the detectors; this is all I do. The data on a magnetic tape are analyzed by a computer, using programs prepared by another individual. I will not believe the results of an experiment based solely on a computer printout. This is one of the features which distinguishes our work from the other experiments. We use a telephone circuit which connects the two detectors so that separate magnetic tape units are not needed, and we always do several experiments, some with real time counting and others with magnetic tape and a computer. Results are discussed or published only if the different methods are in reasonable agreement. Program errors cannot change the character of our results.

Our computing has been criticized and some of the criticism is justified. It is very well documented, and documents have been shown to various members of the panel that the corrected programs support the kind of data reported at Paris in

June 1973, and published in the Proceedings 220 [8]. Furthermore, at Warsaw, I reported the existence of significant numbers of coincidences between detectors in 710 Hertz and 1661 Hertz on a timescale of a few days with statistical significance. In my opinion, this result is very well established.

A list of coincidences employed for the study of the sidereal anisotropy was furnished to Dr. J. A. Tyson. He found evidence for correlations with the very low frequency fluctuations of the earth's magnetic field. Study of the response of the Maryland gravitational radiation detector to magnetic field fluctuations indicated that the response was much too small to account for the earth's magnetic field as the source of the coincidences. If Tyson's correlations are correct, the inference is that the sidereal anisotropy has been correctly observed and that whatever excites the gravitational radiation detectors also excites changes in the earth's magnetic field.

Magnetic tape and computer analyses began supplementing human observer study of pen and ink records in 1970. Outputs of two detectors are recorded on different channels of a magnetic tape. The computer prepares a list of coincidences, and then the accidental coincidences are measured by inserting time delays in one of two channels. The number of real coincidences is then the number at zero delay minus the number of chance coincidences. Mr Brian K. Reid prepared an elaborate computer program which recorded [5] coincident increases of power. He then measured the pulse height after filtering and computed the number of times per day that the given pulse height was exceeded.

Reid left the University of Maryland in 1971. New programs were prepared because it was thought that other groups would have difficulty reproducing the Reid programs.

An incoming signal may or may not increase the power output, depending on the initial phase. To observe effects of both increase and decrease of antenna energy, it was decided to observe some function of the derivative of the amplitude or power. The new programs were based on the following considerations:

Demodulation

Let the output voltage of the gravitational radiation antenna amplifier be given by

$$A = F(t) \sin (\omega_0 t + \phi), \tag{1}$$

where ω_0 is the normal mode angular frequency. The amplitude $F(t)$ and the phase ϕ have values characteristic of signals and noise. It is now common practice to obtain from (1) the amplitude and phase by combining (1) with local reference oscillator voltages $\sin \omega_0 t$ and $\cos \omega_0 t$ to obtain:

$$A \cos \omega_0 t = \tfrac{1}{2} F(t) \left[\sin (2\omega_0 t + \phi) + \sin \phi \right] \tag{2}$$

$$A \sin \omega_0 t = \tfrac{1}{2} F(t) \left[\cos \phi - \cos (2\omega_0 t + \phi) \right] \tag{3}$$

After filtering with a time constant short compared with the antenna relaxation time, (2) and (3) become the averages

$$x = \langle F(t) \cos \phi/2 \rangle \tag{4}$$

$$y = \langle F(t) \sin \phi/2 \rangle \tag{5}$$

An incoming signal may change phase and amplitude of the detector voltage, depending on the initial noise-induced phase relations. The detector output voltage includes narrow band noise of the normal mode of the antenna V_{ANT} and relatively wide band noise V_N from transducers and electronics. To search for sudden changes in amplitude we may observe a function of the derivative of the power P which for convenience is taken as the (positive) quantity:

$$\left(\frac{dP}{dt}\right)^2 = \left[\frac{\Delta(x^2 + y^2)}{\tau}\right]^2 = \left[\frac{\Delta[V_{ANT} + V_N]^2}{\tau}\right]^2 \rightarrow \left[\frac{2\Delta(V_{ANT}V_N)}{\tau}\right]^2 \tag{6}$$

(6) is independent of the phase. Incoming signals which change only the phase would therefore be missed and to include such cases we may search for sudden changes in the quantity [6].

$$\left(\frac{dx}{dt}\right)^2 + \left(\frac{dy}{dt}\right)^2 = \frac{[\Delta V_{ANT} + \Delta V_N]_x^2 + [\Delta V_{ANT} + \Delta V_N]_y^2}{\tau^2} \tag{7}$$

Suppose we insert a sequence of calibration test pulses with the short duration Δt at times $t_1, t_2, t_3 \dots t_n$ and search for the single pulse detector response only at times $t_1 + \Delta t, t_2 + \Delta t, t_3 + \Delta t, \dots t_n + \Delta t$. It is found for pulses which would increase the energy of the normal mode from zero to kT that algorithm (7) gives a larger amount of response pulses exceeding thresholds, than algorithm (6). Perhaps this is the reason that algorithm (7) is preferred by a number of groups.

However, a study of chart records shows that (Figure 1) algorithm (7) produces single response pulses for each test pulse while algorithm (6) may produce a sequence with more than 20 pulses following insertion of a single test pulse, many of them large enough to cross thresholds. This is a consequence of occurrence of the term $\Delta(V_{ANT}V_N)$ in (6). The single pulse excites the antenna and V_{ANT} remains large for the antenna relaxation time. The rapidly varying wide band noise V_N then produces the sequence of large pulses. This does not occur with (7) because ΔV_{ANT} instead of V_{ANT} is combined with ΔV_N. For very weak signals the term $2V_{ANT}\Delta V_{ANT}$ may be important for (6).

In one series of observations 50 single kT pulses were introduced at two-minute intervals. One hundred and ninety-two response pulses exceeding threshold set at five per minute were emitted by the receiver for algorithm (6) in consequence of the proliferation process.

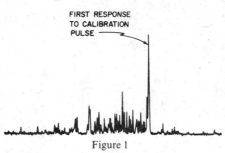

FIRST RESPONSE
TO CALIBRATION
PULSE

Figure 1

Proliferation of pulses for nonlinear algorithm.

We believe that this kind of cascading may result in observation of a larger number of two-detector coincidences for algorithm (6) than for (7), at certain energies.

Furthermore, it seems clear that tests of the single-pulse response of detectors is not sufficient to establish their sensitivity for gravitational radiation experiments.

Observations

The new computer programs did not become available until early 1973. Results were uncertain until late May 1973. Our magnetic tape data processing system converted the detector outputs to digital form with only 6 bit accuracy. To minimize the quantization errors, we employed analogue devices to compute $(dP/dt)^2$ at both Maryland and Argonne installations.

This was then transmitted from Argonne by a telephone line and the digitized $(dP/dt)^2$ written on magnetic tape at Maryland every 0.1 second along with the value from the Maryland detector. Without further processing, the data from both detectors were compared by programs on the Univac 1108 computer. For one kind of analysis [7] thresholds were set and a coincidence defined as a simultaneous crossing (within the 0.1 second writing interval) from low to high values of $(dP/dt)^2$. A histogram for the period June 1–5, 1973, for the threshold crossing algorithm is shown as Figure 2.*

---

* This histogram is for the very controversial tape 217. A copy of this tape together with an unpublished list of coincidences was sent to Professor David Douglass at the University of Rochester. Douglass discovered a program error and incorrect values of the unpublished list of coincidences. Without further processing this tape, he reached the incorrect conclusion that the zero delay excess was one per day. This incorrect information was widely disseminated by him and Dr. R. L. Garwin of the IBM Thomas J. Watson Research Laboratory. After all corrections are applied, the zero delay excess is 8 per day. Subsequently, Douglass reported a zero delay excess of 6 per day for that tape.

   Douglass has also reported computing errors in the data reported at the Warsaw Copernicus Sym-

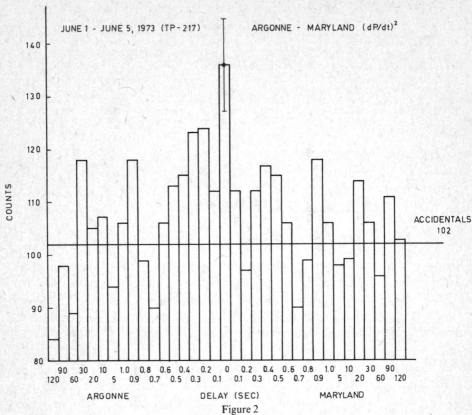

Figure 2

Time delay histogram for simultaneous threshold crossings (June 1–5, 1973), 0.1 second bins.

A second type of analysis [8] defines a coincidence as any pair of points above threshold regardless of previous history. The histogram for the same period as Figure 2 is shown as Figure 3, but the width of the zero delay excess is larger and a higher significance is obtained by grouping together counts in 5 bins. In each instance the standard deviation is computed as the square root of the mean square differences between the mean of the accidental coincidences and the individual delay values.

Figure 3 gives a zero delay excess with a higher level of confidence. For this reason, we decided to process all other data in this report employing the criterion that a coincidence is any pair of points above threshold, without regard to previous history (Figure 3).

posium in September 1973, involving detectors at widely separated frequencies. A 2.6 standard deviation zero delay excess was reported for a six-day period with no claims for it being a positive or negative result. Analyses of other data make it clear that on a time scale at least exceeding four days there is observed a zero delay excess with a level of confidence associated with more than 6 standard deviations for gravitational radiation detectors at frequencies differing by several hundred Hertz.

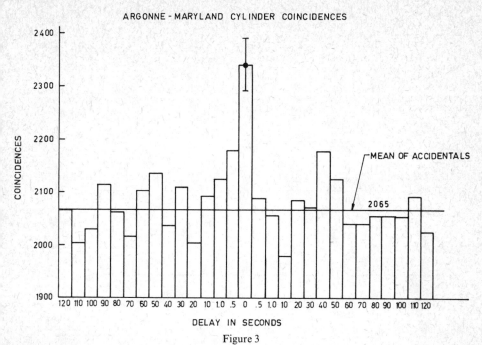

Figure 3

Time delay histogram for coincidence defined as any pair of points above threshold at the same time. Each bin is 0.5 seconds wide.

No significant numbers of coincidences were observed during the latter part of June and July 1973. In July 1973 a new pre-amplifier was installed at the Argonne National Laboratory Gravitational Radiation Detector and the algorithm $(dx/dt)^2 + (dy/dt)^2$ was substituted for $(dP/dt)^2$. No significant numbers of coincidences were observed for the period July–November, 1973, except for one four-day period. In November 1973, parallel experiments were begun employing both $\dot{x}^2 + \dot{y}^2$ and $(dP/dt)^2$ for both the Argonne and Maryland detectors. Again, no significant number of coincidences were observed until mid-December 1973. At that time, it was discovered that the Argonne pre-amplifier had deteriorated in noise performance. A calibration plate was installed at Argonne for introduction of small amounts of energy into the detector by electric fields. For the old (deteriorated) pre-amplifier only one third of the calibration pulses which would couple kT into a cylinder with no lowest mode initial velocity resulted in threshold crossings for the $(dP/dt)^2$ algorithm. The amplifier noise performance was improved to the point were 0.7 of the kT pulses crossed threshold for the $(dP/dt)^2$ algorithm. A significant zero delay excess was observed for the 14-day period immediately following the improvement in sensitivity and shown as the histogram of Figure 4. A histogram for the $\dot{x}^2 + \dot{y}^2$ algorithm for the same period is shown as Figure 5. Clearly these results

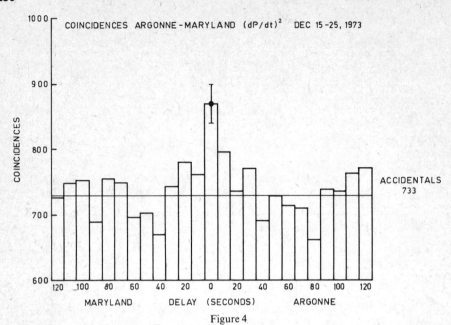

Figure 4

Time delay histogram for December 15–25, 1973 for $(dP/dt)^2$. Each bin is 0.5 seconds wide.

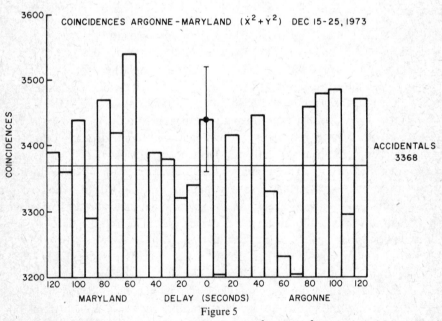

Figure 5

Time delay histogram for December 15–25, 1973 for $(dx/dt)^2 + (dy/dt)^2$. Each bin is 0.5 seconds wide.

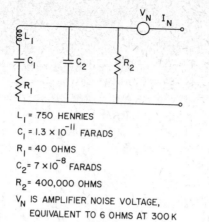

$L_1 = 750$ HENRIES

$C_1 = 1.3 \times 10^{-11}$ FARADS

$R_1 = 40$ OHMS

$C_2 = 7 \times 10^{-8}$ FARADS

$R_2 = 400,000$ OHMS

V_N IS AMPLIFIER NOISE VOLTAGE,
  EQUIVALENT TO 6 OHMS AT 300 K

I_N IS AMPLIFIER NOISE CURRENT,
  EQUIVALENT TO 6×10^7 OHMS AT 300 K

Figure 6

Parameters for Argonne Maryland Experiments, June 1974.

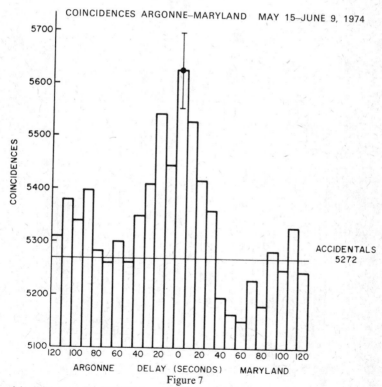

Figure 7

Time delay histogram for $(dP/dt)^2$ for May 15–June 9, 1974. Each bin is 0.5 seconds wide.

are inconsistent with the generally accepted idea that $\dot{x}^2 + \dot{y}^2$ should be the better algorithm.

During the months January–May 1974 results were uncertain—in part because of problems of temperature control with the Argonne installation, in part because a series of experiments was carried out involving new electronics at liquid nitrogen temperatures. Toward the end of April 1974, these experiments were terminated and the parameters of Figure 6 installed. Further improvements in noise performance were made on June 10, 1974, at which time the Butterworth filter bandwidths were further reduced, to 0.8 Hertz for our $\dot{x}^2 + \dot{y}^2$ algorithm. Such reduction was made for $\dot{P}^2$ on July 12, 1974.

For the period May 15–June 9, the zero delay excess characterized by Figure 7 obtained for $\dot{P}^2$, again summing over five bins each 0.1 seconds wide. During this period we were not recording $\dot{x}^2 + \dot{y}^2$. During the period June 18–July 1, 1974, we did not observe a significant result for $\dot{P}^2$ but we did observe a significant zero delay excess for $\dot{x}^2 + \dot{y}^2$, at very low threshold settings as shown in Figure 8.

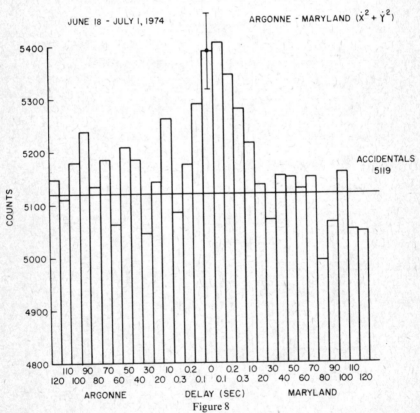

Figure 8

Time delay histogram for $(dx/dt)^2 + (dy/dt)^2$ for June 18–July 1, 1974, 0.1 second bins.

Calibration

As a check on the overall sensitivity, pulsed mechanical forces were applied to both detectors simultaneously, by electric fields at one end of each cylinder. The pulses were sinusoidal wave trains, 50 milliseconds in length, and applied at intervals of two minutes. Outputs of both detectors were recorded on magnetic tape for $(dP/dt)^2$ and $(dx/dt)^2 + (dy/dt)^2$. Our standard programming procedure was then employed to search for these pulses at thirty different thresholds. To smooth the data and minimize effects of real coincidences during that period, delays at multiples of two minutes were applied to one detector output and results averaged over 5 such delays.

For conditions of the operating experiment we find that to obtain a standard deviation excess at zero delay for 150 operating days, summing over 5 bins each 0.1 seconds wide, the experiment could detect the following numbers of pulses per day for the listed increase in energy (assuming the initial energy were zero).

Energy	Number per day For $(dx/dt)^2 + (dy/dt)^2$	Number per day For $(dP/dt)^2$
kT	.07	1.1
kT/3	0.5	14
kT/10	12	160

Conclusion

The past year of observations has been characterized by long periods when there was no significant zero delay excess, and four periods of a few weeks each (in May–June 1973; December 11–25, 1973; May 21–June 9, 1974; June 18–July 1, 1974) when a considerable excess of coincidences was observed, over the accidental ones. As a rule the algorithm $(dP/dt)^2$ gave a larger zero delay excess than $\dot{x}^2 + \dot{y}^2$ except for the period of Figure 8. Each of these periods was characterized by a continuous sequence of tapes, each of which has a zero delay excess. The level of confidence greatly exceeds that which would be expected from simply selecting data which give desired results from a large ensemble, and in each instance the zero delay excess is observed for a range of thresholds. Such results are only observed at zero delay.

We cannot conclude, at this time, that we are dealing with an intermittent source. The attempt to obtain outstandingly good noise performance led to operation of field effect transistors at the large currents recommended by the manufacturer. Under these conditions slow deterioration over a period of about two months was common. Operation now at much lower currents has stabilized the output, and an 8-channel pre-amplifier gives the required noise performance. A further problem has been one of gain adjustment. Our 6-bit recording system, with preprocessing of data before they are written on tape, gives significant errors due to saturation at high gain and significant errors due to quantization noise at low gain. It has been difficult to stabilize the gain for long periods. These difficulties resulted from attempts to improve the experiment so that the detection efficiency would be increased to permit more accurate measurements of anisotropy and polarization. Improved data acquisition systems are now under development in order to give satisfactory solutions for these problems.

On the whole, our results are in fair agreement with the observations of the Munich–Frascati group.

Our recent results suggest a low background event rate with short periods of about two weeks, a few times a year with much larger event rates. Now it is rumored that another group has found the coincidences. I have been waiting five years for someone to find these coincidences and am reminded of the story of Jacob in the Bible. He worked for seven years for a beautiful woman Rachel but received, as a result of trickery, a non-beautiful woman Leah. Then he had to work seven more years for Rachel. I hope the results to be given by the others will not require me to wait five more years. Thank you.

REFERENCES

1. J. WEBER, *Phys. Rev.* **117**, 306 (1960); *General Relativity and Gravitational Waves,* Interscience, New York, 1962, Chapter 8; *Relativity Groups and Topology,* Gordon and Breach, New York, 1964, p. 875; *Phys. Rev. Lett.* **17**, 1228 (1966).
2. J. WEBER, *Phys. Rev. Lett.* **22**, 1320 (1969); *Phys. Rev. Lett.* **24**, 276 (1970).
3. J. WEBER, *Phys. Rev. Lett.* **25**, 180 (1970).
4. J. WEBER, *Lett. Nuovo Cim.* **IV**, 653 (1970).
5. J. WEBER, *Nature* **5375**, 28 (1972).
6. P. KAFKA, *Colloques Internationaux Du Centre National De La Recherche Scientifique Nᵉ 220, Ondes et Radiations Gravitationelles,* p. 181, and Varenna Lecture, International School of Physics, July, 1972.
7. WEBER, GRETZ, LEE, RYDBECK, TRIMBLE and STEPPEL, *Phys. Rev. Lett.* **31**, 779 (1973).
8. WEBER, GRETZ, LEE, RYDBECK, TRIMBLE and STEPPEL, *Colloques Internationaux Du Centre National De La Recherche Scientifique Nᵉ 220, Ondes et Radiations Gravitationelles.*

SCIAMA: Thank you very much Joe and thank you for keeping to your time. I now invite Peter Kafka to continue the discussion.

KAFKA: The experimental work in our coincidence experiment was mainly done by H. Billing and W. Winkler in Munich, and by K. Maischberger in Frascati. The decision to repeat Weber's experiment as closely as possible was made independently by both groups, but the Frascati group was taken over by our institute when ESRIN closed down. The detectors started working in 1972 and we collected some simultaneous data without finding evidence for Weber-type events. Since the experimentalists always wanted to increase sensitivity (like Weber did), a long-term run with stable detectors was started only in July 1973. Due to failures of the tape recording systems we obtained simultaneous data only for 150 out of 300 days until May 1974, but the detectors themselves nearly did not change over that period. This stability is very important, because it guarantees uniform statistical behavior of the noise data, required for a uniform evaluation over the whole year. My colleague F. Meyer and myself have developed the theory of optimal signal detection in such an experiment, and we think there is not much room left for quarreling about the best procedure. The algorithm which Weber now calls the "preferred" one, is close to the one which we have used and recommended for several years. I cannot go into details here, but let me sketch the basic ideas. The detector can be represented by an equivalent circuit as in Figure 9. Its output is a voltage $U(t)$ as a function of time, roughly a sine at $\omega_0 = 1660$ Hz with slowly varying amplitude and phase. We know its statistical behavior in pure noise. The "Gaussian" character defines a metric in the functional space of all $U(t)$ in a neighborhood of a given moment t_0. Using this metric defined by the noise, we can project any observed output on the output which would be produced by a unit signal arriving at time t_0, in the absence of noise. This projection defines the "signal content" λ at time t_0

$$\lambda(t_0) = \int_{-\infty}^{+\infty} G(t)\, U(t_0 + t)\, dt \tag{1}$$

The *optimal filter* $G(t)$ is calculated from the equivalent circuit and the unit signal. (But for short pulses no other property of the signal except the pulse strength comes in!) $\lambda(t)$ is still oscillating at ω_0. Since we are not interested in the exact arrival time or the relative phase between the signal and the detector, we average over the absolute phase (of the 1660 Hz output) and define a squared signal content $\Lambda^2(t_0)$, referring to an arrival time only roughly at t_0. The long term mean value $\overline{\Lambda}^2$ of the optimal signal content function $\Lambda^2(t)$ in pure noise can be used for a definition of optimal sensitivity: We define a dimensionless pulse strength E_g by comparison with

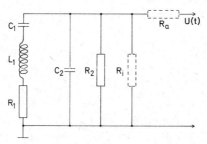

Figure 9

Equivalent circuit of Weber-type detectors, C_1, L_1, R_1 represent the mechanical oscillator; C_2, R_2 the transducers; R_a, R_i the amplifier noise.

a short pulse which would excite the cylinder from emergy zero to 1kT. (Detector dependent unit pulse $E_g = 1$!) Then, optimal sensitivity is defined by

$$\Phi = (\Lambda^2 \text{ in pulse of } E_g = 1)/\bar{\Lambda}^2. \tag{2}$$

The optimal filter involves a certain time-scale of exponential smoothing ($e^{+\mu t}$). The optimal smoothing time $1/\mu$ and the corresponding optimal sensitivity Φ result from the values of the equivalent circuit, and are in good approximation:

$$\left. \begin{aligned} 4\mu^2 &\approx \omega_0^4 C_1^2 R_2'(R_1 + R_a)/(1 + R_a R_2' \omega_0^2 C_2^2) \\ 1/\Phi &\approx 4 \cdot [(R_1 + R_a)/R_2' + R_1 R_a \omega_0^2 C_2^2]^{1/2}, \end{aligned} \right\} \tag{3}$$

where $\omega_0^2 = 1/L_1 C_1$ and $1/R_2' = 1/R_2 + 1/R_i$.

In order to compare the optimal sensitivities of experiments with different detector masses, we had better compute the spectral density (at ω_0) of a short gravitational pulse of $E_g = 1$. One finds

$$Z(\nu) = (c^3/2\pi G) \cdot (\Delta l/l)^2 \approx 1.2 \cdot 10^7 (M_{\text{Munich}}/M) \, \text{erg/cm}^2 \cdot \text{Hz} \tag{4}$$

(G = gravitational constant: l = length of cylinder; M = mass; Δl = amplitude after the pulse, if it was zero initially.) Table 1 compares the equivalent circuits, optimal filter times and optimal sensitivities of the detectors in Munich and Frascati with Tyson's and Weber's of last year. The last line is the relevant one for comparison of optimal sensitivities. It shows the spectral density of the signal which is on the average simulated by pure noise, if one uses the optimal evaluation. Clearly, any deviation from the optimal data processing can only decrease the sensitivity to short pulses.

TABLE 1

Values for the Equivalent Circuits of Some Detectors. M: Munich (1973/74); F: Frascati (Jan. 1974); B: Bell Lab. (1973/74)[10]; W'73: Maryland (1973)[11]; W'70: Maryland (1970)[4]. R_2' Comprises R_2 and R_i in Figure 9. Optimal Filter Time μ^{-1}, Optimal Sensitivity Φ and Spectral Density Z_Φ in a Pulse of Strength 1kT/Q. (1GPU = 10^5 erg/cm²Hz.)

	M	F	B	W'73	W'70
$C_1(10^{-13}F)$	3.6	9.1	100	122	5
$L_1(10^3 H)$	25.7	11	5	.75	20
$R_1(\Omega)$	760	750	82	50	2000
$C_2(10^{-9}F)$	.8	1.2	50	70	100
$R_2'(10^6\Omega)$	44	22	1	.23	.1
$R_a(\Omega)$	120	80	20	15	?
μ^{-1}(sec)	.31	.15	1.4	.69	>2.8
Φ	48	35	18	10	<2
Z_Φ(GPU)	2.5	2.9	23	12	>60

Therefore, we have to ask how good our actual evaluation is, as compared to the theoretical optimum. The actual evaluation is as follows:

Using a reference oscillator near frequency ω_0, we decompose the output $U(t)$ into two slowly varying function $x(t)$ and $y(t)$, the "phase-plane components in a co-rotating frame." Figure 10 shows an example for the output of the Munich detector in the x,y-plane. The 600 data points of 1 minute are connected by a line. The radius r of the circle represents the long-term mean value of $r^2 = x^2 + y^2$. This corresponds to the energy 1kT in the cylinder plus about 2% wide-band noise energy. Near the point marked P in Figure 10, an artificial pulse of $E_g = 0.1$ was applied. One can see that it is nearly drowned by the noise. Let us therefore apply a reasonable approximation of the theoretically optimal filtering process, and determine the signal content function for this minute of data. At the moment t, we define the smoothed future position by $x^+ = \mu \cdot \Delta t \cdot \sum\limits_{n=0}^{\infty} x(t + n \cdot \Delta t) e^{-n,\mu\Delta t}$, the smoothed past by $x^- = \mu\Delta t \cdot \sum\limits_{n=1}^{\infty} x(t - n\Delta t) e^{-n,\mu\Delta t}$ (and correspondingly y^+ and y^-). $\Delta t = 0.1$ sec is the discretization time. Then we approximate the optimal signal content Λ^2 by

$$\Lambda_0^2 = (x^+ - x^-)^2 + (y^+ - y^-)^2 \tag{5}$$

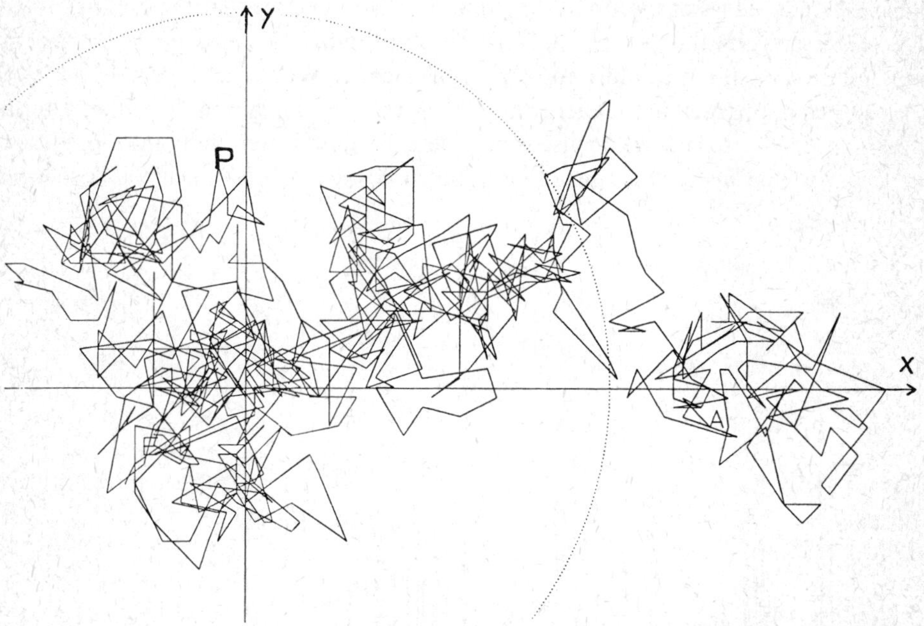

Figure 10

One minute of noise in the x,y-phase plane, in the co-rotating frame. (Munich detector.) The circle represents the long term mean of $r^2 = x^2 + y^2$ (i.e., the cylinder energy 1kT, except for a correction of a few percent). Near the moment marked P an artificial pulse of $E_g = 0.1$ was applied.

Of course, this can only be a good approximation, if the beat between the reference oscillator and the cylinder is always negligible. To achieve this by temperature control would be expensive. Instead, it is much simpler to control the reference oscillator by the detector output, and this is what we do. If there is no beat, the function $\Lambda_0^2(t)$ obeys the same statistical law as the optimal $\Lambda^2(t)$:

If we define the normalized signal content S by

$$S(t) = \Lambda^2(t)/\overline{\Lambda}^2 \tag{6}$$

the probability to find S above a threshold S_0 in pure noise, will be

$$W_0(S_0) = e^{-S_0}. \tag{7}$$

At the arrival of a signal pulse of strength E_g the observed normalized signal content will be found above threshold S_0 with the probability

$$W_1(S_0, \Phi E_g) = \int_{S_0}^{\infty} e^{-S - \Phi E_g} \cdot I_0(2\sqrt{S \Phi E_g})\, dS \tag{8}$$

where I_0 is the modified Besselfunction of order zero.

Let us look at the normalized signal content function $\Lambda_0^2(t)$ for the same minute for which we showed the x,y-data. In the next 3 slides (Figures 11a,b,c) it is shown for 3 different values of the smoothing time $1/\mu$. You can see that the optimal value $1/\mu \approx 0.3$ sec gives the best result. Now, let us be unfair and look at the same artificial pulse with the procedure formerly preferred by Weber. In Figure 12 we plot the squared jump $(\Delta E)^2$ in the energy $E = x^2 + y^2$, also normalized with its long-term mean value. You see that the pulse is not there, because it went fully into the phase. Instead you find many crossings of high thresholds due to noise near the beginning

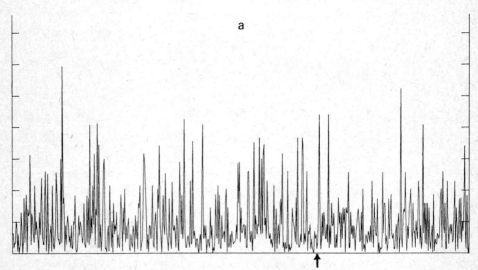

a

b

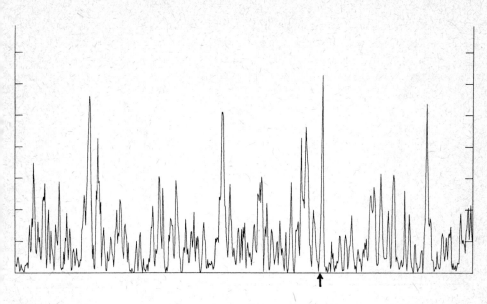

c

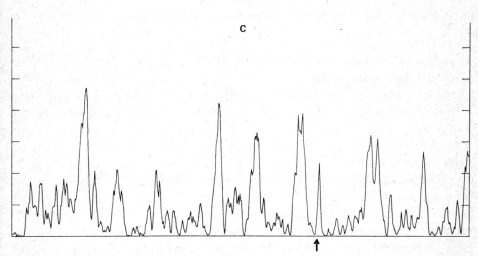

Figure 11

Normalized signal content function $\Lambda_0^2(t)$ for the data from Figure 10. a) With no additional filtering;
b) With optimal filtering ($1/\mu = 0.3$ sec); c) with too much filtering ($1/\mu = 0.7$ sec). The horizontal extension is one minute, the arrow points at the artificial pulse; the vertical scale is in steps of 1 ($=$ long term
mean value of S).

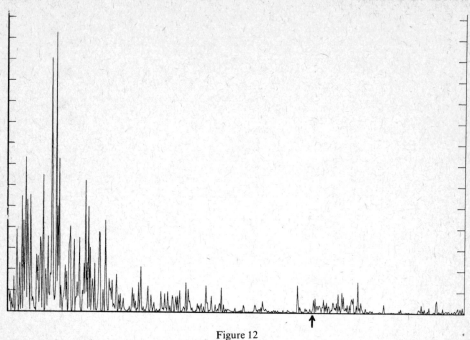

Figure 12

The same as in Figure 3, but with the signal function $(\Delta E)^2$, formerly preferred by Weber.

of the minute, when the energy $x^2 + y^2$ was high. This is just a nasty demonstration of what Weber called "proliferation"—but of course one is not allowed to make a comparison between the procedures from selected data.

Hence, let us now look at the calibration of our procedure in long series of artificial pulses. It does not matter whether the pulses are applied over condensor plates or piezoelectric transducers. (We now choose the latter procedure because it is more stable.) The pulser has to be calibrated by direct observation of strong pulses in the phase plane. For pulses of 5kT, e.g., future and past are sufficiently sharply defined to measure the length of the arrow in the x,y-plane. The unit is supplied by the 1kT circle in this plane (after subtraction of the small contribution of the wide band noise). Since we know the distribution (8) for the observed signal content in pulses of given strength E_g, with a detector and algorithm of sensitivity Φ, we can now determine the actual sensitivity of our evaluation and compare it with the theoretical optimum. Figure 13 shows the theoretical probability $W_1(S, \Phi E_g)$ from equation (8) for various values of the parameter ΦE_g. It also shows the results of two test series in Munich and one in Frascati. We simply plot the fraction of the test pulses which produced S-values above threshold, 0.1 sec after the pulse arrival. The actual sensitivity is then given by the parameter ΦE of the fitting curve, divided by the known pulse strength E_g. We find actual sensitivities of about 40 and 28 for

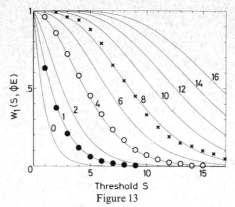

Threshold S

Figure 13

One-detector detection probability $W_1(S, \Phi E_g)$ at threshold S for sensitivity Φ and pulse strength E_g. Theoretical curves for various ΦE_g. Calibration of actual sensitivity Φ from series of artificial pulses with known strength E_g. Filled circles: 3500 pulses with $E_g = 1/40$ in Munich; Open circles: 1000 pulses with $E_g = 0.1$ in Munich; crosses: 100 pulses with $E_g = 0.28$ in Frascati. From the known E_g and the parameter ΦE_g of the best fitting theoretical line follows the actual sensitivity Φ.

Munich and Frascati, respectively, corresponding to 83% and 80% of the corresponding optimal values. This shows that a more sophisticated approximation of the optimal filter (taking into account more detailed spectral features of noise and signal and using better time resolution) is scarcely worthwhile. The Frascati value still has to be multiplied by 1.25 due to a higher mass. Hence we are on the safe side when we take a sensitivity $\Phi \approx 30$ for Frascati (referring to 1kT pulses defined for Munich), in spite of a slow deterioration over the last year.

You have seen that our theory of evaluation is nicely confirmed by the tests with artificial pulses of known strength. I have not developed the theory for Weber's $(\Delta E)^2$-algorithm. Because of its non-linearity it is more complicated. I therefore looked at it only experimentally. For that purpose we simply determine the probability W_1 for the same series of pulses, but with Weber's signal function. The result you see in Figure 14: The solid lines are the same curves as in Figure 13, but this time in a logarithmic scale. The lowest line, for $\Phi E_g = 0$, represents the probability W_0 to be above threshold in pure noise; the other ones are for various pulse strengths. The broken lines I found in the series of artificial pulses, using $(\Delta E)^2$ as the signal function. One might think that the higher values of the broken lines show a superiority of Weber's procedure; however, this "advantage" is more than cancelled by the high values of W_0.

To make this clear, we have to consider the concept of "sensitivity" more carefully. What we need is not a high "detection efficiency" W_1 but a small "*minimal detectable rate.*" A rate R of pulses of strength E_g is detected at threshold S with m standard deviations within an observation time T, if the additional number of crossings

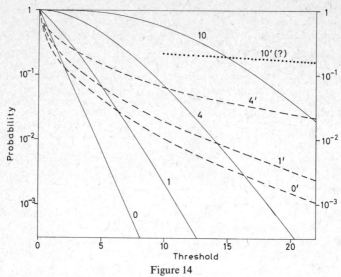

Figure 14

The solid lines show again the theoretical curves for $W_1(S, \Phi E_g)$, however in logarithmic scale. The broken lines represent the result with Weber's $(\Delta E)^2$-procedure, applied to the same series of artificial pulses on the Munich detector. (The line for 1kT is roughly extrapolated.) In order to estimate the analogue of W_1 in Weber's procedure for his own detectors, one has to extrapolate to Weber's smaller value of optimal Φ.

is greater than m times the random fluctuation of the accidental crossing number. Using our probabilities W_0 and W_1, this condition reads

$$R > \frac{\sqrt{W_0(S)}}{W_1(S, \Phi E_g) - W_0(S)} \cdot \frac{m}{\sqrt{T\tau(S)}}$$

(Here $\tau(S)$ is the average peak duration of the signal function above S in noise.) For fixed E_g, the minimum of (9) defines the *optimal threshold* for unlimited observation time T. For ΦE_g between 2 and 15 one finds nearly a linear relation $S_{\mathrm{opt}} \approx 4\Phi E_g - 2.6$. For the Munich detector ($\Phi \approx 40$) and pulses of 0.1 kT this would mean that one should look for crossings at threshold $S \approx 13$ rather than 4 ($= 40 \cdot 0.1$) in order to find such pulses most significantly against the noise. If the observation time is limited, the optimal thresholds for stronger pulses will come down to a value S_T where no accidentals are expected. Then the minimal detectable rate is given by

$$R \approx \frac{m}{T \cdot W_1(S_T, \Phi E_g)} \tag{10}$$

instead of (9).

For two or more detectors with unknown relative phases the optimal signal content function would simply be a sensitivity-weighted sum of single detector functions.

However, since we want to discriminate against local disturbances, we will require that both detectors cross threshold in a real event. The detection probability in coincidence is then given by a product of two values of W_1. Figure 15 shows this probability for three pairs of thresholds in our coincidence experiment.

With the same arguments which led to equations (9) and (10), we can now derive the optimal pairs of thresholds and minimal detectable rates for given pulse strength and observation time. The plot of minimal detectable pulse rate versus pulse strength for given observation time is the most reasonable representation of sensitivity. It is shown in Figure 16 for $T = 150$ days with the actual evaluation procedure in the Munich–Frascati experiment, for the optimal evaluation in Weber's experiment, and for the $(\Delta E)^2$-evaluation which Weber preferred last year. The latter curve is a rough estimate because extrapolation from our sensitivity to Weber's was involved.

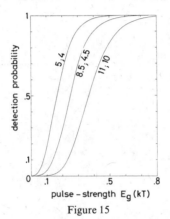

Figure 15

Detection probability in coincidence in the Munich–Frascati experiment, for 3 pairs of thresholds. (This shows why strong pulses are excluded by our null result at thresholds 11; 10!)

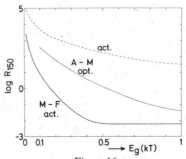

Figure 16

Detectability of short pulses in 150 days. Daily rate R_{150} of pulses of strength E_g (unit $1kT = 120$ GPU), detectable with 3 sigma in Munich–Frascati (actual evaluation) and in Argonne-Maryland 1973 (estimates for optimal and actual evaluation). Weber's present detectors and procedures may be a bit improved, and thus the corresponding curves may be a bit lower now.

(Of course, Weber should determine these curves experimentally for his detectors.) As you see, we have to find Weber's events with extreme significance if they are short gravitational pulses. (A threatening step forward of the moderator. . . .)

What we have found, I haven't shown yet. Shall I show the last slide? (Laughter in the audience, moderator steps back in resignation.)

Figure 17 shows the results over the whole period of 150 days at 3 pairs of thresh-

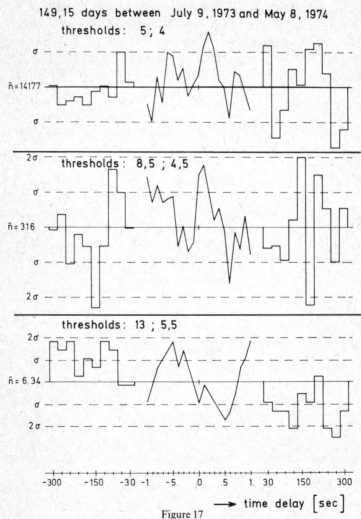

Figure 17

Some results of 150 days. Number of "peaks" above thresholds in coincidence and with 40 values of relative delay, at 3 pairs of thresholds. $\bar{n}$ is the mean of the outer 22 bins plus the zero bin. σ marks the square root of the mean. This negative overall results allows to set upper limits to the average rate of short pulses in the 150 days. These limits are given by the lowest curve for the minimal detectable rates in Figure 16.

olds. (Evaluation was done for many more pairs!) You see that the number of coincidences without time delay is not significantly higher than at arbitrary time delay. There was no problem with synchronization of the data tapes, because time is written every minute, and simultaneous artificial pulses were detected in the predicted way. Our negative result allows us to set upper limits to the average daily rate of short pulses arriving at the earth (with favorable direction and polarization) within the observation time. These limits are given by the lowest curve in the last picture (Figure 16). You see that they are far below the rates stated formerly by Joe Weber.

The very last picture (Figure 18) is the one in which Joe Weber thinks we have discovered something, too. This is for 16 days out of the 150. There is a 3.6σ peak at zero time delay, but you must not be too much impressed by that. It is one out of 13 pieces for which the evaluation was done, and I looked at least at 7 pairs of thresholds. Taking into account selection we can estimate the probability to find such a peak accidentally to be of the order of 1%. Hence, before I say definitely that we have not found any evidence for Weber pulses, I shall study carefully the fluctuations in time. This analysis can be done quickly from the stored results of a long period of evaluation, and I will start it when I am back in Munich.

SCIAMA: Thank you very much, Peter. And now I call on Ron Drever.

DREVER: I plan to talk about two different experiments here; one which has lasted about two years, and one which has lasted two weeks so far. Perhaps to give a little background, I should explain how we got to be doing these experiments. I think it was about three or four years ago when Joseph Weber produced the sidereal distribution for his events, and this result, at the time, seemed to me to be very significant and extremely exciting. In fact, so exciting that I felt impelled to do something about it; we could not just let something like that lie there. What could we do to find out more about these exciting signals, to find out more about gravitational waves and about the sources? So, thinking about it a bit, I decided to develop a slightly different type of detector from Weber's one, with the aim of finding out more information. Now, my aim was *not* to try to find out if Weber's work was right or wrong, but to find out more about gravitational waves. Therefore, our devices were developed to be rather different from the ones which he has built.

In this talk I will first go through the experiments in which we searched for pulses, some of which have been published already, and at the end I will go on to the new experiment which we have just done in the last two weeks.

The detectors which we have built are as shown schematically in Figure 19. They differ from Weber's ones in that they have two separate masses linked by piezo-electric transducers. The point of this arrangement is that the coupling between the mechanical and electrical systems is very much stronger than in other detectors—

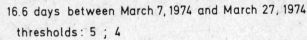

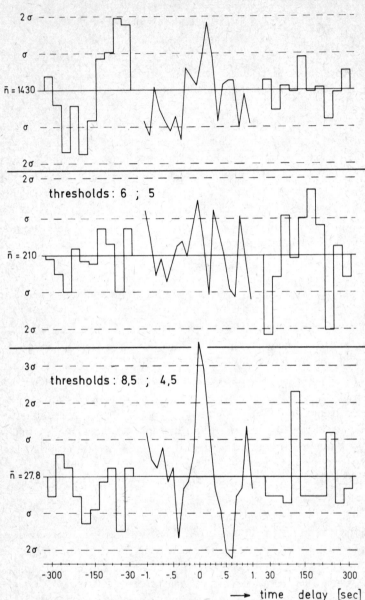

Figure 18

As in Figure 17, but for the piece with the largest positive peak observed in one out of 13 pieces (at several pairs of thresholds), which were either about 8 or about 16 days long. The probabilities that this is due to chance or due to real pulses, both seem to be of the order of 1 percent.

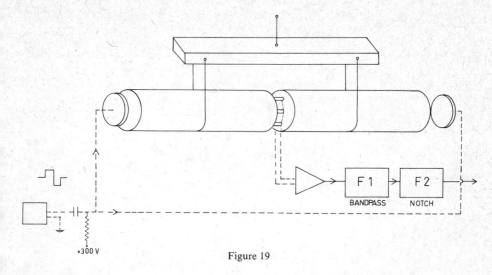

Figure 19

by a factor of the order of 10^4 to 10^5 —and thus for the same energy in the mechanical system we get very much larger electrical signals. This means that with amplifiers of similar quality to those of other workers we can have a very much wider band-width. In fact the bandwidth is so large (it covers about a kilohertz) that we can see the individual cycles of the bar vibrations. Thus, we have the possibility of studying any signals in very much more detail than with other detectors.

Because of this wide bandwidth we process the signals on line; we have a filter that defines the bandwidth, and we have another filter (the "notch filter") which is a little analog computer which solves the equation of motion of the bar and gives an output signal which represents the force acting. Thus, the latter device will tell us the waveform of any gravitational force acting on the system. What we do then is to look for interesting waveforms appearing in the system. An important point to stress is that we have a calibration device for pulses. We have plates near the ends of the detector, and we can test the response to pulses of various kinds by applying electrostatic pulses to the detector systems.

Figure 20 shows one of the detectors; they are not terribly big, but as we shall see their sensitivity is comparable to that of most of the other detectors. The masses are about 300 kilograms. Figure 21 shows the electronic system. I will not go into details, but merely make the point that, as we have such a wide bandwidth, it is not practi-cable for us to record all the signals coming in; it would involve too much magnetic tape. We have to process the signals on line, and we have electronics connected to the two detectors (which are, in fact, in the same building), which selects interesting events and records them in detail by an oscilloscope and photographic system. There is also a digital computer system which examines the signals at quite a high

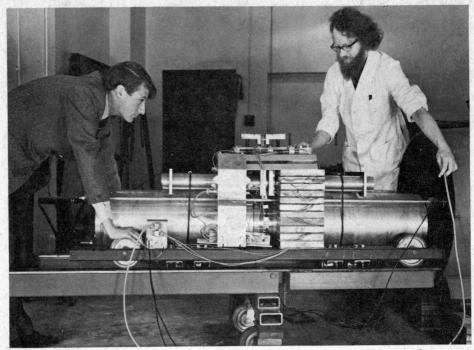

Figure 20

TIME DELAY BETWEEN PULSES FROM DETECTOR 1   AND DETECTOR 2
(SEPTEMBER 1972 – APRIL 1973)

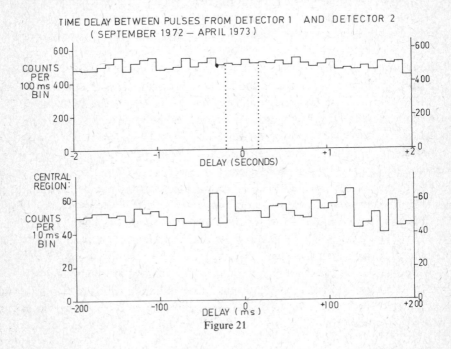

Figure 21

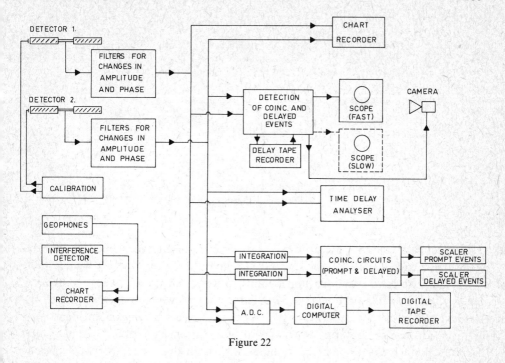

Figure 22

rate and makes quick decisions about whether they are interesting or not; it records any interesting signals.

I will summarize some of the results; we have a large mass of data, most of which have been published. Figure 22 shows the time delay distribution for a particular seven-month run. It is the same type of plot that Joseph Weber was showing us. It covers the delay range −2 to +2 seconds. In fact we have no sign of a peak at zero delay, unfortunately. When we first plotted this we were very saddened indeed, because this is a good, long run. The central part of the same distribution is shown below, with the timescale magnified by a factor of 10. There is still no peak. This was rather saddening because we felt that the sensitivity is at least as good as that of Weber's experiments.

Now it is necessary to try to make some comparison of sensitivities, because our detectors are so different. We can certainly find out how sensitive our detector is for any arbitrary waveform, for we can apply that force waveform and see how it is detected. But there is some difficulty in knowing how sensitive Weber's experiments were, so what we have had to do is estimate his sensitivity. We have tried to make it an upper limit estimate neglecting his amplifier noise. In the table are shown the results of this comparison.

TABLE

Assumed Pulse Energy (Units of kT given to a Weber bar initially at rest.)	Detection Efficiency in coincidence expt.		Deduced Pulse Rate per day	
	Expt. of Ref. 1 (estimated)	Present Expt.	Expt. of Ref. 1.	Present Expt.
3.0	<0.10	0.8	>7.5	0.04 ± 0.09
1.0	<0.0056	0.09	>130	0.39 ± 0.79
0.3	<0.0004	0.0004	>1800	89 ± 180

(Ref: 1—J. Weber, *Nuovo Cim.* **4B**, 201 (August 1971), Figure 3b.)

(Present Experiment: R. W. P. Drever, J. Hough, R. Bland and G. W. Lessnoff.)

We postulate gravitational wave pulses of various energies, the values taken corresponding to energies of 3 kT, 1 kT, and 0.3 kT given to Weber's bar if it were initially at rest, and we have estimated upper limits to the efficiency of Weber's system for these pulses. For one particular experiment by Weber we have then estimated the incident rate which the experiment has to imply. These are *minimum* incident rates, and range from 7 to 1800 pulses per day for gravitational waves of these particular energies. If we then compare these figures with what our own experiment gives— and our experiment has generally much higher efficiency, we think—we find the numbers are completely discordant. (This is shown in the last two columns of Figure 23). Again, this is a saddening thing, and we find it hard to understand.

I should put in a little warning here. Our experiment is most efficient for short pulses. If there were longer pulses we would not be quite so efficient and therefore we do not completely discount the possibility of long pulses. In most of our runs the experimental data are recorded in several parallel modes at the same time; by photographic recordings, by an on-line computer, and by an automatic time analyzer. The photographs give the most interesting data because you can get the full detail there. In seven-month period we only had one good candidate for a gravitational wave: it is one we have mentioned before. Figure 23 is the record for that event. The top trace shows sine wave which is, in fact, the Brownian motion of the detector,

5   ·1   4   ·0   7   ·2   8   ·7   7   3   1

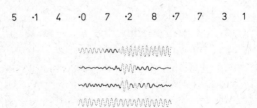

Figure 23

and shows you *can* see it, cycle by cycle. The scale here is a millisecond for each wiggle of the sine wave, and here we see something has certainly caused the amplitude to increase. The other detector was giving us the bottom trace, and there is no apparent change of amplitude there; but if you measure the photograph carefully you find the phase has changed. That has been detected correctly by our electronic processing system. This system displays a signal related to the force acting on the first detector in the second trace, and on the second detector in the third trace. The interesting thing about this event is that these forces are in phase, and are very similar in waveform; and that is a good candidate for a gravitational wave. But of course we cannot prove that it is a gravitational wave. However, an important point is that this was unfortunately the *only* candidate we had in that seven-month period.

I should just mention too, to update these results a little, that we also do look for long pulses using the computer. Figure 24 shows time-delay histograms for these at various threshold levels, which do not in fact show any positive effect; but the sensitivity in this case is not so good, and I will not say much about these results.

To update the results of this whole experiment let me conclude by saying that the thing has been running for about 16 months altogether; we have found no more good candidates at all. We had two slightly interesting pulses which I may talk about later, but we do not think they are very good candidates and so our results are still negative.

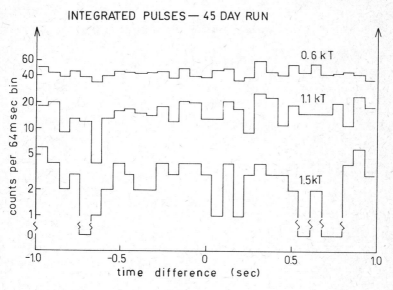

INTEGRATED PULSES — 45 DAY RUN

0.6 kT :	−8 ±11 per month
1.1 kT :	5 ±7 per month
1.5 kT :	1 ±3 per month

Figure 24

Having had this experiment going for about 16 months now we thought, well, it is a bit disappointing all this; we were hoping to get these wonderful gravitational waves, with a certain amount of excitement. However, under the stimulus of Joseph Weber's work, we have built these detectors, we have them in our hands now, can we do anything else with them? And, we thought, yes we could. What about looking for continuous background gravitational waves?

Of course we have not got the sensitivity to detect the predicted levels of flux which you might really expect to be present. But even if our detectors are limited in this sense, it seems worth while looking anyway just to check if there is anything there. We have done an experiment in the last few weeks just to do that. What we have done is to carry out a cross-correlation between the outputs from the two detectors.

I should mention that Dr. James Hough was the person principally responsible for this particular experiment; several people were involved in the two experiments discussed here.* Neither of these are one-man experiments.

For the correlation experiment we have slightly modified the electronic system, as shown in Figure 25. We have our detectors, the filters defining the system bandwidth, the inverse filters; and the signals then come out transformed to represent the forces acting on the bars. We then use the digital computer, on-line, to cross-correlate these two signals. In fact, we do not multiply them together; we use the simpler technique of one-bit correlation. This is only a technical matter; it is easier for the computer to do. So we look for correlated forces acting on the two detectors.

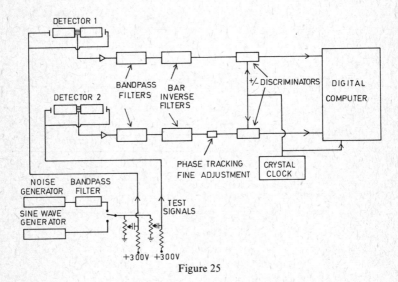

Figure 25

* Pulse experiment: R. W. P. Drever, J. Hough, R. Bland and G. W. Lessnoff. Continuous-wave experiment: J. Hough, R. Bland, R. W. P. Drever and J. Pugh.

It is important to test such a system, to show that it is working. We have applied two forms of signals to test. it We had in mind two types of signals. One was a sine wave signal such as might have been produced by vibrations of a neutron star, perhaps relatively near. The other one was a wide-band signal, perhaps something like a black-body spectrum, so we used a noise generator to simulate that. We could apply either of these test signals to the detectors, and, in fact, detect them by the computer.

First, I must show you the frequency response of the system, because an experiment like this, particularly for sine waves, is only reasonable if you have enough bandwidth to have a fair chance of finding something. Figure 26 shows the frequency response found by exerting a sinusoidal force on the detector and sweeping it through the bandwidth. It indicates that the detector is effective from about 900 to 1100 Hz. The little dip is where the resonance frequency of the bar was—it is a dip in this particular measurement because the person doing the testing did not want to hold the oscillator too long on that frequency in case oscillations built up and damaged the system. We had an unfortunate accident that way before; a test detector broke at the cemented joints when a large sine wave was being applied.

The arrow indicates the frequency of a particular sine wave test, in which the computer extracted the correlation shown in Figure 27. This is the correlation function produced by the computer in a short run, lasting about half an hour, with an applied force of 4×10^{-8} Newtons at a frequency of 910 Hz. This corresponds to a very small fraction of kT in the bar (something like 10^{-4} kT), and the computer has produced a sine wave correlation curve just as it should do. That shows the system works. Let me also show it works for a continuous wide-band spectrum. Figure 28 shows a similar curve produced by applying a random test force, and we get a decaying correlation function; the decay constant corresponds to the band-

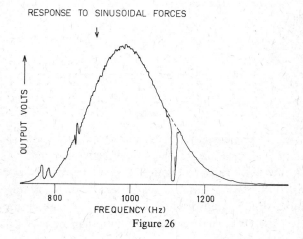

RESPONSE TO SINUSOIDAL FORCES

OUTPUT VOLTS

FREQUENCY (Hz)

800   1000   1200

Figure 26

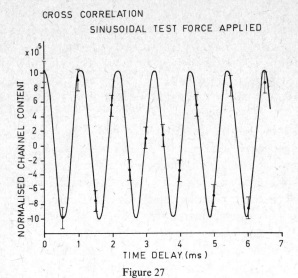

Figure 27

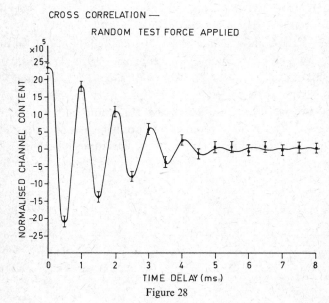

Figure 28

width of the system. That works too; again the signal corresponds to a very small fraction of kT, but the correlation pulls it out perfectly well.

Now we come to the real runs. In Figure 29 we have some data obtained only in the last two weeks. (Since the Cambridge, Massachusetts, meeting where I gave some earlier results, we have improved the equipment by making the computer run four times as fast, thus giving us more data points than before.) The figure shows

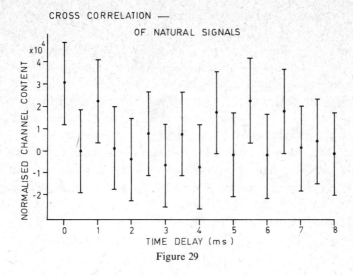

CROSS CORRELATION —
OF NATURAL SIGNALS

Figure 29

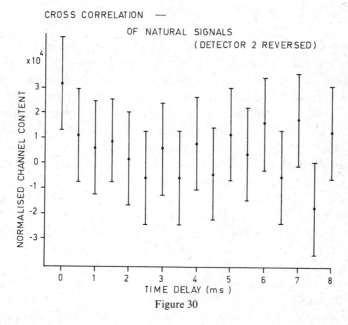

CROSS CORRELATION —
OF NATURAL SIGNALS
(DETECTOR 2 REVERSED)

Figure 30

the correlation curve from one run, with the errors marked in. Essentially there is no effect. A signal might have produced a rise at zero time delay. Well, you might say there is a very slight indication of a very small effect there, less than the errors. In case this was something real we tested it by reversing the polarity of one of the pre-amplifiers, and we get another correlation curve with the polarity reversed. This is shown in Figure 30; it has the same shape, which shows that any effect that may have been present was a spurious one, produced perhaps by mains pickup. The conclusion is that there is no sign of any real effect at all. Let me now try to tell you what that corresponds to in terms of numbers. We can express it in various ways. We might set a limit to the component of the temperature of the bar due to correlated forces of about 230 millidegrees Kelvin. Rather a small temperature. We've set a limit to the gravitational sine wave flux of any origin, in the range 900 Hz to 1100 Hz, of about 1.7 w/cm^2; we have set a limit to a flat frequency spectrum of broadband noise in the region of 0.38 w/cm^{2}Hz. These limits are not really interesting astronomical ones; they are much larger than you could expect from reasonable sources. However, we might have been lucky; there might have been a close-by neutron star, or there might have been a strong broadband local source. We feel it was worthwhile looking anyway. Maybe I should remark that these are preliminary data; we may change our minds on the precise numbers in a short time. Perhaps I will come back afterward and talk on the relevance of these results to the other problem. Thank you.

SCIAMA: Thank you. I now call on Tony Tyson.

TYSON: The people involved in this high sensitivity search are Dave Douglass and Roger Gram at the University of Rochester, Ron Decker and myself at Bell Laboratories and Bob Lee at Stanford University, who does the domputer analysis of most of our data. The experiment, as you know, consists of searching for an increase in the amount of energy in the lowest longitudinal mode of the bar. In a minute I will show you the equivalent circuit of that electrically, but physically this is what is going on. There is a certain amount of energy E deposited in the bar by the gravitational wave of flux density F_v per unit frequency. If $\sigma(v)$ is the differential absorption cross-section, the integral over frequency of the flux density times the differential cross-section for absorption is then the amount of energy that the bar absorbs from the gravitational wave: $E_{abs} = \int F_v \sigma(v) \, dv$. For a resonant detector we have a Breit–Wigner response and we can integrate over the absorption cross-section, and consider just the flux at the resonant frequency of the bar. What these resonant bars are doing then is sampling the power spectrum of gravitational radiation at one frequency. The integrated absorption cross-sections for these detectors are: for our detectors 1.5×10^{-21}cm^{2}Hz, and for Weber's detectors it works out to be 0.61×10^{-21}cm^{2}Hz.

The reason why we did this experiment was to try to examine the possibility that there really were bursts of gravitational waves. You have just heard about a couple of very beautiful experiments done by Ron Drever, in which he has looked cycle by cycle in the gravity. We decided to forgo obtaining that type of information, and trade off that information for increased sensitivity. There is therefore a considerable difference between the kinds of experiments Weber, Kafka and myself have been doing and the experiments that you have just heard about, in the sense that the

Weber-type experiment is one done with a bar of aluminum which acts, more or less, as an integrating calorimeter. It asks the question "was there an increase or a decrease in the amount of energy in one of the longitudinal modes in the last tenth of a second?" (We also search for a sudden phase change.) If there was, we get some output. The sensitivity therefore is going to be a function of how we analyze the data, the kind of integration times that are used in the experiment, the size of the bar, etc. I think you can appreciate the fact that we can trade off a lot of wide-band information about the detailed short time nature of the pulse for increased narrow-band sensitivity. What we have out of these bars, then, is a signal with a characteristic signature as a function of time, if something came along and excited the bar. The interesting feature of these antennas is that the filtered signal coming out of the electronics is pretty much insensitive to the detailed nature of the original pulse in the metric. Figure 31 is a picture of the large bar we constructed at Holmdel three years ago. This bar, which is larger than the Weber antennas, is 3.5 meters long, 68 cm in diameter, and weighs 3720 Kg. The electromechanical transducers are strain gauges symmetrically placed around the center. Figure 32 shows the detection scheme. You see the sine wave from the bar changing in amplitude slowly as the Brownian motion of the bar changes with time. This is followed by a four-quadrant multiplier which puts you into the rotating frame at the bar frequency. The coordinates of this co-rotating frame are X and Y,

Figure 31

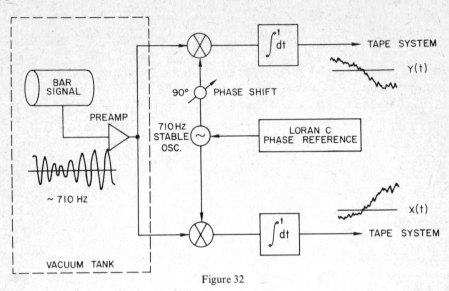

Figure 32

which you have heard about before, which are the outputs from the electronics and are recorded on the magnetic tape. In our system, each one of the two antennas is like this; it has a stabilized local oscillator, and these outputs X and Y are sampled every tenth of a second by a magnetic tape system, with two entirely separate tape systems, one for each antenna. One antenna is at Bell Laboratories, the other is at the University of Rochester about 300 miles north. There is no physical connection between the two antennas.

However, in order to really believe in an experiment of this type, it is necessary, as you have heard from Kafka and Drever, to be able to put fake gravitational waves into such a pair of antennas. For that purpose, four years ago, Laurie Miller and I designed an absolute calibration system in which we put capacitive end-plates up next to the end of the bar. By putting voltages between the end-plate and the bar, we were able to excite the bar in any way we wanted to; we could imitate any shape of the metric as a function of time. For the data which I report here we had for several months a calibration process in which we introduced small calibration pulses locally at both antennas, at each end of the two-antenna array. Then we looked at the computer analysis to see if the computer found these calibration pulses, and we measured the efficiency with which these have been uncovered. These pulses were not put in at exactly the same time, otherwise we would obscure the question of whether gravitational waves were exciting the antennas. We ask the clocks at each station to insert the pulses exactly two seconds apart in real time, so that in our time lag plots we will see a peak at $+2$ seconds from the calibration pulses if there were any during that period of time.

Figure 33 shows what the power output of this kind of antenna looks like as a

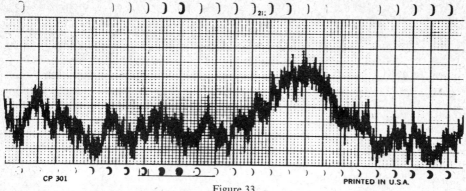

Figure 33

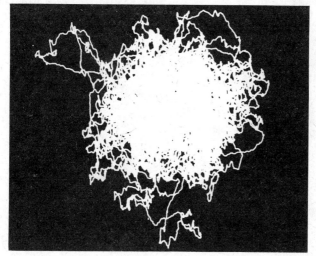

Figure 34

function of time. This is just the square root of $X^2 + Y^2$ (the distance from the origin in the complex plane) as a function of time. This very slowly varying signal is the Brownian motion, and the very fast noise you see on top of it is the residual wide-band noise of the pre-amplifier. Figure 34 shows what the output looks like in the complex plane of one antenna. The computer looks at each antenna's complex plane. This is a whole day's integration of the behavior of the antenna in the complex plane. This is the same thing as integrating an optimally filtered output over the recent past, present and immediate future. There exists an optimal algorithm for this which we have computed and have also tested experimentally. The nice thing about doing the experiment using our autocalibration technique is that we can introduce any sort of force on the two antennas, we can calibrate the entire system in this way and we can find out how effective our filter is. We can change the filter to find out if we can get a better filter for extracting these calibration signals from the noise. Figure 35 shows the kind of signal we are looking for as a function of time. Again, we are sacrificing information about millisecond structure in the curvature— we are monitoring the phase and amplitude changes in the bar oscillation, and we do not know anything about the detailed shape of the metric itself. Any burst of gravitational radiation which has measurable power at the bar resonant frequency will give us this output from the electronics. It jumps up in a very short time compared to the ringdown time τ which is 100 seconds. The rise time of the electronics t is a tenth of a second, and we wish to use an optimal algorithm. I will not go through this. It is possible to obtain an optimal algorithm mathematically that agrees closely with the optimal algorithm that we discovered experimentally. Any detector/electronics system has a *unique* optimal algorithm.

Figure 36 is the result of looking with an optimal algorithm and the output of

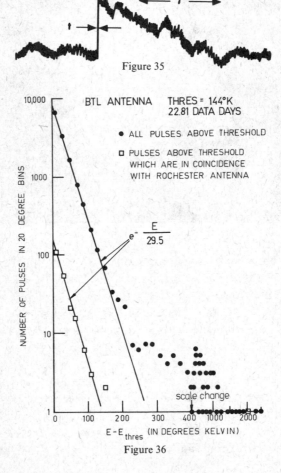

Figure 35

Figure 36

our antenna at Bell Laboratories. Plotted here is the log of the number of pulses above a certain threshold as a function of the energy of that pulse above the threshold. You see here, in the solid line strong evidence for Boltzmann statistics. It is very necessary to have plots like this to convince oneself that the antenna is running correctly. There are a few excess pulses at high energy. These could be gravitational waves (with such a short integration time, we have some excess pulses due to local excitation). However, when we go into coincidence with the other nearly identical antenna at the University of Rochester, there is no evidence for simultaneous excitation. In Figure 37 again we have the number of simultaneous excitations above threshold plotted vertically as a function of the energy of that excitation. We are looking at the energy distribution of coincidences between these two antennas 300 miles apart on a tenth-second timescale. Here again, Boltzmann statistics. No evidence for any excess at either high or low energy. We have extended this, by the way, all the way down to zero energy.

Now to the time lag plots. Figure 38 is a typical time lag plot: vertically the number of coincidences between these antennas is a function of time lag between the two. There is no evidence at zero time lag for any excess. We have many such lag plots. But are we using the correct algorithm? Is there perhaps an even better one? We can compare the type of derived optimal algorithm which we use with other algorithms that other people use. In Figure 39 are the distributions for noise and calibration pulses for the linear algorithm that we are using. What we have done here in this experiment is to apply 93 calibration pulses at 1 kT and look at the computer output and see how many have been discovered. It finds practically all of them

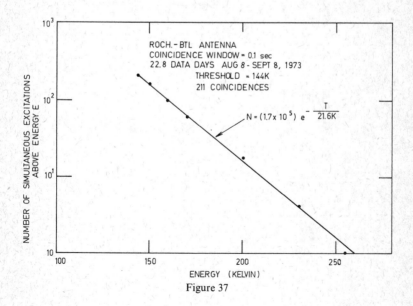

Figure 37

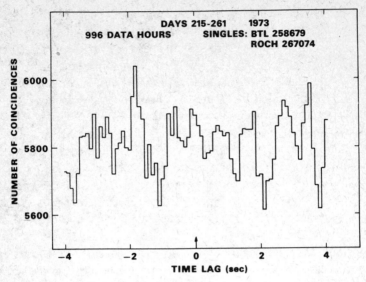

Figure 38

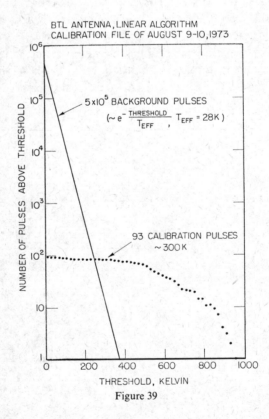

Figure 39

as you can see here. The dotted lines are the calibration pulses and the solid lines the background, again Boltzmann distributed. Now bear in mind how this looked, and now if we look at Figure 40 this will indicate the performance of the non-linear algorithm. The algorithm is $\dot{P}^2$ where P is $X^2 + Y^2$. $\dot{P}^2$ is what you heard about from Weber just a few moments ago. With the same data analyzed with the $\dot{P}^2$ algorithm, you can see that the signal to noise ratio is substantially worse. To see how this affects the search for gravitational radiation we look at the next two figures. Figure 41 shows the non-linear algorithm in the form of a time lag plot. During the time that these data were being taken, many calibration pulses were inserted, but we do not see any excess at plus two seconds where they should appear. In Figure 42, however, we have the result of using our own optimal algorithm, on the same data, and here are the calibration pulses that were inserted at $+2$ seconds.

In the search for gravitational radiation, we must do this experiment as a function of threshold: we do not know if there are just a few pulses of high amplitude or many pulses of very low amplitude. Figure 43 is a very low threshold time lag plot. Again there is no excess at zero time lag. (During this period of time there were no calibra-

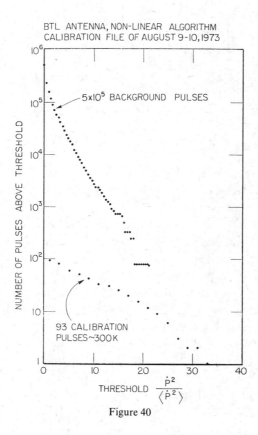

Figure 40

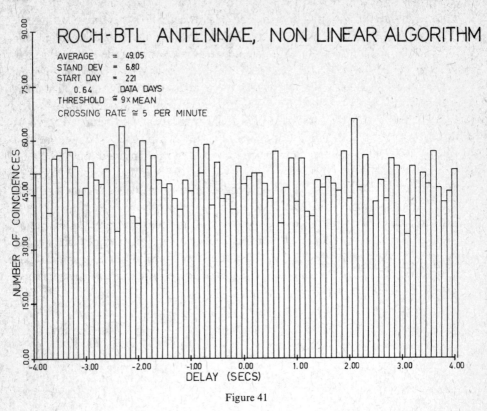

Figure 41

tion pulses.) We have many lag plots like this for different groups of 4 data days and for many thresholds. We can plot all of this data, as observed coincidence rate normalized per day, as a function of the energy deposited in the bar. Figure 44 shows the results of the searches that we have analyzed to date. The lower dashed line here is the most significant limit. The scale on top assumes unity efficiency of detecting bursts in noise. In fact, this efficiency falls off at lower energies. This is from all the time lag plots. The single triangle point is from a null experiment that was done with my antenna at Holmdel, in which we looked with the optimal single antenna integration time for any excitation that occurred in three months. The Munich–Frascati results at 1600 Hertz are the open rectangles. I should point out that the bars we are using are resonant at 710 Hertz and sample the gravity at that frequency. There are three reasons why the sensitivity of these detectors is very much greater than other presently operating detectors: 1) increased mass; 2) better impedance match and higher coupling and Q; and 3) optimal signal filtering. All these measurements taken together constitute a strong null result.

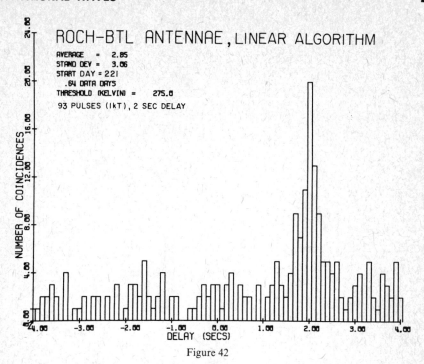

Figure 42

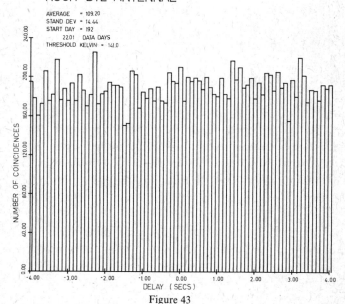

Figure 43

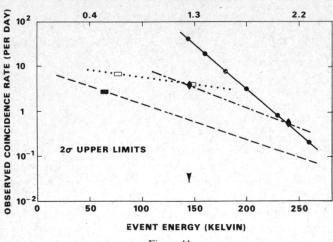

Figure 44

SCIAMA: Thank you very much Tony. I do apologize for being ruthless, but we also want the opportunity for a discussion of all these contributions. First I want to ask the panelists to discuss for a few minutes what we have just heard and then we can have an open discussion with the audience. Perhaps Joe, would you like to begin?

WEBER: Yes, I do not accept the sensitivity analyses made by others of the Maryland array. To really discuss this fully would require having Maischberger, who is concerned with the electronics engineering of the Munich–Frascati array, here and would involve issues of electronics engineering which do not concern this conference, let me therefore indicate my disagreement and let the matter rest here. An important question is "Do the artificial pulses really simulate gravity waves?" I do not know whether they do or not. The evidence from the histograms is that they probably do not because often we get better results from the histogram which is not preferred. There are some data where the preferred algorithm did give better results. I would like to ask Dr. Kafka two questions. One is, "How do you compute your number of coincidences; are they just the pairs of points above threshold, or is there some rejection of pairs of points which appear close together?"

KAFKA: No, there is no rejection, but I plot histograms of two different kinds. In one I count all pairs of points where both detectors were above threshold. But those are not all independent events and the observed fluctuation will not follow a $\sqrt{N}$-law. Therefore, in a different kind of histogram I also count the "number of peaks." If a peak contains more than 1 point I count only one event.

WEBER: For the histograms which you show, the one with the 3.6 sigma peak, what was then on that one?

KAFKA: I am not even sure at the moment. But there is not much difference between the two procedures if one uses our signal content function. [Added in proof: The picture was for the count of "peaks." When all pairs of points above threshold are counted, the zero delay peak is only 2.8 times the observed standard deviation. I should also say that the probability that the peak was caused by real pulses can also be estimated to be small, if one takes into account the result at various thresholds.]

WEBER: Well, there would be in our case. The second question I will ask is "How do you compute the standard deviation?" You showed a sigma, two sigma, three sigma. How do you compute this?

KAFKA: I just mentioned it. The theoretical value depends on the definition of the events. For independent peaks the observed sigma coincides quite well with the value given by the $\sqrt{N}$-argument. For the other kind of histogram a correction comes in for the average number of points per independent peak.

WEBER: Right. But the object displayed on your slide, was that the measured value or the theoretical value?

KAFKA: With our signal content function the agreement between predicted and observed fluctuation is always good. [Added in proof: For my last picture the theoretical fluctuation was used for the definition of σ. The observed fluctuation was slightly larger—hence, the significance of the peak would be judged to be a bit lower.]

WEBER: My own analysis of Kafka's data and system suggests that the sensitivities of the two installations are comparable and this difference he talks about is not large. Thank you.

SCIAMA: Any other comments from the panelists?

KAFKA: I should answer to Joe Weber that there is always doubt in everything and that we certainly try and exclude such doubts as well as we are able to. I have shown you how we calibrated our sensitivity. We have made lots of independent tests, and everything seemed to be consistent. I think our sensitivity is reliable, and then ours should be the most sensitive coincidence experiment working.

DREVER: Perhaps I might just express a personal opinion on the situation because you have heard about Joseph Weber's experiments getting positive results, you have heard about three other experiments getting negative results and there are others too getting negative results, and what does all this mean? Now, at its face value there is obviously a strong discrepancy but I think it is worth trying hard to see if there is any way to fit all of these apparently discordant results together. I have thought about this very hard, and my conclusion is that in any one of these experiments relating to Joe's one, there is always a loophole. It is a different loophole from one experiment to the next. In the case of our own experiments, for example, they are not very sensitive for long pulses. In the case of the experiments described by Peter Kafka and Tony Tyson, they used a slightly different algorithm which you would expect to be the most sensitive, but it is only most sensitive for a certain kind of waveform. In fact, the most probable waveforms. But you can, if you try very hard, invent rather artificial wave forms for which this algorithm is not quite so sensitive. So it is not beyond the bounds of possibility that the gravitational waves have that particular kind of waveform. However, our own experiment would detect that type of waveform; in fact, as efficiently as it would the more usually expected ones, and so I think we close that loophole. I think that when you put all these different experiments together, because they are different, most loopholes are closed. It becomes rather difficult now, I think, to try and find a consistent answer. But still not impossible, in my opinion. One cannot reach a really definite conclusion, but it is rather difficult, I think, to understand how all the experimental data can fit together.

TYSON: I would merely like to comment that all the experiments of the Weber type, where you have an integrated calorimeter which asks the question: "Did the energy increase or decrease in the last tenth of a second?"—all those experiments, of which my own, Weber's and Kafka's are an example—would respond in a similar manner to a given pulse shape in the metric given the same algorithm. I think it must be something which only your detector is sensitive to and not ours. And that is the conclusion I have to draw.

WEBER: It can be seen from the calibration data which have been shown to a number of the panelists that at the level kT/3 there is a substantial amount of proliferation. It appears to me that it would be worthwhile for the Munich–Frascati group to look at their data for May 21 to June 15, 1974 with both algorithms.

KAFKA: I will do that. But I must say that in the 150 days, we did not have any pulses stronger than half a kT. This remark, I think, excludes any influence of "proliferation" (which, by the way, would produce a wide peak and not a narrow one).

TYSON: I wish to say a few words here about proliferation. Before I do so, I would like to say that we, as a matter of course, over the last year, have been analyzing our data not only with the optimal algorithm but also with the algorithms $\dot{P}^2$ and $\dot{X}^2 + \dot{Y}^2$ and we have not seen any excesses at zero time lag on any of these. Of course, we go down to a low enough threshold, so we have all sorts of candidates. Figure 45 is an example here of what happens when we calibrate an antenna of the Weber type with artificial gravitational waves, and, as you see here, we have two kinds of filters. I am plotting the log of the number of events counted above threshold as a function of the threshold, and we put in 10^4 calibration pulses. During that same period of time there were a million background pulses at zero threshold. You see here for a nonlinear filter (and this nonlinear filter is the $\dot{P}^2$ one) that the background is indeed very high. Not only does the signal proliferate, not only do the signal pulses increase, but the background increases enormously over the linear filter. In this case the linear filter was $\dot{X}^2 + \dot{Y}^2$, and so the proliferation indeed does occur but because you cannot invoke a Maxwell Demon in the system, not only does the signal itself proliferate, but the noise grows. And as you can see here, the noise grows even more, so that the signal to noise ratio goes down when one goes from a linear to a nonlinear filter.

SCIAMA: I think this is a good moment to throw the discussion open to the floor. If anyone would like to ask a question of any of the panelists or make a statement, now is the time to do it.

QUESTION: Joe, at a Niels Bohr Institute given about three and a half years ago,

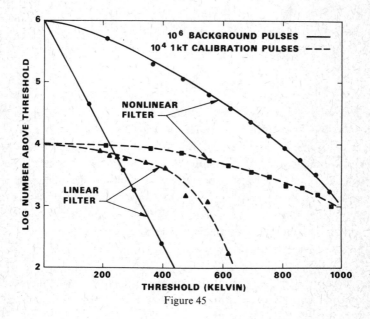

Figure 45

you saw, I believe, about one event per day, something like that, and you had a strong sidereal correlation. Have you seen the same thing in the runs since then?

WEBER: I have not looked. To do the sidereal anisotropy, we need about six months of data. I did not think we could do it any better now. In view of the intense criticism of the experiment, it seemed most important to change our procedures, make them more like those of other groups, and try to find coincidences using the procedures and algorithms of others.

We believe that the old algorithms which we used to generate the histograms displayed in *Nature* in 1972 are correct, but they involved 100 pages of Fortran and we did not think anyone else would be interested in reproducing them. The sidereal anisotropy will have to wait until the present controversy is somewhat better resolved.

QUESTION: Would one of the panel members be willing to comment on experiments that are not represented on the platform?

TYSON: I would like to comment on Braginsky's experiment. He came to the conclusion, after running it for several months and not obtaining any evidence for excess at zero time lag, that there were no gravitational waves of the intensity that was originally reported by Weber in 1970, present at Braginsky's frequency which was about 1600 Hertz. And so he is not running his experiment any longer. He had two detectors which had the same mass as Weber's bars. The electronic pick-off technique was different; it had capacitive sensitive instead of piezo-electric sensors. But I would rate the sensitivity of Braginsky's experiment, considering the way he analyzed the data, as about equal to Weber's sensitivity in 1971.

WEBER: I do not agree.

SCIAMA: Thank you for that comment Joe.

TYSON: There was a question "How about Garwin?" The question there is complicated because in the I.B.M. experiment it was assumed that there is a complete distribution of signal intensities, the Boltzmann distribution, if you will, of the intensities of signals such that occasionally you find a big one. With that assumption, however, they can set very good limits and exclude the large flux implied by the observations of Weber just by using that assumption. Their bar, of course, is smaller than Weber's by roughly a factor of five or ten, and therefore the absorption cross-section is down by that amount. But their analysis is quite superior to Weber's, I think, and their sensitivity is nearly as good, as a result.

QUESTION: There is something that has been puzzling me for quite a while. Joe Weber reported on some experiments with a disc antenna that linked to a cylinder antenna and, as I understand it, he said that when he had the disc wired up to respond to tensor waves he got coincidences. When he wired it to respond to scalar waves, the coincidences went away. It seems to me that if you can make coincidences go away, they must have been there, or you could not make them go away. Could someone comment on this?

WEBER: I believe those experimental results are correct.

QUESTION: But if you stand by them, and no one wants to object to them, does not that mean that you are the winner?

WEBER: No, I do not think that. Well, if you look at my wife, you will see that I have won!

QUESTION: We have heard from two sides statements about the optimal evaluation procedures. As far as I got it from the course, these procedures are indeed essentially the same. However, if two different people use the same word "optimal," then one would have to conclude that they are essentially the same. Could one on the spur of the moment conclude that they are essentially the same? Or, are there still disagreements between these two so-called optimal procedures?

TYSON: I can comment on the process by which we obtained our filter and, as far as I understand, it is in essence the same as the type of filter that is used by the Kafka group. The filter which we have looks like this. What we are interested in doing is obtaining the convolution integral of the output, say, $X(t)$ or, for example $Y(t)$ with the filter kernel $H(t)$ over some period of time long enough so that we can define the event:

$$\bar{X}(t) = \int_{-\Delta t}^{+\Delta t} X(t - \tau) H(\tau) \, d\tau.$$

This convolution is some statistic which you may then threshold. In fact, what one generally does is threshold the quantity $\bar{X}^2 + \bar{Y}^2$. The way that we obtain the optimal algorithm is twofold. First of all, I did a calculation of the shape of the algorithm. In optimal signal theory the optimal algorithm in frequency space is given by the complex conjugate of the signal, before you filter the signal, divided by the noise power spectral density. This gave the shape, say, in the time domain. This told us that, first of all, the optimal algorithm had to be an odd function of time. And secondly, it had to be zero area. We then went ahead and developed an optimal algorithm $H(t)$ for two antennas in coincidence. We did this by trial and error on various types of signals applied to the end plates. The final algorithm was independent of the shape of signal applied to the end plates for all types of signals which we used. The optimal

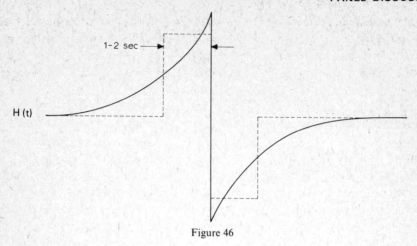

1-2 sec

H (t)

Figure 46

filter has a sharp transition with 0.1 second rise. This is somewhat sharper than for the single antenna optimal filter: for two antennas in coincidence you want better time resolution than you have with just one antenna, in order to get the coincidence real-to-random rate up. The filter decay time (exponential decay on both sides) is roughly two seconds, and as far as I know, that is more or less the filter that Kafka used.

KAFKA: I think you have done the same calculations which we did. We have not yet pared our results, but I suppose they agree in principle if they are found from the same principles of signal detection. The numbers for the optimal filters and corresponding optimal sensitivities do of course depend on the different equivalent circuits. As you could see from my table, the filter time scale which is about 0.3 seconds in Munich, would be about 1.5 seconds for Tyson's detector. [Added in proof: This is nice for him, because he is still far beyond the discretization interval, but on the other hand it requires an even tighter limit on the beat between cylinder and reference oscillator!]

QUESTION: This particular session and sessions of the Cambridge conference and at Liverpool in the last two years are of the same order. Namely, there are people who report one result and there are people who report the opposite result. I think the community would be happy to have this situation resolved, to everybody's satisfaction. Now the question is, of course, that different people have performed different experiments with sophisticated individual apparatuses and they all use what they believe to be the best algorithm possible in their analysis. Can you possibly combine some of the people of the various groups and see if they can possibly agree on a reasonable approach to analyzing the data, because I understand the various difficulties and efforts necessary. Perhaps the next step is to get people together from the various groups working around the world and see if they can agree on some reasonable algorithm that in the end will make them all happy. There are variations in the basic experiments but at least the basic algorithm or basic set of algorithms is satisfactory. This seems to be the next step. We always seem to be bantering back and forth and not resolving the issue. There seems to be the same situation in terms of people who firmly believe in their algorithms and maybe it is time for them to get together. Do you agree?

SCIAMA: Thank you very much.

TYSON: I would like to reply to that. There has been a great deal of intercommunication here. Much of the data has been analyzed by other people. Several of us have analyzed each other's data using either our own algorithm or each other's algorithms. And as far as I know the only evidence I have seen so far in the direction of confirmation of Weber's results is Kafka's 3.6 sigma peak, which is the result of a lot of selection, as he mentioned. There has been a lot of analysis by several of us, of other people's data, and as far as I know, there is no confirmation of the result. I should point out that there is a very important difference in essence in the way in which many of us approach this subject and the way Weber approaches it. We have taken the attitude that, since these are integrating calorimeter type experiments which are not too sensitive to the nature of pulses put in, we simply maximize the sensitivity and use the algorithms which we found maximized the signal to noise ratio, as I showed you. Whereas Weber's approach is, he says, as follows: He really does not know what should be happening, and therefore he or his programmer is twisting all the adjustments in the experiment more or less continuously, at every instant in time locally maximizing the excess at zero time delay. I want to point out there is a potentially serious possibility for error in this approach. No longer can you just speak about Poisson statistics. You are biasing yourself to zero time delay, by continuously modifying the experiment on as short a time scale as possible (about four days), to maximize the number of events detected at zero time delay. We are taking the opposite approach, which is to calibrate the antennas with all possible known sources of excitation, see what the result is, and maximize our probability of detection. Then we go through all of the data with that one algorithm and integrate all of them. Weber made the following comment before and I quote out of context: "Results pile up." I agree with Joe. But I think you have to analyze all of the data with one well-understood algorithm.

WEBER: It is not true that we turn our knobs continuously. I have been full time at the University of California at Irvine for the last six months, and have not been turning the knobs by remote control from California. In fact, the parameters have not been changed for almost a year. What we do is write the two algorithms on a tape continuously. The computer varies the thresholds to get a computer printout which is for 31 different thresholds. The data shown are not the result of looking over a lot of possibilities and selecting the most attractive ones. We obtain a result which is more than three standard deviations for an extended period for a wide range of thresholds. I think it is very important to take the point of view that the histogram itself is the final judge of what the sensitivity is.

SCIAMA: The authorities have allowed us to extend this session for 15 minutes and I believe the best thing to do at this point is to go on to the final item on the agenda I referred to at the outset, that is to talk about future possible detectors. It is my impression as a layman that in the past few years there has not been a very great increase in sensitivity of these detectors—as though Joe's designs from the beginning have reached some kind of threshold of possibility. Some of us Astrophysicists, quite apart from the controversy we have been hearing about this morning, would like to be able to detect, let us say, gravitational waves from a supernova explosion in the Virgo cluster of galaxies. Such an event would probably occur several times a year, giving a very acceptable event rate. But, of course, the sensitivity you would need to detect such an event would have to be very great. A very crude estimate would suggest that one would need sensitivities of the order of say a million to 10^8 times greater than the kind of sensitivities we have today. I think we ought to spend a few minutes discussing the question. Is it conceivable that future designs of detectors have any hope, let us say in the next ten years, of achieving an improvement in sensitivity of that kind of order of magnitude? What I have arranged is that, first of all, Tony Tyson will say a few words about the Stanford experiment which is already under construction (Stanford and Rome, I should say) and then a few words about his own vision about detectors of the future; Ron Drever will then say a few words about his vision. We will either end at that point or, if we have a few moments to spare, the other panelists can throw in their visions too. I ask Tony Tyson to take up this topic.

TYSON: Thank you. I would like to do that. But, first I would like to close the previous discussion by saying that I agree with Joe Weber that the histograms are the whole story. However, for the histograms to be the whole story, one has to calibrate both antennas; you have to put signals into both antennas simultaneously; you have to put a known force into both antennas to see what the result is. The one thing that we really need now from Joe Weber more than anything is one of these plots of the number of pulses above threshold as a function of threshold, for simultaneous calibrated excitations of the two antennas. Absolute calibration is absolutely necessary. Now I should say something about the recent Stanford results. It is a very difficult experiment. Let me explain to you what it is. They want to get very high sensitivity. To get very high sensitivity you have to have large coupling to the bar, as Ron Drever pointed out. You also have to have very low noise. To get low noise, you have to eliminate the Brownian motion of the bar. You do that by cooling to low temperatures; that noise goes like kT. But you also have to eliminate the amplifier noise. Unfortunately, you cannot just eliminate amplifier noise by cooling. You cannot dump your hi-fi into a dewar of liquid nitrogen and expect better noise performance. So, the Stanford people have set out to make a completely superconducting detector with a superconducting transducer, a superconducting amplifier first stage called a SQUID. This stands for Superconducting Quantum Interference Device. All of this technology is possible today. Although possible, getting these three new technologies working together at the same time is very difficult indeed. They have the dewars, they have the aluminum, but they do not have many of the other items which they need. The biggest problem right now, as I understand it, is that they are having difficulty floating their aluminum in a noise-free way on a magnetic field. They want to float the aluminum bar in a magnetic field by the Meissner effect. The critical field for aluminum is 100 gauss. They need several thousand gauss to float this large bar of aluminum. So, their approach has been to attempt to coat the bar with a hard superconducting coating which has a high critical field and will float in a large magnetic field. They are now trying to vacuum-sputter on some niobium titanium, which is a very expensive process that I think eventually would work. Another unfortunate delay is that they have designed a transducer which should be resonant near the frequency of the bar. But the bars are resonant up above a kilohertz and the transducers they had working are resonant below 700 hertz, and they cannot seem to get the two together. I suspect that in the next year this will be overcome. I think we estimated three years ago that it would take three years. Perhaps we should give it another three years, but it is a very worthwhile project. The eventual sensitivity should be somewhere around 100,000 times the sensitivity of Weber's experiment of 1970. This will possibly allow them to see supernovae in the Virgo cluster, if the stars have enough angular momentum to start with, but it is on the edge of detection. As for my own hopes for the future, I am now most excited about the short range. I can well sympathize with the Fairbank group at Stanford and with Bill Hamilton's group at L.S.U. and with the Rome group in their difficulties in cooling large bars. I think, however, that it is possible to get noise temperatures well below what we have achieved at room temperature. With our room-temperature bar we have about 20 degrees Kelvin effective noise temperature now for the detection of the signature which we are searching for in the presence of noise. One ought to be able to do even better than that. Fundamental limitations set in strongly at about a half a degrees Kelvin. We have designed and are now testing a new kind of transducer which, when applied to the large 300-degree Kelvin bar, should give a noise temperature around one degree Kelvin or lower.

SCIAMA: Thank you very much; Ron?

DREVER: Perhaps it might be useful if I were to give a slightly broader view on ideas about the future development of the field. I should commence by saying that before Joseph Weber started his experiments nobody believed we had a hope of detecting gravitational waves. If you worked out the fluxes expected they were incredibly small. Then Joe found his pulses, apparently. These pulses were, in fact, ten times bigger than anybody might have expected, as well as being far more frequent. This focused attention on that region of the spectrum and all experimental groups have therefore been studying that same region: for pulses lasting some milliseconds, and of the same magnitude as Joe's. In my opinion, I think it now seems probable that Joe has been mistaken. We are not yet certain but it is probable. If one accepts as a hypothesis for the moment that he is mistaken, then the field has in a sense rather broadened. Perhaps one should see what other things can be done as well. One is not necessarily just trying to search in that particular region of sensitivity for those par-ticular kinds of pulses. This broadens one's viewpoint. One might consider both pulses and experiments for continuous radiation. In the case of pulses, I think that one obviously needs a huge improvement in sensitivity in order to see predicted effects. Tony Tyson has told us about one obvious direction in which to go—low-temperature large detectors. In the last two years or so, I think two other possibilities have become apparent. One is the technique being developed in Russia by Braginskii and his group in Moscow, who have been investigating the possibility of building rather small detectors of single crystals which have exceedingly high Q when operated at low temperatures. If these experiments are successful, and it is too early to say if they will be yet, this may provide an alternative to the technique being developed by Fairbank and the Rome and Louisiana groups. Another technique which is coming into view now is the quite different possibility of having separate masses which are a long distance apart, so that you get motions of absolute magnitude bigger than you would get with two masses at the separation of about two meters available in a bar. One may monitor the separations using laser techniques in a kind of Michelson inteferometer arrangement. This, I think, is a very promising technique in the long term and I can see important advances in that direction in the next few years. These two methods apply of course to pulses. Supposing one considers the question of continuous waves. We can see several sources from which we would expect to get signals, for example, pulsars (as long as the cores are not symmetrical) and binary stars. If one calculates the fluxes from these, as Bertotti showed in some of these cases earlier in this meeting, they are very small indeed. Can one hope to detect them? Well, I think one can. And I can see now three ways in which this might be done, perhaps. Using resonant detectors of very high Q, such as the techniques being developed by Braginskii, is one possibility. Another is to use the laser interferometer technique with a very large base line; that is well adapted to low frequency signals. A third one is the one which Bertotti talked about—the heterodyne technique. These are three new possibilities for looking for low-frequency signals. The experiments are at least as difficult as the past experiments, but I think now they will go on as well, and I think that they may possibly work in a few years. Let me just make a final remark on this. The experiments have much greater potentialities than is apparent to those who are not in the game. If in fact the pulses Joe found were what everyone expected them to be—short pulses, of quite strong intensity—I myself am quite certain that all of the groups would have found them by now. The important thing is that it would have been possible to find the directions and begin to produce maps of the sky of the signals; for by comparing phase information from detectors at different places, or from times of arrival, one can in principle find the directions of individual pulses. Unfortunately, this program has been held up because none of us has found the pulses, with the exception of Joe who knows how to do it rather better, or has been more lucky. However, I still feel strongly that these pulses will eventually be discovered, perhaps at a much lower level. As soon as that happens, I think the field will develop very rapidly. From a confirmed initial discovery that can be reproduced readily, I think the thing could rapidly spread to where we would have a real astronomy and we would be producing maps of the sky of gravitational wave sources.

WEBER: From my understanding of the standards of particle physics, if you get a histogram with a zero time delay excess of 3.6 standard deviations, that is regarded as a positive result. Perhaps Ron Drever has higher standards; it would be nice to have a larger excess with a higher level of confidence, but please be patient.

TYSON: I think these results of Weber's would be convincing if it were not for all the selection and bias to zero time lag.

KAFKA: Since there were many questions from participants about what we would be able to see, I should like to make one last remark about the sensitivity of the present coincidence experiments. Yesterday, Bludman told us about a possible observation of an anti-neutrino evnet. Though it seems unlikely to me, let us speculate with Bludman that it was connected with the collapse of a star of several solar masses, and that it also emitted gravitational radiation, say 2 percent of a solar mass, in a bandwidth of 1kHz, at a distance of 1 kpc. Then it turns out that the gravitational wave pulse would correspond to 1/40 kT in our or Weber's detector—and this is exactly what our noise simulates on the average. Hence, with a sufficient number of such detectors it would be a marginally detectable event. With coincident independent evidence from neutrinos or electromagnetic waves, the existence of the gravitational pulse would be significantly proved, and one could already start an analysis of its polarization properties. Therefore, if one is extremely optimistic about sources (which I am usually not), one has already reached a stage where one should look for something.

TYSON: I would like to second that. I think that every time we have looked into the sky with a new kind of detector, a new black box, we have found something which we did not expect. I do not think we should be discouraged by the fact that one needs a factor of a million or so improvement over Weber's sensitivity of 1970, before one can really expect to see a supernova in the Virgo cluster. I think we are going to be surprised long before then. Right now we have antennas that are perhaps 200 times more sensitive than Weber's experiment. We do not yet see anything, but then antennas are always increasing in sensitivity. I am very excited about the possibilities.

SCIAMA: Ladies and gentlemen, I fear we have to bring this session to a close but I am very glad it has ended on this optimistic note, because I too believe we are going to detect either pulsars or supernovae or both or something else in the next few years. So now I ask you to thank all the panelists for this clarifying discussion.

Conference Summary: More Results than Ever in Gravitation Physics and Relativity

JOHN A. WHEELER

Princeton University, Princeton, New Jersey 08540, U.S.A.

Never was an old saying more relevant than at this conference: The larger the island of the known, the greater the shoreline of the unknown. Would that it were possible to go on consulting with colleagues from near and far on the many fascinating issues that came up at this conference, and go on doing this under such favorable arrangements! Did we sometimes in imagination see Einstein at these discussions? We knew that for him this was a favorite subject, in a favorite country. How could he have failed to smile and to enjoy these meetings? For the hospitality, for the organization, and for ever so much more, I know that the thanks of all present go to the agencies that sponsored this meeting, to the officers of the local organizing committee, including not least President Ne'eman, Professor N. Rosen, Professor J. Rosen, and Professor Shaviv, and to many other devoted hands and hearts.

The number of able young men at this conference and the caliber of their contributions provide living evidence for the extraordinary upsurge of activity in gravitation physics and relativity in the past decade. Another index of activity is the number of contributions at GR7, so great that often three simultaneous sessions were necessary during this week-long meeting.

It would be out of place to attempt here any detailed summary of all the individual research reports. Moreover, excellent review papers were given at the conference. They survey the state of our knowledge in many of the major divisions of our field. In view of these circumstances, this closing report may be most useful if it takes a sample participant and asks what seemed to him eight highlights or perspectives opened up by the meeting. Naturally there will be as many answers to such a question as there are participants. Moreover, any such arbitrary selection of eight topics necessarily leaves out a large number of important contributions, possibly even what in retrospect might prove to be the most important single finding reported at this meeting. Therefore, for a proper appreciation of what went on, the student must turn to the conference proceedings themselves. They provide the only true summary of this meeting.

1. THE QUANTUM RADIANCE OF A BLACK HOLE

No topic attracted more interest than the quantum radiance of a black hole. There is a standard expression (Bardeen, Carter and Hawking, 1973) for the change in mass-energy of a black hole when additions are made to its charge Q, its angular momentum J, and the proper surface A, of its horizon:

$$dM = \frac{\text{acceleration of gravity of horizon}}{2\pi} \cdot \frac{dA}{4} + (\text{angular velocity of horizon}) \cdot dJ +$$

$$+ (\text{electrostatic potential at horizon}) \cdot dQ = \frac{g}{2\pi} \frac{dA}{4} + \Omega dJ + \Phi dQ \qquad (1)$$

(geometric units; $G = 1$, $c = 1$). This result recalls the formula for the change in mass-energy in a thermodynamic transformation,

$$dM = T dS + \Omega dJ + \Phi dQ \qquad (2)$$

(see, for example, Landau and Lifshitz, 1958). Bekenstein (1972, 1973, 1974) argued that the connection between the two formulas is more than an analogy; it is an identity. The area of the horizon of a black hole, divided by the quantum of angular momentum,

$$\hbar(\text{cm}^2) = \hbar_{\text{conv}} G/c^3 = (\text{Planck length})^2 =$$

$$= (1.66 \times 10^{-33} \text{ cm})^2 = 2.612 \times 10^{-66} \text{ cm}^2 \quad (3)$$

is not only analogous to entropy; it *is* entropy, up to a numerical factor of order unity. The acceleration of gravity at the horizon, multiplied by $\hbar$, is not only analogous to temperature; it *is* temperature, again up to a numerical factor of order unity, according to Bekenstein. Hawking (1974a, b) provided a deeper mathematical foundation for these considerations of Bekenstein, showed why a black hole must radiate, gave a formula for the radiation, and determined the two numerical factors left undetermined by Bekenstein; thus, the temperature is

$$T(\text{cm}) = (G/c^4) k T_{\text{conv}} = \hbar(\text{cm}^2) g(\text{cm}^{-1})/2\pi \qquad (4)$$

and the entropy is

$$S(\text{dimensionless}) = S_{\text{conv}}/k = A(\text{cm}^2)/4\hbar(\text{cm}^2) \qquad (5)$$

For the special case of a non-rotating uncharged (i.e., Schwarzschild) black hole, where horizon gravity is

$$g = M/R_{\text{hor}}^2 = 1/4M \tag{6}$$

the temperature is

$$T_{\text{conv}} = 0.616 \times 10^{-7} \deg(M_\odot/M) \tag{7}$$

No one sees any possibility whatever for directly measuring such a temperature, nor the associated thermal electromagnetic radiation, for any black hole of solar mass or greater. No one sees any possibility whatever for any process of astrophysics or human technology to produce a black hole of quarter solar mass or smaller (Harrison, Thorne, Wakano, and Wheeler, 1965). However, Hawking (1973) points out that quantum fluctuations in geometry at the time of the big bang itself may well give birth to small black holes. Moreover, he notes (1974a,b) that such a primordial black hole, when endowed initially with a suitable mass, of the order of $10^{15}g$, will be able to survive until today, radiating more and more strongly as its mass evaporates away, until in the last 0.1 sec of its life it goes out with a bang equivalent in energy to some millions of hydrogen bombs. Therefore the Bekenstein–Hawking black hole radiance is in principle of the very greatest interest.

No subject has been of more intense concern in informal discussions at this conference than the derivation of Hawking's formula for the quantum radiance. There are as many approaches to the derivation as there are investigators working on the question. More appropriate to recall here than the derivations—fit subject for a future meeting—is the final result, on which most workers, however different their methods, nevertheless agree. This result is conveniently stated in the form of a "doctor's prescription". This prescription is most quickly arrived at by considering an old and simple problem of physics that has in it no reference whatever to a black hole or to gravitation: How much radiation does a spherical black body (not a black hole!) put out? The answer is expressed as the product of four factors:

The area of the radiating surface

$$4\pi R^2$$

The average component of the velocity of a photon normal to the surface, $c/4$, when one deals with electromagnetic radiation; or, when one deals more generally with thermal emission of any kind of particle

$$v/4$$

The number of independent modes of oscillation of the electromagnetic field or other field per unit volume that lie in the interval of circular wavenumber of interest, dk, and have the state of polarization of interest

$$4\pi k^2 dk/(2\pi)^3$$

The average number of photons per mode of oscillation, when one is dealing with radiated *number*; or the average energy per mode of oscillation, when one is dealing with radiated *energy*; with a minus in the denominator when one is dealing with bosons and a plus sign for mermions

$$\frac{1 \text{ or } \hbar\omega}{e^{\hbar\omega/T} \pm 1} \tag{8}$$

Multiply to obtain the total radiation in the specified frequency interval.

1.1. The "access factor"

Now subdivide this radiation into classes, each class characterized by the total number of units of angular momentum j that the representative outgoing quantum carries away, as well as its angular momentum m around some preferred axis, and its polarization π. Thus arrive at the "doctor's prescription",

$$\begin{matrix} \text{number of quanta or amount} \\ \text{of energy emitted per unit} \\ \text{time in specified interval } dk \end{matrix} = \frac{v\,dk}{2\pi} \sum_{j,m,\pi} \Gamma_{\omega,j,m,\pi} \frac{1 \text{ or } \hbar\omega}{e^{C/T} \pm 1} \tag{9}$$

Here $C = \hbar\omega$ is an abbreviation for the "cost of emission" in the numerator in the Boltzmann exponent. Nowhere in the formula is there any explicit reference to the radius R of the radiating sphere. Instead there appears the "access factor" Γ. It depends on the angular momentum of the quantum, or, equivalently in the standard semiclassical approximation, on the impact parameter b of the quantum (Fig. 1):

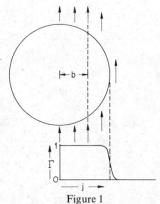

Figure 1

Access factor Γ as a function of impact parameter b or angular momentum quantum number j for the impact of a photon on, or the emission of a photon from, a sphere of radius R. The sphere above (dimension cm) and the range of j values below (integers; dimensionless) are so scaled that point of transition from full access to zero access, $b = R$ in the upper diagram, matches with the corresponding point of transition, $j = kR$, in the lower diagram.

$$b = \left(\begin{array}{c} \text{impact} \\ \text{parameter} \end{array} \right) = \left(\begin{array}{c} \text{classical distance} \\ \text{of closest approach} \end{array} \right) = \frac{\text{(angular momentum)}}{\text{(linear momentum)}} = \frac{j\hbar}{k\hbar} = \frac{j}{k} \quad (10)$$

It also depends on the size of the emitting object. Again in the semiclassical approximation, all those quanta make contact with the completely absorbing surface for which the impact parameter is less than the radius; and all those quanta miss the surface for which the impact parameter is greater than the radius; that is,

$$\Gamma \simeq 1 \quad \text{for} \quad j < kR$$

$$\Gamma \simeq 0 \quad \text{for} \quad j > kR \quad (11)$$

In actuality the access factor Γ falls rapidly from unity to zero in the neighborhood of $j \sim kR$ provided that the value of kR is large compared to one. Under these circumstances, to carry out the sum over all modes of emission is to make the first step in recovering from the "doctor's prescription" (9) the standard surface–area proportionality and all the other features of the standard formula,

$$\left(\begin{array}{c} \text{energy emitted} \\ \text{per unit time} \end{array} \right) = 4\pi R^2 \cdot \frac{c}{4} \cdot \frac{\pi^2}{15} \frac{T^4}{\hbar^3 c^3} \quad (12)$$

for the total blackbody radiation of the sphere; thus,

$$\sum_{j,m,\pi} \Gamma_{\omega,j,m,\pi} \simeq 2 \sum_{j=0}^{kR} \sum_{m=-j}^{j} 1 = 2 \sum_{j=0}^{kR} (2j + 1) \simeq 2k^2 R^2 \quad (13)$$

Here the factor 2 presupposes a radiation with two independent states of polarization.

When the patient is not a blackbody but a black hole, the doctor, in this Stephen–Hawking case, employs the standard tried and true prescription (9) for the quantum radiance, with the following modifications:

1. Replace the "cost of emission" $C = \hbar\omega$ (rest mass plus kinetic energy) in the Boltzmann exponent in (9) by the "corrected cost of emission" as given in standard treatises on statistical mechanics (see, for example, Landau and Lifschitz, 1958) for a rotating charged body,

$$C = \hbar(\omega - m\Omega) - e_i \Phi \quad (14a)$$

where mh is the azimuthal angular momentum of the emitted quantum and e_i is its charge.

2. Insert for the temperature T in the Boltzmann exponent (constant over the radiating surface) the value

$$T = gh/2\pi \quad (15)$$

where g, the horizon value of the gravity, is constant over the horizon (the "zeroth law of black hole physics", the history of which is summarized by Carter (1973)).

3. Recognize that the total angular momentum of a quantum in the nonspherically-symmetric geometry of a rotating black hole is not a constant of the motion. Therefore do not identify the label j of the access factor $\Gamma_{\omega,j,m,\pi}$ with total angular momentum. Instead, identify j with the index number of the appropriate spheroidal harmonic when one separates the wave equation for the radiation in question in Boyer–Lindquist coordinates. Table 1, adapted from Press (1973), recalls what one knows about this separation.

TABLE 1

Analysis into Harmonic Components of (1) a Field Propagating in the Geometry of a Black Hole, (2) a Perturbation in This Geometry Itself, or (3) a Coupled Combination of the Two. Also the Representation of the Harmonic Component in Terms of a Single Scalar Function of Position, and the Separation of the Partial Differential Equations for This Function into Four Ordinary Differential Equations in the Four Separated Variables t, r, θ and ϕ (Table taken from Press (1973) with rearrangements and supplements)

Geometry	Field	References
Schwarzschild	Scalar	Bel (1963); Price (1971, 1972a)
	Neutrino, electron	Brill and Wheeler (1957); Wheeler (1971b)
	Electromagnetic	In terms of vector harmonics, Wheeler (1955); in terms of spinorial field components, Price (1971, 1972a) and Bardeen and Press (1973)
	This geometry itself	In terms of tensor harmonics, Regge and Wheeler (1957), Vishveshwara (1968, 1970), Zerilli (1970a, 1970b); in terms of spinorial field components, Bardeen and Press (1973).
Reissner–Nordstrøm	Scalar	Carter (1968a, 1968b)
	Neutrino, electron	Brill and Wheeler (1957); Wheeler (1971b)
	Coupled EM–geometrodynamic	Zerilli (1974), Moncrief (1974a, 1974b), Chitre (1975)
Kerr	Scalar	Carter (1968a, 1968b)
	Neutrino	Teukolsky (1972, 1973, 1974)
	Electromagnetic	Teukolsky (1972, 1973, 1974); Fackerell and Ipser (1972).
	Geometrodynamic	Teukolsky (1972, 1973, 1974); Wald (1973)

4. Solve the "radial" wave equation for the spheroidal harmonic in question to determine the access factor Γ. This is the point where the size of the black hole enters. So does the effective potential that runs from the horizon of the black hole to infinity. It measures the combined effect of "gravitational", "centrifugal", and electromagnetic forces.

To define the access factor, let a wave of the given (ω, j, m, π) start outward from just outside the emitter. Then the fraction of the intensity of this wave that arrives at infinity measures Γ, and the fraction of the intensity that gets reflected back down the black hole, is $1 - \Gamma$. As an alternative, equivalent, and often more convenient way to define Γ, let a wave of unit strength be envisaged as running inward from infinity, and *identify the fraction of the intensity reflected back out by the potential barrier as* $(1 - \Gamma)$. In this case all comparisons of intensity are made at infinity.

1.2. Superradiance

Distant from the spontaneous quantum radiation considered by Hawking is the classical superradiance whose existence was first pointed out by Zel'dovich (1971, 1972) and by Misner (1972), whose meaning was clarified by Bekenstein (1973), and whose magnitude has since been calculated: (a) analytically in appropriate limiting cases by Starobinsky (1973) and Starobinsky and Churilov (1973), and (b) numerically for the entire frequency range by Press and Teukolsky (1972) and Teukolsky and Press (1974). In superradiance, a wave of given (ω, j, m, π) runs in from infinity toward a black hole and is returned back to infinity with augmented strength. The increase in the energy of the wave comes at the expense of the rotational energy of the black hole. It is for waves what the Penrose process is for particles, an instance of *activity*, the ability of a rotating black hole to communicate energy to its surroundings (Fig. 2). It manifests itself in a value of the reflection coefficient $(1 - \Gamma)$ greater than unity; that is to say, in *a negative value for the access factor* Γ.

A negative value of Γ makes no difficulties for the quantum radiance prescription (9), for a simple reason. The access factor Γ in the numerator of (9) becomes negative then and only then, and the black hole becomes superradiant then and only then, when the denominator

$$e^{C/T} \pm 1 \tag{16}$$

in (9) is also negative.

In contrast to the quantum radiance of a black hole, governed by the ratio

$$\frac{\Gamma}{e^{C/T} \pm 1} \tag{17}$$

and always positive, superradiance is entirely classical, but conditional. For it to occur, two conditions must be met. First, the radiation in question has to have Bose–Einstein character (minus sign in (16)). Second, the "corrected cost of emission" C in (14a) has to be negative.

1.3. Quantum radiance as affected by the number of types of neutrinos

Nothing is more impressive about black hole radiance than the blackness of the emitter. A star, opaque to photons, is normally almost perfectly transparent to neutrinos and to gravitational radiation, and is therefore a significant radiator of

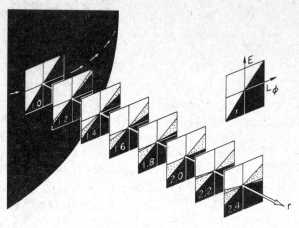

Figure 2

Tilt of the light cone on approach to the horizon of an extreme Kerr rotating black hole ($J = M^2$). Interior of backward light cone, black; interior of forward light cone, white; "neutral region" (spacelike intervals), dotted. For simplicity only the equatorial plane of the black hole is considered, and only photons which are directed exclusively in the azimuthal direction (ϕ changing; no change in r or θ). L_ϕ represents the angular momentum of the photon in the direction of increasing ϕ; E is its energy. Far away ($r \to \infty$) the forward light cone has its usual form, $E = |L_\phi|/r$, and a photon has positive energy. Within the ergosphere (from $r = M$ to $r = 2M$), a photon still has positive energy, as that energy is judged in a local Lorentz reference system. In contrast, the energy as defined by E, like the angular momentum L_ϕ, (1) is an "integral of" or "constant of" the equations of motion of the photon on its way from r to ∞ (when it *can* get to infinity!); (2) thus provides a measure of the energy of the same photon as judged by the "common-market standard of energy as measured at infinity"; and (3) can be negative for a photon trapped inside the ergosphere. *Example*: For $r = 1.4M$ (inside ergosphere), and for a photon with negative L_ϕ (traveling against the direction of the rotation of the black hole), the energy is $E = 0.234L_\phi/M$. Such a photon, though locally endowed with positive energy, has negative common-market energy. It *reduces* the mass energy of the black hole when it falls into it (Penrose–Christodoulou process). In contrast, a photon going with the rotation (positive L_ϕ) has positive common-market energy ($E = 0.417\ L_\phi/M$). The tilted double light cone that one has for a photon splits up for a particle into two hyperboloids, also tilted; but otherwise the situation for particles is similar to that for photons. If a particle or a photon in an allowable state (on upper sheet of hyperboloid, or on forward light cone, or endowed with positive energy as seen in a local Lorentz frame) can have a common-market energy E that is negative, can it not likewise have a direction of progress, $dt/d\tau$, in common-market time t that is also negative? If this idea were correct, it would open up the possibility to travel back into one's own past; but it is not correct. The 4-vector (E, L_ϕ) and the 4-vector $\big(\mu(dt/d\tau), \mu(d\phi/d\tau)\big)$ are the same vector in covariant ("*l*-form") and contravariant ("tangent vector") representations; but the metric is non-diagonal (non-zero $g_{t\phi}$). The coordinates t and ϕ are not orthogonal. The common-market energy E can be negative, but the rate of advance of common-market time t with particle proper time can never be negative.

neither. In contrast, a "hot" black hole, that is, a small black hole, is as good at emitting neutrinos and gravitons as at emitting photons; and as effective in sending out mu-neutrinos as electron-neutrinos. Thus any estimate of the life of a small black hole against Hawking's quantum evaporation has to depend in an important way on the number of different kinds of neutrinos that there are in nature. On that one knows as little as one does about the number of particles beyond the electron and the mu-meson: Zero? Infinity? A finite but small number? Or a finite but large number?

1.4. Cosmological limit on types of neutrinos

If this question evidences one link between gravitation and neutrino physics, cosmology recalls another. Were the density of mass-energy in the universe greater by one or more factors of ten than the so-called critical cosmological density,

$$\rho_{cos} = \left(\frac{3}{8\pi}\right)\frac{1}{a}\left(\frac{da}{dt}\right)^2 \sim (3/8\pi)(49\text{km/sec Megaparsec})^2$$

$$= (3/8\pi)(20 \times 10^9 \text{lyr})^{-2} = 3 \times 10^{-58}\text{cm}^{-2} \quad \text{or} \quad 4 \times 10^{-30}\text{g/cm}^3 \quad (18)$$

it would have greatly shortened the time-scale of the universe. There would not have been time enough since the big bang for stars and star clusters to have evolved so far as we see them to be evolved. Therefore (18), or ten times it, provides an upper limit on the density. One neutrino family contributes to this density the amount

$$\rho_v(\text{one family}) = \frac{7\pi^2}{120}\frac{T^4}{\hbar^3 c^5} \quad (19)$$

unless a chemical potential shifts the equilibrium away from the equality of numbers of neutrinos and antineutrinos presupposed in (19). In this event the contribution of neutrinos and antineutrinos together is only increased above the value (19). Moreover, in the absence of other considerations, it is reasonable to assume for neutrinos roughly the same temperature today as the primordial cosmic fireball radiation, 3 deg K, on the view that both were once in equilibrium with matter and both have been cooled together by the expansion of the universe.*

---

* In actuality neutrinos are believed to have broken their thermal link with matter at a time when matter was very hot, and (e^+e^-) pairs were overwhelmingly more numerous than the electrons seen today. Under these conditions the calculated energy present in electron pairs was about 7/4 of the energy present in the form of electromagnetic radiation. On further cooling the pairs must have been frozen out, dumping their energy into the electromagnetic field. Thereupon the T^4-proportional energy of EM radiation must have risen to $\sim 11/4$ of the value otherwise to have been anticipated, bringing it to this extent out of alignment with the energy of any single neutrino family, though otherwise both energies decrease alike with time during the expansion of the universe. On this account, in the calculation from

This reasoning leads to a lower limit of the order

$$\rho_\nu(\text{one family, 3 deg K}) \sim 6 \times 10^{-34}\,\text{g/cm}^3 \tag{20}$$

for the contribution of a single family of neutrinos to the cosmological density; and to an upper limit for the allowable number of families of neutrinos of the order of

$$N_\nu(\text{max}) \sim \rho_{\text{cos}}/\rho(\text{one family}) \sim (4 \times 10^{-30}\,\text{g/cm}^3)/(6 \times 10^{-34}\,\text{g/cm}^3) \sim 10^4 \tag{21a}$$

Subsequent to the conference, Martin Rees has pointed out (Rees, 1975) that a number of varieties of neutrons of the order of 10–100 times that already recognized would so raise the pressure in the earliest days of the universe and so alter the time scale of expansion that great difficulties would arise in accounting for the observed ratio of helium to hydrogen. From this argument one arrives at an upper limit of the order of

$$N_\nu(\text{max}) \sim 10^2 \tag{21b}$$

In conclusion, no way is evident to exclude the possibility that a black hole evaporates via quantum radiance $\sim 10^2$ times as fast as one has previously supposed. Such a neutrino-enhanced radiation would lead to no observable effect whatsoever for a black hole of solar mass or greater, because its previously calculated life was already greater than the estimated age of the universe by so many more than two powers of ten. However, a small black hole of primordial origin would have to be more massive than previously supposed by a factor of the order of $N_\nu^{1/3}$ in order to survive from big bang to today. The mass converted into *electromagnetic* energy in the last 0.1 sec of the life of the object would be

$$\Delta M(\text{EM, last 0.1 sec}) \sim 0.7 \times 10^8 \text{g}/N_\nu^{2/3} \tag{22}$$

dependent in an important way on the number of families of neutrinos. To observe the final flash of such an object would therefore not only confirm the existence of primordial black holes, but also tell one something new about elementary particles.

(19) of the energy in a single family of neutrinos, it would be reasonable to take for T_{conv}^4, not the value (3 deg K)^4 obtained directly from measurements on the primordial cosmic fireball radiation, but rather $(4/11)(3 \text{ deg K})^4$. This correction is neglected for simplicity in the present order-of-magnitude discussion. It would not be possible to ignore such a correction if in addition to the energy dumped into the EM field by freezout of (e^+e^-) pairs one also had energy dumped into the EM field *after* neutrino decoupling by freezout of $(\mu^+\mu^-)$ pairs and the many many hadron and other particle pairs that one could list. The actual order of events is not known. Until it is known, an important element of uncertainty will reside in the present analysis, which is based on the assumption that almost all of these other particle pairs froze out *before* neutrinos decoupled from matter. Appreciation is expressed here to William H. Press for a discussion in which he pointed out that cosmological density puts a tighter limit on the number of families of neutrinos than does the rate of evaporation of black holes.

1.5. Chemical potentials and the laws of conservation of baryon number and lepton number

This discussion presupposes that the horizon of a black hole is not endowed with a chemical potential. But a chemical potential μ does appear in the statistical-mechanics theory of the equilibrium of a rotating charged body in every other normal context. The "corrected cost of emission" in the Boltzmann exponent contains a term additional to that shown in (14a), and reads (see, for example, Landau and Lifshitz, 1958)

$$C = \hbar(\omega - m\Omega) - e_i\Phi - \mu_i \tag{23}$$

where the subscript i refers to the type of particle being emitted. Moreover, a positive chemical potential μ_i for a given particle favors the emission of that particle and inhibits the emission of the corresponding antiparticle. Therefore the chemical potentials for the various emissions can be imagined to be so adjusted, and in normal evaporation processes automatically will so adjust themselves, that what is given off is identical in content of conserved components with what went in to form the object in question in the first place.

The theory of the quantum radiance of a black hole is not sufficiently developed to allow a definitive judgment whether a black hole is characterized by chemical potentials. If it is, these potentials would be automatically fixed by the requirement that baryon number and lepton number be conserved in a cycle in which a black hole is first formed, and then evaporates away (Wheeler, 1974).

Whatever may be the final conclusion on this question of black-hole chemical potentials, three comments are appropriate. First, it is interesting in principle to conceive of a cycle in which one puts together matter to make a black hole, and then lets that black hole evaporate. However, it is questionable whether such a cycle is achievable except in imagination. It may well be that the beginning of the cycle can be followed, or the end, but not both. Either the star can be followed from formation through its collapse (then one can imagine counting the number of baryons that go in, and the number of leptons; however, the resulting black hole is too big, and lives too long to evaporate. Therefore one cannot count what comes *out*) or the black hole is small enough to evaporate before the Einstein-predicted contraction and collapse of the universe. Then the ejected particles can be counted. However, such a black hole has to be primordial, formed in the big bang. Then no way offers itself to tell what goes *in*. In neither case is it evident how to make any direct check whatsoever on the laws of conservation of particle number. This difficulty of checking may rule out even in principle any definition of any effective baryon number and lepton number for a black hole. However, the alternative is also conceivable, that one should forget the Einstein-argued finiteness of the time scale of the universe in considering black-hole physics—in which case total evaporation can be imagined eventually to occur, and total output of particles will be as well defined as total input.

Second, as far back as the 1930's, one learned to understand in easily visualizable terms the production of pairs of electrons in a sufficiently strong electric field. One had to deal with a tunneling between one region where locally-negative-energy states are occupied and another region where locally-positive-energy states of the same common-market energy are empty. No equally simple picture has yet been developed for the quantum radiance of a black hole. For someday filling this gap in our understanding it is therefore useful that G. W. Gibbons has explained to us at this meeting the mechanism of loss of charge by a black hole, a process where there is some overlap between the old picture of pair creation and the new aspects of quantum radiance. In an important preprint, Brandon Carter (Carter, 1974) has given additional insight into this process (see also Zel'dovich, 1972a; Bekenstein, 1973).

Third, by any method of analyzing properties of a black hole, whether from the study of a test object around it, or by the electrical repulsion it exerts on a test charge or by the superradiance to which it gives rise, one sees not the slightest possibility to distinguish this black hole from another black hole of the same mass, charge, and angular momentum formed from a quite different mix of baryons and anti-baryons, leptons and antileptons, and radiation. In this sense it is as true as ever that the laws of conservation of particle number are "transcended" in black hole physics (Wheeler, 1971b).

2. ANALYSIS OF REALISTIC SCENARIOS OF GRAVITATIONAL COLLAPSE

Three signals come from a black hole: gravitational radiation during formation, X-ray from the accretion disk after formation, and "activity"—transfer of energy from a rotating black hole to its surroundings. For the detection of these three signals there is much observational work in progress, and much progress in the instruments of observation. In contrast, predictions lag behind on what is to be observed, and nowhere more so than on gravitational radiation. Detailed calculations of great difficulty and beauty have been made on the collapse of systems endowed with spherical symmetry, but unhappily they provide the one exception to the otherwise universal rule of gravitational radiation during collapse. Allowance for rotation —and rotation-induced fragmentation, when it occurs—however difficult, is essential for any proper prediction of the radiation.

For the prediction of the astrophysical evolution up to the point of collapse and radiation, one supposes that rotation is without great influence. On this basis Schwarzschild, Hayashi, Paczynski and others calculate the scale of time for the successive steps in the evolution of a single star. Others, not least van den Heuvel (1974), analyze the great effect upon the evolution of one star produced by the companion star in a close binary. At this meeting Martin Rees has given a most useful survey of what we know about the "natural economy" of black holes—how many new ones are produced per year, how long one is "live", and how many "dead" black holes one can reasonably believe our galaxy to contain today.

For an analysis of the very difficult hydrodynamics of collapse nothing would be more useful for a first orientation than exact results for a sufficiently wide variety of idealized situations to allow one to extrapolate or interpolate to the scenarios of actual physical interest. A model need not even have rotation to be of interest, provided only that it has a deformation comparable to that produced by rotation. Among models for which some hydrodynamic analysis has been carried out, and much more would be appropriate, are these:

1. Newtonian gravitational collapse of a spheroidal cloud of dust of uniform density.

2. The same for an ellipsoidal cloud of dust (Mestel, 1965; Lin, Mestel, and Shu, 1965).

3. Initial stages of the fragmentation of an infinite plane slab of nuclear fluid (Renfrew, 1972).

4. Final stages of the approach of a black hole to ideality (Price, 1972a,b; Thorne, 1972; Fackerell, 1971; Press and Teukolsky, 1973).

Interesting and important as it is to treat idealized models in all detail, they represent only a first step on a long road, culminating in the detailed hydrodynamics analysis of realistic scenarios. What a moving picture of one such scenario might look like can at most be suggested in a highly schematized and symbolic way by Fig. 3. Only when the several stages of the hydrodynamics are worked out will one be able to calculate within anything like 20 percent latitude the intensity and other features of the gravitational radiation emitted in the successive steps of collapse:

1. Continuous radiation of ever-increasing frequency given off during the contraction of a slowly rotating and slightly non-axisymmetric white dwarf core to a rapidly rotating and still slightly non-axisymmetric neutron-star pancake.

2. The pulse of quadrupole gravitational radiation given off during the last rush of this approach to pancake shape even when any departure from axisymmetry is zero or negligible.

3. The pulse or pulses emitted as this pancake fragments—a fragmentation conditional upon the hydrodynamic system having a sufficient angular momentum.

4. Continuous radiation of changing frequency or frequencies given off as a fragmented system gradually loses angular momentum.

5. A pulse of radiation emitted each time that two fragments amalgamate to make a more massive fragment.

6. A pulse given off each time that a neutron star collapses to a black hole.

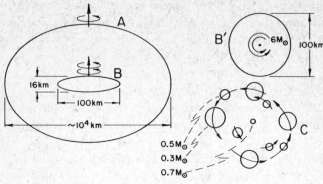

Figure 3

Collapse (A to B or B′) of the slowly rotating white dwarf core of a star to a rapidly rotating neutron star pancake; fragmentation (B to C) of this pancake into separate neutron stars; and slow dissipation (C) of the angular momentum of this revolving system by gravitational radiation, allowing the separate pieces one by one to amalgamate into a single neutron star or black hole. This diagram of a "collapse, pursuit, and plunge scenario", taken from Ruffini and Wheeler (1971) is, as they emphasize, schematic only, and must undergo great alterations for modest changes in the values of the mass and angular momentum of the original white dwarf core. For the very different scenario for the formation of a neutron star or a black hole by accretion in a double-star system, see van den Heuvel (1974).

A great difference will be expected between one scenario and another as to whether fragmentation will or will not occur, as to the number of fragments when fragmentation does occur, and as to the number of distinct stages run through by the system in settling down to a single final compact object. In selecting between one such motion picture run and another the decisive parameters are two in number: the mass M and angular momentum J of the system before collapse. May the day soon come when for at least three values of M, and three values of J (or J/M^2), or a total of nine representative cases, one can give to 20 percent accuracy predicted curves for gravitational wave amplitude and polarization in their dependence upon viewing angle and time. Not until then will one know what is the characteristic "signature" that gravitational collapse writes with the pen of gravitational radiation. Not until then will one have the information required to design a detector of this radiation of maximal astrophysical relevance.

Weber bar detectors of present sensitivity we expect to detect collapse processes that take place within this galaxy, with an average time between events of the order of fifty years. The second generation of detectors, now under construction at Stanford, Baton Rouge, and Rome (Everitt, Fairbank, and Hamilton, 1972; Boughn et al., 1973), can be envisaged to be followed by a third generation of even greater Q-value and sensitivity, able to detect collapse anywhere in the Virgo Cluster of galaxies, perhaps of the order of once a month (MTW, 1973). The signal comes through all overlying layers of matter and all surrounding clouds of gas and dust. No other observational means offers itself to acquire more decisive insights into the dynamics of collapse, except neutrinos (Chudakov, Ryajskaya and Zatsepin, 1973; Kotzer, Lord, and Reines, 1975) and gravitational radiation. Only by continuing and intensifying present work, as well on collapse hydrodynamics as on gravitational wave detectors, can we win that new understanding of collapse, and that new power over gravitational radiation, that stand within our grasp.

Mathematical tools are under development that promise new power to analyze hydrodynamic and geometrodynamic complexity individually and in combination. The work of DeWitt and Smarr, reported by L. L. Smarr, on computer analysis of the collision between two black holes, step by step identifies and solves problems of stability, of accuracy, of checks, of the best balance between elliptic initial value equations and hyperbolic dynamic equations, of optimal "slicing" of spacetime, and —perhaps most difficult of all—optimal display of the results. Preparing the algebra prior to computer coding, whether for hydrodynamics or relativity or both, is a heavy task in itself, towards the lightening of which much has already been done (summarized, for example, in MTW (1973, p. 342)), and more has been reported at this meeting by H. I. Cohen, O. Leringe and Y. Sundblad in one paper, and by I. Frick in another.

3. BLACK-HOLE PHYSICS

3.1. The X-ray "signature" of a black hole

Calculation of gravitational radiation at the moment of formation of a black hole, and construction of detectors, constitute a continuing investment with a great payoff expected around the corner. In contrast, calculation of the X-radiation from the accretion disk around an already existing black hole, and lofting X-ray telescopes to detect this X-radiation, make up an investment of equal vision, whose first marvellous payoff is already in hand. This is not the place to review the evidence that the X-ray object Cyg X−1 is a black hole. For that it is enough to refer to the report of M. Rees at this conference, and reports by him and by R. Giacconi at another conference (chapter in Debever (1974)), as well as the book of Giacconi and Gursky (1974). It is more useful here to recall a central objective still to be achieved. Can one by calculation or observation or both establish the X-ray "signature" of a black hole? Can one identify some characteristic feature of the spectral distribution, or of the fluctuations in time, or of the radiation from some revolving and then inward plunging hot spot, or of the polarization, that will distinguish easily and unambiguously between a black hole and a neutron star, without need for a discrimination by way of mass? A. P. Lightman and D. M. Eardley, by analyzing in their report some of the instabilities expected in the inner part of the accretion disk, give us a most helpful impression of the problems that will have to be solved before we can arrive at a reliable X-ray "signature" for a black hole.

3.2. Superradiance

The order of history was not first the quantum radiance of a black hole and then superradiance, but first superradiance (Zeldovich, 1971, 1972b; Misner, 1972) and then quantum temperature and quantum radiance (Bekenstein, 1972, 1973, 1974; Hawking, 1974, 1975). Therefore it is of interest that much more detailed calculations are now available, as reported here by Press in his review and Teukolsky in his written report (see also Press and Teukolsky, 1973; Teukolsky and Press, 1974), than one ever had before. In this superradiance, one is concerned with the comparison between the radiation that comes out and that which went in. For this reason it is especially interesting to know that a charged black hole will send out electromagnetic radiation when illuminated with pure gravitational radiation. The original static electric lines of force (nonrotating black hole) or static electric and magnetic lines or force (rotating black hole) are set avibrating. Thereby electromagnetic waves are generated and run out into space. It is evident that the black hole serves as a convertor to transform incoming gravitational radiation to outgoing electromagnetical radiation. The cross-section for this process will evidently be of the order of the geometric dimensions of the black hole itself in the extreme case—very far from astrophysical realization—of a black hole with charge comparable to the Misner–Nordstrøm limit and angular momentum comparable to the Kerr limit. Several papers in the literature deal with aspects of this conversion process (see Table 1; also Johnston, Ruffini, and Zerilli, 1973, 1974; Chitre, Price, and Sandberg, 1973, 1975; Johnston, 1974; Boughn, 1975; Chitre, 1975; Weinstein, 1975). Reports given at this conference add to our understanding. However, we are still far removed from a detailed knowledge of the conversion cross-section in its dependence upon angle and frequency.

3.3. Scattering by a black hole

For the simpler process of scattering of radiation by a noncharged rotating black hole, or the even simpler case where there is no rotation (Matzner, 1968), one has reasonably detailed information on the angular distribution in the geometrical optics limit of high frequency. A special case of the geometrical optics analysis requires special considerations—that in which one looks at radiation scattered straight back toward the source. When close up, the observer will see a whole series of rings of brightness corresponding to photons that have made half a loop, one and a half loops, two and a half loops, and so on, around the black hole. However, Mashhoon (1973a, 1974) showed that for the ideal case of the 180° observer, like the source, removed to infinity, the contributions to the back scattered amplitude from the several parts of the several rings all add together destructively and one gets zero back scattering.

In the opposite limiting case of low frequency and long wavelength, the Rayleigh analysis of scattering of light by a small particle invites us to consider the scattering cross-section as proportional to the fourth power of the frequency and the square of the static polarizability. What is the polarizability of a black hole? Or is this quantity zero? In that event one would expect to have to determine the next relevant coefficient beyond the static polarizability in an expansion of the radiation moment in powers of the frequency. Or is the polarizability of a black hole even a well-defined quantity, when due account is taken of the slow fall-off of the gravitational influence at great distances? Much is to be found on the response of a black hole to static electric fields in the analyses of Cohen and Wald (1971) and Hanni and Ruffini (1973); but to answer these questions about polarizability and scattering would seem to require an additional investigation.

The reports that we have heard at this conference of new analytical methods and new computer approaches increases our assurance that we will soon know enormously more than we do today about the scattering power of black holes of a variety of types and for a variety of radiations for the whole range of frequencies and angles.

3.4. "Activity": tidal kick

The transfer of energy from a rotating black hole to its surroundings, otherwise known as "activity", is not limited to superradiance. Transfer of energy by the Penrose–Christodoulou process (Penrose, 1960; Christodoulou, 1970) was the original example of activity and remains an instructive mechanism. In its extreme form, it envisages the split up of an incoming mass into two fragments propelled apart at substantial velocity, one going into a "positive root state" of negative energy (Fig. 2) ("indebtedness" or "due-bill" state) from which it is captured into the black hole thereby decreasing the total mass energy of the black hole. The other fragment flies off to infinity with a greater content of mass-energy than the incoming object brought with it in the first place.

Fascinating as the Penrose–Christodoulou "due-bill" process is as a matter of principle, does it have any significant physical application? One can imagine the decay of an incoming elementary particle so well timed and so energetic that the particle flies apart at the right place and in the right direction. However no one sees the slightest prospect of a real astrophysical mechanism that would lead to any significant number of such carefully tailored elementary particle transformations. Nobody has ever come up with a process of other than elementary particle origin that would drive apart two fragments at a speed that is a substantial fraction of the speed of light, yet it has been pointed out more than once (Bardean, Press, and Teukolsky, 1972; Starobinsky and Churilov, 1973; Wald, 1974) that it is necessary and sufficient to have a separation velocity of half the speed of light or greater to make the "due-bill" process work in its extreme form (rest *plus* kinetic energy out greater than rest *plus* kinetic energy in).

Do the jets shot out in opposite directions from quasistellar objects signal the existence of a new and exciting process that would meet this $v > c/2$ criterion for the "due-bill" process? The two centers of radio-emission at the ends of the two jets are known to move apart with a substantial fraction of the velocity of light. Moreover, Blandford and Rees (1974, 1975) have given a perfectly straightforward astrophysical explanation for the jet ejection process. They consider relativistic plasma generated in an active galactic nucleus. They envisage this plasma as confined by a surrounding cloud of gas. This cloud has some rotation. By reason of this rotation it is thinner at the poles than elsewhere. In these two domains of lowered tamping the plasma pushes its way through the confining gas by the much-studied "snow-plow" mechanism. The plasma emerges in "twin-exhaust" jets. They have relativistic velocity. This twin-exhaust process is remarkable in the magnitude of the velocities it imparts and the masses it propels ($10^5 M_\odot$ to $10^{10} M_\odot$). It can supply enough *velocity* for the "due-bill" mechanism, but in two other respects it appears unworkable:

1. The twin-exhaust leaves behind the bulk of the mass of galactic-nucleus-plus-confining gas. In this respect it gives something far short of the proper *two*-body explosion that was envisaged by Penrose and Christodoulou.

2. A galactic scale of distances marks the separation of the two blobs of matter emerging in the double exhaust; yet this gigantic machine has to fit into the close neighborhood, the ergosphere, of a rotating black hole. Such a black hole would have to be enormous in size (10^5lyr or 10^{23}cm or more) and preposterous in mass (10^{23}cm or 10^{51}g or $10^{18} M_\odot$ or more) to accommodate the twin-exhaust machine. This is not the way to make the due-bill process astrophysically relevant!

To make break-up in the ergosphere of a black hole astrophysically relevant, it would seem necessary to give up the "all or nothing" philosophy. It is not required that one should gain everything in order to gain something. It is not required that the rest plus kinetic energy of what goes out should exceed the rest plus kinetic energy of what goes in. It is enough to impart astrophysical interest to the process that the kinetic energy of what goes out should exceed by a good margin the kinetic energy of what goes in. Moreover, for this gain to occur, it is not at all required that the two fragments separate with a velocity equal to any substantial fraction of the speed of light, nor is it even necessary for the fragment left behind to go into a state which has negative energy as judged by the common market standard of energy as measured at infinity. Neither is it necessary to invent a mechanism to drive the fragments apart. One already exists; tidal disruption (Wheeler, 1971a; Mashhoon, 1971, 1973b, 1975).

It might at first sight seem that the velocity of separation brought about by tidal disruption is too small to bring about ejection of a fragment with any substantial velocity. This would certainly be an appropriate assessment of the situation if one depended on the Penrose-Christodoulou mechanism in its extreme form for propulsion. However, in actuality, one depends neither on the due-bill mechanism nor on being inside the ergosphere nor even upon having a black hole. By way of illustration it is enough to recall the familiar example of a spaceship that swoops in hyperbolic orbits with velocity V close to a planet and blows apart into two modules with separation velocity v. The gain in kinetic energy-per-unit-of-mass of the emergent module is of the order Vv and can be very big even when v itself is modest. Of course, either V or v or both must be relativistic if the module is to emerge relativistic (Wald, 1974); the law of conservation of energy is not violated. The gain of the one fragment comes primarily at the expense of the energy of the residual module, which does not retain enough velocity to escape.

Wide though the scope is of this familiar planetary "kick process" it clearly operates at its best when the speed V at closest approach is maximal; that is, as close to the speed of light as one can arrange. No better way to achieve such velocities has ever been proposed than entry of the incoming object into an orbit that takes it in a close loop around a black hole. The conditions achieved are still more extreme when the black hole is rotating at the Kerr maximal rate or close to it, for then, with the appropriate direction of the incoming orbit, the object acquires a substantial additional increment in the effective velocity V from the frame-dragging. The only additional requirement is breakup in the most favorable direction. Mashhoon's calculations on the total deformation of a rotating star falling into the ergosphere of a sufficiently large black hole show that this directionality does not have to be engineered. It can come about automatically. In an appreciable fraction of all infall scenarios, the calculated kinetic energy imparted in the form of increasing tidal deformation would suggest velocities of ejection (not yet calculated!) of hundreds of km/sec.

3.5. White holes

At a meeting, less than a year ago (International Astronomical Union Symposium No. 64, Warsaw, September 1973; DeWitt-Morette, 1974) Zel'dovich marshalled the long known arguments against the existence of any such object as a "white hole" and pointed out that any such object, for example, one of solar mass, even if formed in the primordial big bang, will be expected to break up at a time of the order of 10^{-5} sec. At the present conference, Eardley has given us a new reason to drop the idea of a "white hole" and with it the word itself. His considerations focus on the energy that would be transported out of such an object if it existed and on the drastic modification that would be produced by this energy flow in the originally presupposed geometry.

This review of some of the aspects of black hole physics including X-ray signature, "activity" in general and superradiance in particular, "conversion" of gravitational radiation into electromagnetic radiation, simple scattering, and "tidal kick", is enough to suggest that 90% of the detailed work of analysis of this physics, in all its ramifications, still lies ahead of us—a reminder, if one were needed, of the exciting future of this topic.

4. STATISTICS

A review by Professor Ehlers of progress in relativistic statistical mechanics, thermodynamics, and continuum mechanics recalls anew how successful and how comprehensive are the well-established principles of these long-cultivated fields, even
under relativistic conditions. We received a fresh impression of the strength of the
principles of statistical mechanics and the law of increase of entropy with time.
Nevertheless the foundation for this "onesidedness in time" remains to us as mysterious as ever. A wireless antenna loses energy. An atom drops from a higher state
to a lower state. A star pours out radiation. A fast electron that is deviated sends out
an electromagnetic wave and suffers a loss of energy. Heat flows from a hot body to
a cold body. Biological reproduction proceeds forward in time. Memory reaches
backward in time. Moreover, nowhere that we look do we see the arrow of time
reversed. The onesidedness of time is, so far as we can tell, cosmological in its scope.
This circumstance, far from explaining the onesidedness of statistics, has in the
minds of many colleagues even raised the question whether entropy will forever
continue to increase. Yes, today the arrow of time points to the future; but also today,
the universe is expanding. If Einstein is right and the universe is closed, there will
come a time, general relativity tells us, when the universe will stop expanding and
start contracting. Will the arrow of statistical time (as contrasted to the continuous
advance of dynamical time) then turn around and point to the past? Will entropy
then decrease? Stars sop up radiation? All who are then alive have a sense of time
opposite to our own? In this event, far from seeing the universe contracting, they
will see it expanding and see entropy increasing (Wheeler, 1962; Gold, 1967; Davies,
1974).

It is difficult to name a topic of principle in physics on which views are more
divided than on this: Does the statistical arrow of time always point the same way?
Or does a moment come of "dead center" or of "turning of the tide" when this arrow
reverses sense? The liveliness of the issue is perhaps best shown by the fact it is not
lively. Some respected colleagues who suppose entropy forever to increase find
the idea of a turning of the tide impossible to consider let alone accept. Other respected colleagues regard the concept of a tie between statistics and cosmology as
so natural that they find it equally impossible to contemplate any other foundation
for the arrow of time.

Happily there are two signs that the question of a cosmological reverse in the
arrow of time is passing from the domain of opinion to the domain of science. One
is the paper of W. J. Cocke (1967) on the mathematical foundations and logical self-
consistency of "double-ended statistics". The other is the measurement of radiation
reaction carried out by Partridge (1973) and subsequently analyzed by Davies
(1974).

The idea is simple behind the Partridge experiment and most other proposed experiments to look for a premonitory evidence for a turn-around in the arrow of time. The exponential decay with time of a temperature difference or the population of atoms in an excited state or number of uranium nuclei that have not yet undergone decay is replaced according to the calculation of Cocke by a hyperbolic cosine. The minimum in the number occurs at the time of "turning of the tide". The characteristic time scale for the turn-around from exponential decay to exponential increase is governed in the highly idealized and necessarily simplified model of Cocke by the characteristic time of the single elementary process envisaged in his analysis. For a process of long characteristic time scale, the calculated departure from the normally assumed simple exponential decay shows up long before the minimum. To search for this departure or to set an upper limit on it is the objective.

Not radioactivity but radiation damping is the process on which Partridge focused attention. In effect, he ran a telescope backward. He put a source of radiation at the focus and measured the power consumption. He found the drain of radiation reaction to be the same within a fraction of a percent whether he directed the telescope at a nearby absorber or at the darkness of space. Light travelling out into space, simple estimates indicate, will go several 10^{10} years or more before it is absorbed. At that time, according to Einstein's picture of a closed universe, the universe will be in its contracting phase. It is often reasoned that damping here and now depends on the back reaction of the absorber there and then. On this view one might suppose that a reversal in the statistics at the time of absorption would alter the force of reaction observed today from today's standard value. This expectation is contrary to what Partridge observed. A closer examination (Davies, 1974, based on Wheeler and Feynman, 1945) shows that no alteration would be expected unless each atom of the source could see out into the emptiness of space both forward and backward, more after the fashion of something like a bazooka than a traditional telescope. This two-way radiation experiment is yet to be done. Also to be done is any real calculation of what is to be expected on the basis of double-ended statistics, either in this experiment or in observations (Dicke and Peebles, 1962) comparing the effective lifetime for the same decay processes several 10^9 years in the past.

In brief, the cosmological domain of statistics confronts us with fascinating and perplexing issues. It presents us with a deep question of principle. It calls for new observations. It demands penetrating theoretical analysis to plan decisive measurements. It is virgin territory.

5. COSMOLOGY

Matzner's report gave an impressive overview of many of the chief results of cosmology, both observational and theoretical, as we see them today. It ranged from the primordial cosmic fire ball radiation and the "ring of fire" or bright halo that one might expect to see in the otherwise isotropic (Boynton and Partridge, 1973) distribution of this radiation on some models of cosmology (unpublished 1970 lecture of Matzner and other references given in the report of Matzner; see also Ryan and Shepley, 1974; Ellis and King, 1974) to the age-old problem (Peebles, 1971, 1974a) what primordial pattern is required in the original disturbance in density and flow of matter to account for the distribution of galaxies as we see it today (Peebles and Hauser, 1974; Peebles, 1974b,c; Peebles and Groth, 1975); and reaching from the formation of the elements (Fowler and Hoyle, 1964; Wagoner, 1973) to the question whether the universe is open or closed (Gott, Gunn, Schramm, and Tinsley, 1974).

Is the universe open or closed? On no central issue of cosmology is there greater divergence of evidence today. Einstein's philosophical arguments speak for closure. An appreciable body of astrophysical evidence speaks against it.

To determine the so-called deceleration parameter q_0 from source counts is the goal of some of the greatest and most skilful observers of our times. This important measurement nevertheless requires such care in interpretation, demands so many corrections, and is afflicted with such uncertainties that the final number still today leaves the door open to either cosmology.

The quickest way to see that the expansion may be slowing down is still the most elementary. One has only to compare the actual time back to the start of the expansion, a time of the order of 10×10^9 years, as judged from the rate of evolution of stars and clusters of stars, with the apparent, or extrapolated, or Hubble time of $\sim 20 \times 10^9$ years. This is the time it would have taken galaxies to get to their present separations from us, moving with their present separation velocities, with no allowance for the greater velocity in times past. Of course, considerable uncertainties attend both numbers, uncertainties of the order of 30 percent or, conceivably, even more. Even so, it is difficult to find evidence more impressive anywhere else in cosmology for the predicted slowing down of the expansion.

If to fix ideas we take the two numbers, 10×10^9 years and 20×10^9 years, as 100 percent accurate and assume a homogeneous isotropic spherical universe and neglect the pressure and energy content of radiation in comparison to the mass energy of inchoate material ("dust") then Einstein's theory straightforwardly gives all the other illustrative numbers of Table 2. The 30-fold discrepancy between the density of the universe today as called for by these calculations and the density estimated by Oort (1958) gives rise to the well-known "mystery of the missing mass" to which Matzner has already alluded. Of all the evidence for a low density cited by Gott, Gunn, Schramm, and Tinsley (1974) and by Gunn and Oke (1975), none is more impressive than the abundance of primordial deuterium. The sensitivity of the deuterium abundance to density arises as we know from the dependence of the expansion rate on density and from the fact that only a few minutes are required for

TABLE 2

Major Features of the Universe According to Einstein's Theory, as Normalized by Two Key Astrophysical Data, Each believed Uncertain by an Amount of the Order of 30%: (1) the Actual Time, $\sim 10 \times 10^9$ Yr, Back to the Start of the Expansion, as Determined From the Evolution of the Stars and the Elements, and (2) the "Hubble Time", or Time Linearly Extrapolated back to the Start of the Expansion, $\sim 20 \times 10^9$ Yr, that is, the Time Needed for Galaxies to Reach Their Present Distances if They Had Always Been Receding From Us with Their Present Velocities (Adapted from MTW, 1973)

Illustrative values all derived from	
Time from start to now	10×10^9 yr
Hubble time now	20×10^9 yr
Hubble expansion rate now	$49.0 \dfrac{\text{km/sec}}{\text{megaparsec}}$
Rate of increase of radius now	0.66 lyr/yr
Radius now	13.19×10^9 lyr
Radius at maximum	18.94×10^9 lyr
Time, start to end	59.52×10^9 yr
Density now	14.8×10^{-30} g/cm^3
Amount of matter	5.68×10^{56} g
Equivalent number of baryons	3.39×10^{80}

primordial neutrons to decay to protons. Unhappily less satisfactory than this theoretical side of the story is the observational evidence. Determinations of deuterium abundance are made by looking at the absorption of light in interstellar space on its way from a star to the telescope. Only a few such determinations have been made. No one knows how representative are the samples of gas intervening nor how much they have been altered between primordial times and today by cosmic ray impacts and contaminated by ejecta from stars and supernovae.

New light on missing mass comes from the recent work of Ostriker and Peebles (1973) and Ostriker, Peebles, and Yahil (1974). They give arguments from galactic stability that the mass of the typical galaxy must be of the order of 3 to 20 times as great as one has previously estimated. They give reasons to believe that this matter is in the form of stars of modest mass and very low luminosity. Happily for the subject, the direct observational search for this "halo" is now underway. It is difficult to name any single issue in all of astrophysics which draws together a wider variety of important investigations than those going on today concerning in one way or another the mystery of the missing mass.

It has often been suggested that one should make a direct geometrical determination of the curvature of space in the large. In this way, the hope has been expressed, one could find out whether the universe is closed or open even prior to a reliable determination of the average mass density of the universe. More than one calculation has been made and reported (Misner, Thorne, and Wheeler, 1973) of the apparent angular diameter of an object of standard dimensions (if there be any such) as a function of distance (as defined by red shift). In Euclidean space, a "standard" object has an apparent angular diameter which decreases in inverse proportion to distance. However, when the object is far enough away in an ideal spherical space, it is magnified by a kind of lens effect. Then the apparent angular diameter, rather than decreasing, increases with distance. Moreover, the double radio sources associated with quasistellar objects offer a conspicuous "ruler". If anything, the length of this "ruler" will be shortened in early double radio sources as compared to more recent ones by the greater density at early times of the matter through which the "twin exhausts" (Blandford and Rees, 1974, 1975) have to plough their way. Thus if double radio sources of a sufficiently great red shift were to begin to show an increase in apparent angular diameter, one could hardly do anything but regard this effect as evidence for the predicted lens effect.

A closer consideration shows that the situation is by no means as simple as would be indicated by these elementary considerations. It was already pointed out by Zel'dovich (1964) and by Dashevsky and Zel'dovich (1964) (references to this and the subsequent literature in Press and Gunn (1973)) that the clustering of matter into galaxies, deviation from uniformity unimportant for the question of openness or closure, is vitally important for the focusing process. A spray of light rays that starts at a point, and spreads out as it goes, *continues* to spread out as it travels through matter-free interstellar space, even though the universe itself is contracting. Nothing like the elementary focusing effect takes place. We are indebted to R. C. Roeder in his report at this conference for stressing the difficulties posed by this circumstance for any proposed cosmological test of closure, via measurement of apparent angular diameters as a function of red shift. However, if one hope fades, another brightens. Press and Gunn (1973) show that a cosmologically significant density of condensed objects has high probability to cause a distant point source to be gravitationally imaged into two roughly equal images—an effect with testable consequences.

Nothing is more entrancing to the viewer of scenery than the moment when the mist dissipates and he can see whether the hidden feature was a mountain or a lake. Nothing is more tantalizing today than the mist, here and there thinning out but not yet dissipated, that conceals whether the density of space has the cosmological value or is far less; and whether the universe is closed or is open.

However much these two issues belong to science and however important general relativity is in dealing with them, one cannot forget that this science and tool took its birth in philosophy. On this account, I will be remiss if I do not quote what Einstein himself (1950, pp. 107–108) says about closure,

> "Thus we may present the following arguments against the conception of a space-infinite, and for the conception of a space-bounded, universe: (1) From the standpoint of the theory of relativity, the condition for a closed surface is very much simpler than the corresponding boundary condition at infinity of the quasi-Euclidean structure of the universe. (2) The idea that Mach expressed, that inertia depends upon the mutual action of bodies, is contained, to a first approximation, in the equations of the theory of relativity;... But this idea of Mach's corresponds only to a finite universe, bounded in space, and not to a quasi-Euclidean infinite universe."

In another place Einstein (1934, p. 52) states,

> "In my opinion the general theory of relativity can only solve this problem [of inertia, for a recent survey of which see for example MTW (1973), §21.12] satisfactorily if it regards the world as spatially self-enclosed."

How are we to look at Einstein's arguments today? One view is "empirical": "These are outdated considerations of purely historical interest. One should pay attention to the field equations alone. One should forget everything else as 'theology'. An open universe is just as conceivable as a closed universe. Only observation can decide".

Another view takes Einstein's considereations quite seriously: "It is only a matter of historical accident that the demand for closure was not stated in an equation as mathematical in appearance as the field equation itself". On this view, the condition for closure is essential in formulating the "initial value data"; and the "initial value data" are essential in formulating what general relativity is all about. There are alternatives to closure as part of the formulation of the initial value data but no alternative so simple as closure. It is one alternative to postulate asymptotic flatness at infinity. It is another alternative to postulate more particularistic data on some closed 2-surface that bounds the 3-geometry embraced in the "initial value problem". What kind of data should be given on such a 2-surface? Mathematical tools we have on hand to try to answer such a question, but no slightest hint of any physical consideration that would make this a reasonable route to follow. And asymptotic flatness (see, for example, the "heirarchical cosmology" of Alfvén and Klein (1962) and De Vaucouleurs (1971)) makes double difficulties. First, it takes the geometry of faraway space out of physics and makes it part of theology, to be discovered by reading Euclid's bible. It puts us back to the days before Riemann, days when, as Einstein (1934, p. 68) puts it,

> "... space was still, for them [physicists], a rigid, homogeneous something, susceptible of no change or conditions. Only the genius of Riemann, solitary and uncomprehended, had already won its way by the middle of the last century to

a new conception of space, in which space was deprived of its rigidity, and in which its power to take part in physical events was recognized as possible."

Why accept this advance for near space and undo it for faraway space? Second, "asymptotic flatness" leaves one lost. How can anyone even define the idea of asymptotic flatness? According to the most elementary considerations of quantum theory there is no such thing as *the* geometry of space. Geometry is not deterministic, it is probabilistic. There is a probability amplitude $\psi(^{(3)}\mathcal{G})$ for this, that, and the other 3-geometry. If a given 3-geometry occurs with appreciable probability amplitude so does any other 3-geometry that differs from the first by an amount of the order $\Delta g \sim L^*/L$ in a region of order L. Thus, no matter how "far away" one goes, one can never arrive at a place where the fluctuations have less than standard strength. Difficult as it is under these circumstances to define "far away", it is even more difficult to define "asymptotic flatness". No one has ever proposed a reasonable definition of "average geometry" such as would seem to be a prerequisite for defining asymptotic flatness. It would be contrary to the whole concept of general relativity as well as to its mathematical methodology, to try to "step out of the manifold" to do any averaging. Under these circumstances, it is difficult not to rate the concept of "asymptotic flatness" as a "concept without a concept"; and difficult to see where else one can turn for a satisfactorily sharp boundary condition compatible with quantum fluctuations, except to *closure*.

Other cosmological issues besides closure attracted attention in this conference, among them homogeneity. Liang discussed the dynamical evolution of primordial inhomogeneities in a model universe. We have only to look at the collapse of a star to a black hole to see one ultimate in this evolution; but Liang considers the problem more generally. King reported on considerations of his own and earlier considerations by Matzner, and by Ellis (1973, 1974) and by Ellis and King (1974), summarized in the question, "Was the big bang a whimper?". The Taub–Misner transition from the closed Friedmann-like Taub geometry to the open NUT geometry has long been known (Misner and Taub, 1968) to take place in a continuous way. However, its continuity is achieved only at the cost of having certain classes of world lines spiral round the universe in the final stages of its collapse to tighter and tighter packing. Thus the presence of the slightest "real matter" builds up an ever-increasing density (in this connection, see also Penrose (1973)). As it goes to infinity, this density destroys the relevance of the model with which one started. One returns to something closer to a Friedmann cosmology with a Friedmann singularity. Friedmann singularity; Misner–Taub or whimper beginning or ending; singularity characterized by the general mixmaster oscillation, whose phase, amplitude and orientation of principal axes varies from point to point (Lifshitz and Khalatnikov, 1963a,b, 1970; Belinsky and Khalatnikov, 1969a,b, 1970; Khalatnikov and Lifshitz, 1970; Belinsky, Khalatnikov and Lifshitz, 1970; Belinsky, Lifshitz, and Khalatnikov, 1971; see also Eardley, Liang and Sach, 1972 for relevant considerations); analytic behavior of singularities (Schmidt, 1971, 1974); causal analysis of singularities (Hawking and Ellis, 1973); when will all these points of view be assimilated into a larger, unified, picture of the earliest and the latest phases of cosmology? Not until then, and perhaps not even then, we can believe, will we be able to state with confidence which kind of cosmological singularity makes physical sense and which does not.

5.1. Is the universe Copernican?

Is the universe Copernican? Are conditions in the vicinity of this galaxy quite similar to those in galaxies elsewhere in the universe? Can we exclude the possibility that densities, for example, are several-fold lower—or several-fold higher—in our cluster of galaxies than in the average of galaxies at greater distances? This is a cosmological issue of great interest; but on it, unhappily, the evidence is too scanty to permit a reliable conclusion today. The isotropy of the distribution of galaxies as seen from our own vantage point is well known and carefully documented (Peebles and Hauser, 1974; Peebles, 1974b,c; Peebles and Groth, 1975). However, to see that the universe is isotropic from our vantage point by no means allows one to conclude that the universe is homogeneous. One has to add, and one is accustomed to adding, the Copernican assumption that things would not look very different from any other vantage point of the universe; only then does it follow that the universe is homogeneous; thus

$$\begin{pmatrix} \text{isotropy} \\ \text{here} \end{pmatrix} + \begin{pmatrix} \text{Copernican} \\ \text{assumption} \end{pmatrix} \rightarrow \begin{pmatrix} \text{isotropy} \\ \text{everywhere} \end{pmatrix}$$

$$\begin{pmatrix} \text{isotropy} \\ \text{everywhere} \end{pmatrix} \rightarrow \begin{pmatrix} \text{homo-} \\ \text{geneity} \end{pmatrix} \tag{24}$$

It is disturbing to be reminded how much of astrophysics depends on pure assumption—the Copernican assumption. The Copernican assumption as it is understood today is very different from what was given to us by the great pioneer whose 500th birthday we celebrated last year. Then, it took man out of the center of the universe and installed the sun at the center of the solar system. Today it rules out anything special about the physics in this part of the universe. One check on this point has been carried through to marvelous precision; Bahcall and Schmidt (1967) conclude that the fine structure constant for galaxies 2×10^9 light years away ($z = 0.17$ to 0.26) agrees with the fine structure constant here to three parts in 10^3 or better. But the Copernican assumption as commonly accepted today carried with it certain overtones and assumptions which are such common property that they are not always explicit stated. Among these is the idea that the universe is a gigantic machine and life on this planet of this solar system of this Milky Way is, by comparison, an accident and an unimportant accident at that. From this assumption it then follows that there is nothing special about this galaxy or this galactic cluster. From saying that this is no special place, it is a small step to say that isotropy of the heavens as seen from here implies isotropy of the universe as seen from any other location—and hence the universe on the scale in question must be homogeneous.

Life, instead of being accidental, is central according to a very different view of the universe which goes back to the Greeks and has received renewed attention in our own time. In the thought of Paramenides of Elea ($\sim$ 500 B.C.), "What is, ..., is identical with the thought that recognizes it"; or in the later formulation of Berkeley (ca. 1710), "No object exists apart from mind"; or, as tentatively restated for the purposes of a recent reanalysis (Patton and Wheeler, 1975), "[We look for] a guiding principle of 'wiring together' past, present, and future that does not even let the universe come into being unless and until the blind accidents of evolution are guaranteed to produce, for some non-zero stretch of time in its history-to-be, the consciousness, and consciousness of consciousness, and communicating community, that will give *meaning* to that universe from start to finish." It has no sense even to talk of a universe, Dicke (1961) in effect argues, unless there is a mind to be aware of it. But to have a mind requires life. And to have life requires heavy elements. To produce heavy elements, in turn, requires thermonuclear combustion. Thermonuclear combustion demands $\sim 10^9$ years of cooking time in the interior of a star. But to have a universe that lives $\sim 10^9$ years in time, according to general relativity, requires a universe with $\sim 10^9$ light years of space. So why, on this view, is the universe as big as it is? Because we are here!

The ridiculous disproportion between the means employed, the universe, and the end achieved, life and mind in one place in that universe, is ridiculous, on this view, only because attention is focused on distance and on volume instead of the relevant parameter, time. The important point is *time* enough for the evolution of consciousness. One goes back again to time as a central point when one speaks of improving the "cost effectiveness" of the process. Look at the waste: $\sim 10^{11}$ stars per galaxy, and $\sim 10^{11}$ galaxies in the universe and all that matter to permit the development of consciousness of that universe on one planet! Institute some savings! Don't chop the project too drastically. Cut down the amount of matter and the size of the universe, not to the amount required for one star, but to the amount required for one galaxy, or $\sim 10^{11}$ stars. Why is not that still more than enough? In amount of matter, yes! In number of stars, yes! But in amount of time, no! The total time from big bang to collapse is cut down from 10^{11} years to 1 year. There is not time enough to produce even one star let alone a planet or life or mind. The purported improvement in the "cost effectiveness" of the program has ended up with the collapse of the program. From this point of view it is far from obvious that there is any extravagance in the scale of the universe.

It would be out of place here even to mention these cosmogonic issues if they did not bear on the cosmologic issue of homogeneity. Homogeneity would not necessarily seem natural if the sole requirement is that the universe give rise to consciousness, and consciousness of consciousness and communication and meaning, in some limited place and for some limited time in its history-to-be. More probable would seem a distribution of density which, by favoring evolution in one region, reduced the "cost" of the system as a whole. It is not necessary to know how to define "cost" nor how to calculate it in order to raise this question. Nor is it necessary to raise this question in order to agree that the Copernican assumption of "no special location for us" as it is employed in cosmology is, so far as density is concerned, very far indeed from having been verified or even tested. We can hope that advances in astrophysics will suggest some observational means to get at this fascinating question.

5.2. How do non-communicating domains of the universe communicate?

Turn from the question of homogeneity of density to the identity of physical laws in the various parts of the universe for another cosmological issue. In light that has undergone a red shift of

$$z = \frac{\lambda_{\text{reception}} - \lambda_{\text{emission}}}{\lambda_{\text{emission}}} = 3 \tag{25}$$

on its way to us from a distant galaxy, we see evidence for the same atoms and the same spectrum as well as the same fine structure constant that we have on earth. Yet on the basis (Patton and Wheeler, 1975) of the illustrative cosmological parameters of Table 2, the light received now, 10×10^9 years after the big bang, was given off by that galaxy only 1×10^9 years after the big bang. At that time it had not yet received a single signal from us. It was destined to have to wait another 1×10^9 years (2×10^9 years after the big bang) before it could get this first information from us about what the laws and constants of physics are here. How did it find out? On this great question, we are still in the dark. Misner (1968, 1969) made a valiant attempt to provide an explanation with his picture of the primordial "Mixmaster oscillations" of the universe which provided opportunity for signals to travel all around at the very earliest days. That picture has had to be abandoned (Doroshkevich, Lukash, and Novikov, 1971; Chitre, 1972) and we have nothing to take its place nor any suggestion of any replacement at our conference.

5.3. How do galaxies form?

What initiates galaxy formation? On few puzzles has the intensity of effort risen more strikingly. Few puzzles have so long continued to defy explanation. Neither in the observational data on the distribution of galaxies nor in the theory of gravitational agglomeration is there any evidence for a well defined length with the dimensions of a galaxy, a cluster of galaxies, or a star cluster (Peebles, 1974a,b). What then governs the size of these large scale features of the universe? It is no explanation to "feed in by hand" some assumed initial spectrum of perturbations in density and in flow. This circumstance is a renewed reminder that physics deals with laws of motion and initial conditions. We have gone far toward understanding the laws. We know no more than Laplace what sets the initial conditions.

Cosmology is laced with great issues, but also lighted up with great achievements, ranging from quasistellar objects to neutron stars and from radio sources to Cyg X-1. No report at this conference does more than Matzner's to convince one that physical cosmology is a great and growing subject, the present scope and solidity of which were beyond imagination 25 years ago. The splendid present-day orchestration of new technology, new observing techniques, new young talent, and new powers of analysis surely promises even greater advances in the coming decades.

6. TESTS OF RELATIVITY

Nowhere has there been a greater increase in precision of observational tests of general relativity than in the bending of "light" by the sun, thanks especially to radiowave interferometry. Nevertheless, we may have still to wait for several more years of this ongoing enterprise of successive improvements before we achieve the better than 1 percent precision that we all hope for so much to resolve questions now unresolved.

Other ways to test theories of gravitation in the solar system have been reported here by Richard. We have also heard at this conference new proposals to improve traditional tests and proposals for new kinds of tests. Among these are radar search for the Lense–Thirring rotation (Jaffe, Miller, and Shapiro), search for the second-order gravitational red shift (Jaffe), search for the gravitational spin-spin interaction (Jaffe), the influence of earth structure on the Lense-Thirring precession of a gyroscope (Teyssandier), and ideas about interaction of gravitation with superconducting matter (Halpern).

If we look upon the bending of "light" by the sun as a measurement whose precision is just at present increasing dramatically, we also have to note that the significance of that measurement is subject to real question because of the possibility that the sun is oblate (Dicke, 1964). On this account, great interest attaches to the independent method of getting at the solar oblateness recorded at this conference by Hill. In whatever way the disagreement is ultimately resolved between the two very different results for the flattening of the sun, it is clear that the work of Dicke and Hill promises to open a new chapter in our understanding of the sun. Is it too much to hope that out of this work we will someday come to understand why the high-energy, temperature-sensitive, fraction of the neutrinos from the sun carries a so much lower flux than we had previously expected (Davis, 1972; Bahcall and Sears, 1972; Hill, McCullen, Brown, and Stebbins, 1975)?

7. FROM EXACT SOLUTIONS TO ALL SOLUTIONS?

Kinnersley's comprehensive and systematic survey of what we know about exact solutions of the equations of general relativity invites the question whether there does not exist some magic algorithm that will give *all* of the solutions of the vacuum field equations. Why do I still have hope that such an algorithm will be discovered? Did I not bet Rainer K. Sachs $5 at Les Houches in 1963 that by 1973 one would know how to get all vacuum solutions? And did I not send a last minute cable to Roger Penrose asking if he could see a way to save me? And in the end did I not have to pay?

It may give some impression of the continuing allure of the hoped-for algorithm to point to what success meant in the theory of minimal surfaces. There, as in general relativity, the equation is nonlinear:

$$\left[1+\left(\frac{\partial z}{\partial x}\right)^2\right]\frac{\partial^2 z}{\partial y^2} + \left[1+\left(\frac{\partial z}{\partial y}\right)^2\right]\frac{\partial^2 z}{\partial x^2} - 2\frac{\partial z}{\partial x}\frac{\partial z}{\partial y}\frac{\partial^2 z}{\partial x\,\partial y} = 0 \qquad (26)$$

The solution proceeds as follows (Douglass, as summarized by Bers (1952); see also Darboux, 1941): Take any real plane algebraic curve C in the (s, t) plane, $s = s(\lambda)$ and $t = t(\lambda)$, and construct the Abelian integrals

$$x = \text{Re} \int \frac{2s}{1+s^2}\, itds \qquad (27a)$$

$$y = \text{Re} \int \frac{1-s^2}{1+s^2}\, itds \qquad (27b)$$

$$z = \text{Re} \int t\, ds \qquad (27c)$$

Now treat s as a complex number, $s = \alpha + i\beta$. Then the functions

$$x = x(\alpha, \beta)$$

$$y = y(\alpha, \beta)$$

$$z = z(\alpha, \beta) \qquad (28)$$

satisfy the original minimal surface equation.

The intrinsic simplicity of the geometry of a minimal surface does not show in the non-linear partial differential equation (26), but does come through in the simplicity of the final solution (27). Einstein's equation is also nonlinear, but it expresses even greater geometric simplicity (Cartan 1928, 1946; MTW, 1973). That this simplicity shows up more conspicuously when we go from the realm of real numbers to the realm of complex numbers is becoming more and more apparent. In only a few years we have progressed to seeing the Kerr solution derived from the Schwarzschild solution by replacing the real mass m by the complex quantity $m + ia$, and from there to the still more comprehensive picture summarized for us here by Newman. We also have more and more results from Penrose (up-to-date summary in Penrose (1975)) and his colleagues in the theory of twistors which suggest that some great new insight lies around the corner. Are we indeed destined someday to have an automatic way to construct all the solutions to Einstein's vacuum field equation?

8. THE DYNAMICS OF GEOMETRY

Each passing year instructs us both how much we learn about physics from general relativity, and how little. Physical law expresses itself through group theory and symmetry; but group theory and symmetry hide the machinery beneath that physical law.

The search for the most illuminating statement and derivation of Einstein's field equation did not end with his death. No statement, we know, shortens more the jump from the classical theory to the quantum theory than the Hamilton–Jacobi formulation. It deals with the simplest object that one can have, short of the wave function itself: the phase of the wave function. For a particle moving in one dimension under the influence of a potential $V(x)$ with the specified energy E, the phase is

$$S(x, t) = -Et + \int_0^x \left[2m\bigl(E - V(x)\bigr)\right]^{1/2} dx. \tag{29}$$

Of course, this phase, like the wave itself, is spread all over (x, t) space. Not the slightest resemblance of anything like the classical path is to be seen. For that, one considers not one wave but the superposition of two waves of slightly different energies. The two waves interfere destructively almost everywhere. The region of constructive interference is marked by the equality.

$$S_E(x, t) = S_{E+\Delta E}(x, t) \tag{30}$$

(condition to determine the world line $x = x(t)$); or, more compactly,

$$\frac{\delta S_E(x, t)}{\delta E} = 0. \tag{31}$$

How this condition of "constructive interference" yields the full story of classical motion is too well known to bear spelling out here. Neither is this the place to recall in detail the analogous discussion of the condition of "constructive interference" as applied (Gerlach, 1969) to determine the dynamics of geometry in all its fullness.

It is enough to recall that the phase of any individual wave is expressed in the form:

$$S = S(^{(3)}\mathscr{G}) \tag{32}$$

This phase is defined all over "superspace" (Wheeler, 1964, 1968; Fischer, 1970; Bers, 1970; Wheeler, 1970; MTW, 1973), the infinite-dimensional manifold, each point of which represents one spacelike 3-geometry $^{(3)}\mathscr{G}$. Thus the whole story of the dynamics of geometry of geometry follows from the simple Hamilton–Jacobi equation satisfied by S (Peres, 1962),

$$g^{-1/2} \left[\tfrac{1}{2} g_{pq} g_{rs} - g_{pr} g_{qs} \right] \frac{\delta S}{\delta g_{pq}} \frac{\delta S}{\delta g_{rs}} + g^{1/2\,(3)} R = 0. \tag{33}$$

It is all very well to have the dynamics of geometry thus briefly epitomized; but why this law and why this Hamilton–Jacobi equation? Why not some other law, some other Hamilton–Jacobi equation? Deser and Goldberg have given us accounts of many of the many ways in which we can look at the dynamics of geometry; but we come back always in the end to the question, why this law, why not some other? It provides only a partial answer to our question to turn back to Hilbert's famous paper (Hilbert, 1915). He derived electrodynamics and vacuum geometrodynamics and the combined theory by postulating the simplest action principle that depends on a 4-dimensional vector field,

$$A_\alpha \tag{34}$$

or on a 4-dimensional metric field,

$$g_{\mu\nu} \tag{35}$$

or on the combination of the two. But why the simplest action principle? Why not some one of the thousand and one alternative action principles that contain these two fields in some other invariant combination? "Imbeddability" is the new and magic and beautiful answer that Hojman, Kuchař, and Teitelboim (1973) (see also Teitelboim (1973a,b) and Kuchař (1973, 1974)) give to this old question. They invisage a 3-vector field

$$^{(3)}A_i \text{ (and its conjugate momentum)} \tag{36}$$

or a 3-tensor field

$$^{(3)}g_{jk} \text{ (and its conjugate momentum)} \tag{37}$$

or both (and their conjugate momenta). Whatever the dynamic law that governs the evolution of these fields with time, as time is pushed forward from the spacelike hypersurface σ_1 to the spacelike hypersurface σ_2 that law must give the same result for the dynamic variables whether this hypersurface is pushed forward first more rapidly on the "right" and then more rapidly on the "left"; or first more rapidly on

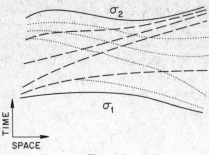

Figure 4

The "history of deformation" indicated by the dashed hypersurface leads from initial-value hypersurface σ_1 to final-value hypersurface σ_2. So does the history indicated by the dotted hypersurfaces. The physics on σ_2 resulting from a complete specification of the initial value data on σ_1 must be independent of the history one chooses to integrate along in passing from σ_1 to σ_2 via the Hamiltonian equations of motion. This heavy but simple requirement suffices to fix the form of the Hamiltonian both for the dynamics of a vector field (giving Maxwell theory) and for the dynamics of the 3-geometry itself (giving Einstein's geometrodynamics) (Hojman, Kuchař, and Teitelboim, 1973).

the "left" and then more rapidly on the "right". If the conditions obtained at σ_2 (by forward integration of the field equations) depended upon the choice of "history" adopted in proceeding from σ_1 to σ_2, then the history of the fields could not be imbedded in any single space-time manifold. Imbeddability would be lost. No one has ever seen a simpler or more compelling way to state the requirements that lead to physics as we know it.

The "group" of deformations of a spacelike hypersurface provides the mathematical framework for the physical considerations of Hojman, Kuchař, and Teitelboim. This is in keeping with the spirit of physics as it has developed in our time. Whether one is concerned with solids or elementary particles, with molecules or nuclei, one finds that the simplest formulation of physical law is one that draws most directly on group theoretic and symmetry considerations (see, e.g., Dyson, 1966; Michel, 1970, 1974).

Unhappily, it is also true that group theoretic and symmetry considerations hide from view any sight of the underlying machinery (see, e.g., Dürr, 1969). The very arguments of group theory that ascribe two, and only two, elastic constants to a homogeneous isotropic medium, give not the faintest clue that these elastic constants are to be found by adding the second derivatives of a multitude of very complicated potential energy curves that describe 100 different atomic and molecular bonds. Even more deeply hidden from view is the fantastic underlying simplicity of the solid: A system of electrons and positively charged nuclei governed by Coulomb's law and Schrödinger's equation and nothing more. Sakharov, following the analogy of elasticity (Sakharov, 1967; developed more fully in Zel'dovich and Novikov (1971)) proposes to regard the constant of gravitation as measuring the "metric elasticity of space", the resistance that space puts up to being bent again following the model of the solid. He proposes to regard the source of this resistance to bending as the zero point energy of particles and fields. However, can we not look in our mind's eye beyond Sakharov to a day when we shall see both geometry and particles as built on something far simpler than either, call it "pregeometry" or call it what one will?

How shall we make progress into this unknown territory? Perhaps it is better to ask how not to make progress! A hundred years of the study of elasticity revealed nothing about atomic and molecular bonds. A hundred years of study of the chemistry of atoms and molecules revealed nothing of Schrödinger's equation. The direction of progress was not "down" but "up", from Schrödinger's equation to atomic and molecular potential energy curves; and from these potentials to elasticity. The hundred striking regularities of chemistry did not require for their explanation a hundred laws of physics. One was enough. Likewise 59 years of study of the dynamics of curved space geometry has not explained the existence of particles; and 79 years of the study of particles has not taught what underlies the structure of particles. Not from geometry and particles to an understanding of "pregeometry" is the direction of progress, if history is any guide, but from "pregeometry" to particles and geometry.

What kind of "structure" is one looking for when he speaks of pregeometry?

One view has it that we will work down from cells to molecules, from molecules to atoms, from atoms to nuclei, from nuclei to elementary particles, from elementary particles to partons or quarks, and so on, level after level, world without end! Another view thinks of this uncovering of layers of structure as bottoming out at some finite nth level. On a third view, the layering of structure neither leads on forever nor terminates but of necessity must lead back, by a kind of "Leibnitz logic loop" to the observer himself, seen at last to be involved in an inescapable way in the very structure of that which he observes.

Gravitational collapse, as evidenced in the big bang and in the collapse of a star to a black hole and in the Einstein-Friedmann-predicted collapse of the universe itself, is the crowning argument for the mutability of the physical world (Fig. 5). Nothing argues more strongly than collapse against the existence of an iron framework of laws, structure, and constants that goes on from everlasting to everlasting. And what is everlasting? Even the very ideas of spacetime and time have to be regarded, not as basic ideas but only as semi-classical approximations, we know, when we turn from a classical description of the dynamics of geometry to the superspace quantum description. "Before" and "after" make good sense in everyday discussions but have to be rejected as meaningless terms in the legalistic quantum description of what is going on in physics. As if this were not shock enough for our long established ways of looking at the world, the quantum principle tells us that the observer is more than an observer, he is a participator. In some strange sense this is a participatory universe.

Much as we admire Einstein for what he did to give us the two overarching principles

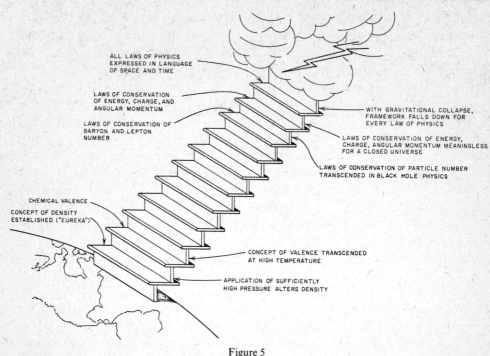

ALL LAWS OF PHYSICS
EXPRESSED IN LANGUAGE
OF SPACE AND TIME

LAWS OF CONSERVATION
OF ENERGY, CHARGE, AND
ANGULAR MOMENTUM

LAWS OF CONSERVATION OF
BARYON AND LEPTON
NUMBER

WITH GRAVITATIONAL COLLAPSE,
FRAMEWORK FALLS DOWN FOR
EVERY LAW OF PHYSICS

LAWS OF CONSERVATION OF ENERGY,
CHARGE, ANGULAR MOMENTUM MEANINGLESS
FOR A CLOSED UNIVERSE

LAWS OF CONSERVATION OF PARTICLE NUMBER
TRANSCENDED IN BLACK HOLE PHYSICS

CHEMICAL VALENCE
CONCEPT OF DENSITY
ESTABLISHED ("EUREKA")

CONCEPT OF VALENCE TRANSCENDED
AT HIGH TEMPERATURE

APPLICATION OF SUFFICIENTLY
HIGH PRESSURE ALTERS DENSITY

Figure 5

"Mutability". Each tread of the staircase of physics registers a law (for example, the fixity of nuclear charge number and mass number). Each riser marks the transcendence of that law (nuclear transmutation in this example), the imposition of conditions so extreme that that law no longer applies. The condition most extreme of all is gravitational collapse. There is no law of physics that has not required space and time for its statement. With the complete collapse of space, the framework falls down for everything one ever called a law (adapted from Wheeler (1973)).

of 20th century physics, relativity and quantum, we have had to abandon his view of reality, "Physics is an attempt conceptually to grasp reality as it is thought independently of its being observed". In direct contrast to that view, Bohr teaches us that an observation is only then complete when it can be communicated to another in plain language; and Wigner (1973) tells us that, "No measurement is complete until the result has entered the consciousness". If the quantum principle has become so much a part of our everyday technology that we have forgotten to be astounded by it, we can always return to the words of Bohr, "If a man does not feel dizzy when he first learns about the quantum of action, he has not understood a word". We can agree that we have not yet seen a point of view that makes clear the necessity of the quantum principle in the construction of the world. Until we do, who can say that he has grasped the first thing about it?

The mystery of collapse and the mystery of "participation" summarize the greatest crisis that physics has ever faced. Once murmuring voices, they grow louder and

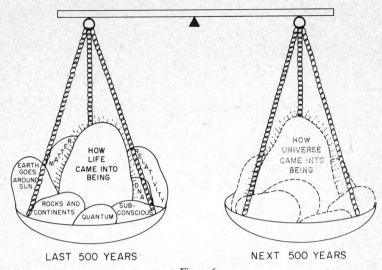

Figure 6

That the discoveries of the next 500 years will outweigh those of the last 500 years is suggested by nothing so much as the magnitude of the great question that still confronts us, "How did the universe come into being?"

louder in our ears. Soon, we can believe, they will unite to thrust an imperative upon us: "Accept a drastically new view of man's relation to the physical universe—or understand nothing."

We used to say, "No theory of particles that deals only with particles will ever explain particles". Today we say, "No theory of physics that deals only with physics will ever explain physics". Tomorrow, who knows what young colleague, inspired by this meeting and the problems we have taken up here, will explain to us the greatest puzzle of all (Fig. 6), how this mysterious universe came into being.

Acknowledgment

Appreciation is expressed to many kind colleagues for assistance and to the University of Washington for hospitality in the period of revision of this report for publication.

REFERENCES

ALFVÉN, H. and KLEIN, O., "Matter-Antimatter Annihilation and Cosmology," *Ark. Fys.* **23**:187–194 (1962).

BAHCALL, J. N. and SCHMIDT, M., "Does the Fine-Structure Constant Vary with Cosmic Time?," *Phys. Rev. Lett.* **19**:1294–1295 (1967).

BAHCALL, J. N. and SEARS, R. L., "Solar Neutrinos," *Ann. Rev. Astron. and Astrophys.* **10**:25–44 (1972).

BARDEEN, J. M., CARTER, B. and HAWKING, S. W., "The Four Laws of Black Hole Mechanics," *Commun. Math. Phys.* **31**:161–170 (1973).

BARDEEN, J. M. and PRESS, W. H., "Radiation Fields in Schwarzschild Background," *J. Math. Phys.* **14**:7–19 (1973).

BARDEEN, J. M., PRESS, W. H. and TEUKOLSKY, S. A., "Rotating Black Holes: Locally Non-rotating Frames, Energy Extraction and Scalar Synchrotron Radiation," *Astrophys. J.* **178**:347–369 (1972).

BARGMANN, V., ed., *Group Representations in Mathematics and Physics: Battelle Seattle 1969 Rencontres,* Springer, Berlin, 1970.

BEKENSTEIN, J. D., "Baryon Number, Entropy, and Black-Hole Physics," Doctoral Thesis, Dept. of Physics, Princeton University, 1972.

BEKENSTEIN, J. D., "Black Holes and Entropy," *Phys. Rev.* **D7**:2333–2346 (1973).

BEKENSTEIN, J. D., "Generalized Second Law of Thermodynamics in Black-Hole Physics," *Phys. Rev.* **D9**:3292–3300 (1974).

BEL, L., unpublished, 1963; cited by BONAZZOLA, S. and PACINI, F., *Phys. Rev.* **148**, 1269–1270 (1966) as "C.I.M.E. salice d'Ulzio."

BELINSKY, V. A. and KHALATNIKOV, I. M., "On the Nature of the Singularities in the General Solution of Gravitational Equations," *Zh. Eksp. & Teor. Fiz.* **56**:1700–1712 (1969a); English translation in *Sov. Phys.—JETP* **29**:911–917.

BELINSKY, V. A. and KHALATNIKOV, I. M., "General Solution of the Gravitational Equations with a Physical Singularity," *Zh. Eksp. & Teor. Fiz.* **57**:2163–2175 (1969b); English translation in *Sov. Phys.—JETP* **30**, 1174–1180 (1970).

BELINSKY, V. A. and KHALATNIKOV I. M., "General Solution of the Gravitational Equations with a Physical Oscillatory Singularity," *Zh. Eksp. & Teor. Fiz.* **59**:314–321 (1970); English translation in *Sov. Phys.—JETP* **32**:169–172.

BELINSKY, V. A., KHALATNIKOV, I. M. and LIFSHITZ, E. M., "Oscillatory Approach to a Singular Point in the Relativistic Cosmology," *Usp. Nauk* **102**:463–500 (1970); English translation in *Advances in Physics 19*, 525–573.

BELINSKY, V. A., LIFSHITZ, E. M. and KHALATNIKOV, I. M., "Oscillatory Mode of Approach to a Singularity in Homogeneous Cosmological Models with Notating Axes," *Zh. Eksp. & Teor. Fiz.* **60**:1969–1979 (1971); English translation in *Sov. Phys.—JETP* **33**:1061–1066 (1971).

BERKELEY, G., ~ 1710, as summarized in article on him by R. ADAMSON, *Encyclopaedia Britannica,* Chicago, Vol. 3, p. 438, 1959.

BERS, L. F. L., "Singularities of Minimal Surfaces," in *Proceedings of the International Congress of Mathematicians, Cambridge, Massachusetts, 1950,* American Mathematical Society, Providence, Rhode Island, Vol. 2, p. 157, 1952.

BERS, L., 1970, "Universal Teichmüller Space," in *Analytic Methods in Mathematical Physics* (R. P. GILBERT and R. NEWTON, eds.), Gordon and Beach, New York, pp. 65–83, 1970.

BLANDFORD, R. D. and REES, M. J., "A 'Twin-Exhaust' Model for Double Radio Sources," *Mon. Not. R. astr. Soc.* **169**:395–415 (1974).

BLANDFORD, R. D. and REES, M. J., "Extragalactic Double Radio Sources—the Current Observational and Theoretical Position," *Contem. Phys.* **16**:1–16 (1975).

BOUGHN, S., "Electromagnetic Radiation Induced by a Gravitational Wave," *Phys. Rev.* **D11**:248–252 (1975).

BOUGHN, S. P., FAIRBANK, W. M., McASHAN, M. S., PAIK, H. J., TABER, R. C., BERNAT, T. P., BLAIR, D. G. and HAMILTON, W. O., "The Use of Cryogenic Techniques to Achieve High Sensitivity in Gravitational Wave Detectors," in *Gravitational Radiation and Gravitational Collapse: International Astronomical Union, Symposium No. 64, Warsaw, Poland, 5–8 Sept. 1973* (C. DEWITT-MORETTE, ed.), Reidel, Dordrecht, Holland, pp. 40–51, 1974.

BOYNTON, P. E. and PARTRIDGE, R. B., "Fine-Scale Anisotropy of the Microwave Background: An Upper Limit at λ = 3.5 Millimeters," *Astrophys. J.* **181**:243–253 (1973).

BRILL, D. R. and WHEELER, J. A., "Interaction of Neutrinos and Gravitational Fields," *Rev. Mod. Phys.* **29**:465–479 (1957).

CARMELI, M., FICKLER, S. I. and WITTEN, L., eds., *Relativity,* Plenum, New York, 1970.

CARTAN, E., *Lecons sur la Géométrie des Espaces de Riemann,* Gauthier-Villars, Paris, 1928 and 1946.

CARTER, B., "Global Structure of the Kerr Family of Gravitational Fields," *Phys. Rev.* **174**:1559–1571 (1968a).

CARTER, B., "Hamilton–Jacobi and Schrödinger Separable Solutions of Einstein's Equations," *Commun. Math. Phys.* **10**:280–310 (1968b).

CARTER, B., "Black Hole Equilibrium States," in *Black Holes* (C. DEWITT and B. S. DEWITT, eds.), Gordon and Breach, New York, pp. 58–214, 1973.

CARTER, B., "Charge and Particle Conservation in Black Hole Decay," preprint from Observatoire de Paris-Meudon, France, 1974.

CHITRE, D. M., "Investigations of the Vanishing of a Horizon for Bianchi Type IX (Mixmaster) Universe," Doctoral Dissertation, University of Maryland, 1972.

CHITRE, D. M., "Electromagnetic Radiation Generated by Gravitational Perturbations of a [Mildly] Charged Rotating Black Hole," *Phys. Rev.* **D11**:760–762 (1975).

CHITRE, D. M., PRICE, R. H. and SANDBERG, V. D., "Electromagnetic Radiation from an Unmoving Charge," *Phys. Rev. Lett.* **31**:1018–1022 (1973).

CHITRE, D. M., PRICE, R. H. and SANDBERG, V. D., "Electromagnetic Radiation Due to Spacetime Oscillations," *Phys. Rev.* **D11**:747–759 (1975).

CHRISTODOULOU, D., "Reversible and Irreversible Transformations in Black-Hole Physics," *Phys. Rev. Lett.* **25**:1596–1597 (1970).

CHUDAKOV, A. E., RYAJSKAYA, O. G. and ZATSEPIN, G. T., "The Project of an Arrangement for the Detection of Neutrino Radiation from Collapsing Stars," *13th International Cosmic Ray Conference, 17–30 August 1973,* University of Denver, Denver, Colorado, Vol. 3, pp. 2007–2012, 1973.

COCKE, W. J. "Statistical Time Symmetry and Two-Time Boundary Conditions in Physics and Cosmology," *Phys. Rev.* **160**:1165–1170 (1967).

COHEN, J. M. and WALD, R. M., "Point Charge in the Vicinity of a Schwarzschild Black Hole," *J. Math. Phys.* **12**:1845–1849 (1971).

DARBOUX, G., *Lecons sur la théorie générale des surfaces,* Gauthir-Villars, Paris. Part I, 1941.

DASHEVSKY, V. M. and ZEL'DOVICH, YA. B., "The propagation of Light in a Nonhomogeneous Nonflat Universe II," *Astronom. Zh.* **41**:1071–1074 (1964); English translation in *Sov. Astr.—AJ* **8**:854–856 (1965).

DAVIES, P. C. W., ed., *The Physics of Time Asymmetry*, University of California Press, Berkeley and Los Angeles, California, 1974.

DAVIS, R., JR., "A Progress Report on the Brookhaven Solar Neutrino Experiment," *Bull. Amer. Phys. Soc.* **17**:527–528 (1972).

DEBEVER, R., ed., *Astrophysics and Gravitation: Proceedings of the Sixteenth Conference on Physics at the University of Brussels, September 1973*, Editions de l'Universite de Bruxelles, 1040 Bruxelles, Belgium, 1974.

DE VAUCOULEURS, G., "The Large-Scale Distribution of Galaxies and Clusters of Galaxies," *Astron. Soc. Pacific Publ.* **83**:113–143 (1971).

DEWITT, C. and DEWITT, B. S., eds., *Relativity, Groups, and Topology*, Gordon and Breach, New York, 1964.

DEWITT, C. and DEWITT, B. S., eds., *Black Holes*, Proceedings of 1972 session of Ecole d'été de physique théorique, Gordon and Breach, New York, 1973.

DEWITT, C. and WHEELER, J. A., eds., *Battelle Rencontres: 1967 Lectures in Mathematics and Physics*, W. A. Benjamin, New York, 1968.

DEWITT-MORETTE, C., ed., *Gravitational Radiation and Gravitational Collapse: International Astronomical Union, Symposium No. 64, Warsaw, Poland, 5–8 September 1973*, Reidel, Dordrecht, Holland, 1974.

DICKE, R. H., "Dirac's Cosmology and Mach's Principle," *Nature* **192**:440–441 (1961).

DICKE, R. H., "The Sun's Rotation and Relativity," *Nature* **202**:432–435 (1964).

DICKE, R. H. and PEEBLES, P. J. E., "Cosmology and the Radioactive Decay Ages of Terrestrial Rocks and Meteorites," *Phys. Rev.* **128**:2006–2011 (1962).

DOROSHKEVICH, A. G., LUKASH, V. N. and NOVIKOV, I. D., "Impossibility of Mixing in the Bianchi Type IX Cosmological Model," *Zh. Eksp. & Teor. Fiz.* **49**:170–181 (1971); English translation in *Sov. Phys.—JETP* **22**:122–130.

DÜRR, H. P., "Approximate Symmetries in Atomic and Elementary Particle Physics," pp. 301–326, in *Properties of Matter under Unusual Conditions* (H. MARK and S. FERNBACH, eds.), Wiley-Interscience, New York, pp. 301–326, 1969.

DYSON, F. J., ed., *Symmetry Group in Nuclear and Particle Physics*, Benjamin, New York, 1966.

EARDLEY, D., LIANG, E. and SACHS, R., "Velocity-Dominated Singularities in Irrotational Dust Cosmologies," *J. Math. Phys.* **13**:99–107 (1972).

EINSTEIN, A., *Essays in Science*, Philosophical Library, New York, 1934. Translated from *Mein Weltbilde*, Querido Verlag, Amsterdam, 1933.

EINSTEIN, A., *The Meaning of Relativity*, 3rd ed., Princeton University Press, Princeton, New Jersey, 1950.

ELLIS, G. F. R. and KING, A. R., "Was the Big Bang a Whimper?," *Commun. Math. Phys.* **38**:119–156 (1974).

ELLIS, H. G., "Ether Flow through a Drainhole-Particle Model in General Relativity," *J. Math. Phys.* **14**:104–118 (1973). Correction in *J. Math. Phys.* **15**, 520 (1974).

ELLIS, H. G., "Time, Grand Illusion," *Foundations of Physics* **4**:311–319 (1974).

EVERITT, C. W. F., FAIRBANK, W. M. and HAMILTON, W. O., "From Quantized Magnetic Flux in Superconductors to Experiments on Gravitation and Time-Reversal Invariance," in *Magic without Magic: John Archibald Wheeler* (J. R. KLAUDER, ed.), W. H. Freeman, San Francisco, California, pp. 201–223, 1972.

FACKERELL, E. D., "Solutions of Zerilli's Equation for the Even-Party Gravitational Perturbations," *Astrophys. J.* **166**: 197–206 (1971).

FACKERELL, E. D. and IPSER, J. R., "Weak Electronic Fields around a Rotating Black Hole," *Phys. Rev.* **D5**: 2455–2458 (1972).

FISCHER, A. E., "The Theory of Superspace," in *Relativity* (M. CARMELI, S. I. FICKLER and L. WITTEN, eds.), Plenum, New York, 1970.

FOWLER, W. and HOYLE, F., *Nucleosyntheses in Massive Stars and Supernovae*, University of Chicago Press, Chicago, Illinois, 1964.

GERLACH, U., "Derivation of the Ten Einstein Field Equations from the Semiclassical Approximation to Quantum Geometrodynamics," *Phys. Rev.* **177**: 1929–1941 (1969).

GIACCONI, R., 1974, "Observational Results on Compact Galactic X-Ray Sources," in *Astrophysics and Gravitation: Proceedings of the Sixteenth Conference on Physics at the University of Brussels, September 1973* (R. DEBEVER, ed.), Editions de l'Universite de Bruxelles, 1040 Bruxelles, Belgium, 1974.

GIACCONI, R. and GURSKY, H., eds., *X-Ray Astronomy*, Reidel, Dordrecht, Holland, 1974.

GILBERT, R. P. and NEWTON, R., *Analytic Methods in Mathematical Physics*, Gordon and Breach, New York, 1970.

GOLD, T., ed., *The Nature of Time*, Cornell University Press, Ithaca, N.Y., 1967.

GOTT, J. F., GUNN, J. E., SCHRAMM, D. N. and TINSLEY, B. M., "An Unbound Universe?" *Astrophys. J.* **194**: 543–553 (1974).

GUNN, J. E. and OKE, J. B., "Spectrophotometry of Faint Cluster Galaxies and the Hubble Diagram: An Approach to Cosmology," *Astrophys. J.* **195**: 255–268 (1975).

HANNI, R. S. and RUFFINI, R., "Lines of Force of a Point Charge near a Schwarzschild Black Hole," *Phys. Rev.* **D8**: 3259–3265 (1973).

HARRISON, B. K., THORNE, K. S., WAKANG, M. and WHEELER, J. A., *Gravitation Theory and Gravitational Collapse*, University of Chicago Press, Chicago, Illinois, 1965.

HAWKING, S. W., "Gravitationally Collapsed Objects of Very Low Mass," *Mon. Nar. Roy. Astr. Soc.* **152**: 75–78 (1971).

HAWKING, S. W., "Black Hole Explosions?" *Nature* **248**: 30–31 (1974).

HAWKING, S. W., "Particle Creation by Black Holes," *Comm. Math. Phys.*, 1975 (in press).

HAWKING, S. W. and ELLIS, G. F. R., *The Large Scale Structure of Space–Time*, Cambridge University Press, Cambridge, U.K., 1973.

HILBERT, D., "Die Grundlagen der Physik," *Konig,. Gesell. d. Wiss. Göttingen, Nacht., Math.— Phys.* **K2**: 395–407 (1915).

HILL, H. A., McCULLEN, J. D., BROWN, T. M. and STEBBINS, R. T., "Energy Transport by Normal Modes: A Possible Resolution of the Neutrino Paradox," preprint submitted to *Phys. Rev. Lett.*, 2 May 1975.

HOJMAN, S., KUCHAŘ, K. and TEITELBOIM, C., "New Approach to General Relativity," *Nature PS* **245**: 97–98 (1973).

HOOKER, C. A., ed., *Contemporary Research in the Foundations and Philosophy of Quantum Mechanics*, Reidel, Dordrecht, Holland, 1973.

ISHAM, C., PENROSE, R. and SCIAMA, D., eds., *Quantum Gravity*, Clarendon Press, Oxford, 1975.

ISRAEL, W., ed., *Relativity, Astrophysics, and Cosmology*, Reidel, Dordrecht, Holland, 1973.

JOHNSTON, M., 1974, "Spin Coefficients and Electromagnetism in Reissner–Nordstrøm Spacetime," A. B. Senior Thesis, Dept. of Physics, Princeton University, 1974.

JOHNSTON, M., RUFFINI, R. and ZERILLI, F., "Gravitationally Induced Electromagnetic Radiation," *Phys. Rev. Lett.* **31**:1317–1319 (1973).

JOHNSTON, M., RUFFINI, R. and ZERILLI, F., "Electromagnetically Induced Gravitational Radiation," *Phys. Lett.* **49B**:185–188 (1974).

KHALATNIKOV, I. M. and LIFSHITZ, E. M., "General Cosmological Solutions of the Gravitational Equations with a Singularity in Time," *Phys. Lett.* **24**:76–79 (1970).

KLAUDER, J. R., ed., *Magic without Magic: John Archibald Wheeler,* W. H. Freeman, San Francisco, California, 1972.

KOTZER, P., LORD, J. and REINES, F., eds., *Proceedings of the Summer Workshop on Project DUMAND* [Deep Underseas Muon and Neutrino Detector], University of Washington and Western Washington State College, 1975.

KUCHAŘ, K., "Canonical Quantization of Gravity," in Israel 1973, pp. 238–288, 1973.

KUCHAŘ, K., "Geometrodynamics Regained: A Lagrangian Approach," *J. Math. Phys.* **15**:708–715 (1974).

LANDAU, L. D. and LIFSHITZ, E. M., *Statistical Physics,* (translated from the Russian by E. PEIERLS and R. F. PEIERLS), Pergamon, London, also Addison-Wesley, Reading, Massachusetts, 1958.

LIFSHITZ, E. M. and KHALATNIKOV, I. M., "Problems of Relativistic Cosmology," *Usp. Fiz. Nauk* **80**:391–438 (1963a); English translation in *Sov. Phys.—Uspekhi* **6**:495–522 (1964).

LIFSHITZ, E. M. and KHALATNIKOV, I. M., "Investigations in Relativistic Cosmology," *Advances in Physics* **12**:185–249 (1963b); translated from the Russian by I. L. BEEBY.

LIFSHITZ, E. M. and KHALATNIKOV, I. M., "Oscillatory Approach to Singular Point in the Open Cosmological Model," *Zh. Eksp. & Teor. Fiz. Pis'ma* **11**:200–203 (1970); English translation in *Sov. Phys.—JETP Lett.* **11**:123–125 (1971).

LIN, C. C., MESTEL, L. and SHU, F. H., "The Gravitational Collapse of a Uniform Spheriod," *Astrophys. J.* **142**:1431–1446 (1965).

MARK, H. and FERNBACH, S., eds., *Properties of Matter under Unusual Conditions,* Wiley-Interscience, New York, 1969.

MASHHOON, B., "Kerr Black Holes and the Problem of Jet Formation in Radio Galaxies," *Bull. Am. Phys. Soc.* **16**, 34 (1971).

MASHHOON, B., "Scattering of Electromagnetic Radiation from a Black Hole," *Phys. Rev.* **D7**:2807–2814 (1973a).

MASHHOON, B., "Tidal Gravitational Radiation," *Astrophys. J.* **185**:83–86 (1973b).

MASHHOON, B., "Electromagnetic Scattering from a Black Hole and the Glory Effect," *Phys. Rev.* **D10**:1059–1063 (1974).

MASHHOON, B., "On Tidal Phenomena in a Strong Gravitational Field," *Astrophys. J.* **197**:705–716 (1975).

MATZNER, R. A., "Scattering of Massless Scalar Waves by a Schwarzschild Singularity," *J. Math. Phys.* **9**:163–170 (1968).

MEHRA, J., ed., *The Physicist's Conception of Nature,* Reidel, Dordrecht, Holland, 1973.

MESTEL, L., "Problems of Star Formation—I," *Roy. Astro. Soc. Quart. J.* **6**:161–198 (1965).

MICHEL, L., "Applications of Group Theory to Quantum Physics: Algebraic Aspects," in *Group Representations in Mathematics and Physics: Battelle Seattle 1969 Recontres* (V. BARGMANN, ed.), Springer, Berlin, pp. 36–143, 1970.

MICHEL, L., "Simple Mathematical Models of Symmetry Breaking. Application to Particle

Physics," 26 March 1974 lecture at the Warsaw Symposium in Mathematical Physics, issued as a preprint from Institut des Hautes Etudes Scientifiques, 91440 Bures-sur-Yvette, France, May 1974.

Misner, C. W., "The Isotropy of the Universe," *Astrophys. J.* **151**:431–457 (1968).

Misner, C. W., "Mixmaster Universe," *Phys. Rev. Lett.* **22**:1071–1074 (1969).

Misner, C. W., "Stability of Kerr Black Holes against Scalar Perturbations," *Bull. Amer. Phys. Soc.* **17**, 472 (1972).

Misner, C. W. and Taub, A. H., "A Singularity-Free Empty Universe," *Zh. Eksp. & Teor. Fiz.* **55**:233–255 (1968); English original in Sov. Phys.—*JETP* **28**:122–133 (1969).

Misner, C. W., Thorne, K. S. and Wheeler, J. A., *Gravitation,* W. H. Freeman, San Francisco, California, 1973.

Moncrief, V., "Odd Parity Stability of a Reissner–Nordstrøm Black Hole," *Phys. Rev.* **D9**:2707–2709 (1974a).

Moncrief, V., "Stability of Reissner–Nordstrøm Black Holes," *Phys. Rev.* **D10**:1057–1059 (1974b).

O'Connell, D. J. K., ed., *Study Week on Nuclei of Galaxies,* Vatican City, 1970, *Pontificiae Academiae Scientarum Scripta Varia, No. 35,* North Holland, Amsterdam, 1971.

Oort, J. H., "Distribution of Galaxies and the Density of the Universe," in *Onzieme Conseil de Physique Solvay: La Structure et l'évolution de l'univers,* Editions Stoops, Brussels, Belgium, 1958.

Ostriker, J. P. and Peebles, P. J. E., "A Numerical Study of the Stability of Flattened Galaxies; or, Can Cold Galaxies Survive?," *Astrophys. J.* **186**:467–480 (1973).

Ostriker, J. D., Peebles, P. J. E., and Yahil, A., "The Size and Mass of Galaxies and the Mass of the Universe," *Astrophys. J. Lett.* **193**, L1-4 (1974).

Parmenides of Elea, ~ 500 b.c., poem *Nature,* part "Truth" as summarized by A. C. Lloyd in "Parmenides," *Encyclopaedia Britannica,* Chicago, Vol. 17, p. 327, 1959.

Partridge, R. B., "Absorber Theory of Radiation and the Future of the Universe," *Nature* **244**:263–265 (1973).

Patton, C. M. and Wheeler, J. A., "Is Physics Legislated by Cosmogony," in *Quantum Gravity* (C. Isham, R. Penrose and D. Sciama, eds.), Clarendon Press, Oxford, pp. 538–605, 1975.

Peebles, P. J. E., *Physical Cosmology,* Princeton University Press, Princeton, New Jersey, 1971.

Peebles, P. J. E., "The Gravitational Instability Picture and the Nature of the Distribution of Galaxies," *Astrophys. J.* **189**:L51–L53 (1974a).

Peebles, P. J. E., "The Nature of the Distribution of Galaxies," *Astron. and Astrophys.* **32**:197–202 (1974b).

Peebles, P. J. E., "Statistical Analysis of Catalogs of Extragalactic Objects. IV. Cross-Correlation of the Abell and Shane–Wirtanen Catalogs," *Astrophys. J. Supp. No. 253,* **28**:37–50 (1974c).

Peebles, P. J. E., and Groth, E. J., "Statistical Analysis of Catalogs of Extragalactic Objects. V. 3-Point Correlation Function for the Galaxy Distribution in the Zwicky Catalog," *Astrophys. J.* **196**:1–11 (1975).

Peebles, P. J. E. and Hauser, M. G., "Statistical Analysis of Catalogs of Extragalactic Objects. III. Shane–Wirtanen and Zwicky Catalogs," *Astrophys. J. Supp. No. 253,* **28**:19–36 (1974).

PENROSE, R., "Gravitational Collapse: The Role of General Relativity," *Revista de Nuovo Cimento* **1**:252–276 (1969).

PENROSE, R., "Internal Instability in a Reissner–Nordstrøm Black Hole," *Internat. J. Theor. Phys.* **7**:183–197 (1973).

PENROSE, R., "Twistors and Particles: An Outline," preprint from Mathematical Institute, Oxford University, January 1975.

PERES, A., "On the Cauchy Problem in General Relativity II," *Nuovo Cimento* **26**:53–62 (1962).

PRESS, W. H., "Black-Hole Perturbations: An Overview," *Ann. New York Acad. Sci.* **224**: 272–277 (1973).

PRESS, W. H. and GUNN, J. E., "Method for Detecting a Cosmological Density of Condensed Objects," *Astrophys. J.* **185**:397–412 (1973).

PRESS, W. H. and TEUKOLSKY, S. A., "Floating Orbits, Superradiant Scattering and the Black-Hole Bomb," *Nature* **238**:211 212 (1972).

PRESS, W. H. and TEUKOLSKY, S. A., "Perturbations of a Rotating Black Hole. II. Dynamical Stability of the Kerr Metric," *Astrophys. J.* **185**:649–673 (1973).

PRICE, R. H., "Nonsperhical Perturbations of Relativistic Gravitational Collapse," Doctoral Dissertation, California Institute of Technology, 1971.

PRICE, R. H., "Nonsperhical Perturbations of Relativistic Gravitational Collapse. I. Scalar and Gravitational Perturbations," *Phys. Rev.* **D5**:2419–2438 (1972a).

PRICE, R. H., "Nonspherical Perturbations of Relativistic Gravitational Collapse. II. Integer-Spin, Zero-Rest-Mass Fields," *Phys. Rev.* **D5**:2439–2454 (1972b).

REES, M., personal communication, 1975.

REGGE, T. and WHEELER, J. A., "Stability of a Schwarzschild Singularity," *Phys. Rev.* **108**: 1063–1069 (1957).

RENFREW, D., unpublished calculations (as a graduate student at Princeton University) of the initial stages of the fragmentation of an infinite slab of nuclear fluid, 1972.

RUFFINI, R. and WHEELER, J. A., "Relativistic Cosmology and Space Platforms," in *Proceedings of the Conference on Space Physics,* Interlaken 1970, European Space Research Organization, Paris, France, pp. 45–174, 1971; updated version in REES, RUFFINI and WHEELER, 1974.

RYAN, M. P. JR. and SHEPLEY, L., *Homogeneous Cosmologies,* Princeton University Press, Princeton, New Jersey, 1974.

SAKHAROV, A. D., "Vacuum Quantum Fluctuations in Curved Space and the Theory of Gravitation," *Doklady Akad. Nauk S.S.S.R.* **177**:70–71 (1967); English translation in *Sov. Phys. Doklady* **12**:1040–1041 (1968).

SCHMIDT, B. G., "A New Definition of Singular Points in General Relativity," *Gen. Rel. Grav.* **1**:269–280 (1971).

SCHMIDT, B. G., "A New Definition of Conformal and Projective Infinity of Spacetimes," *Comm. Math. Phys.* **36**:73–90 (1974).

STAROBINSKY, A. A., "Amplification of Waves during Reflection from a Rotating 'Black Hole'," *Zh. Eksp. & Teor. Fiz.* **64**:48–57 (1973); English translation in *Sov. Phys.—JETP* **37**:28–32 (1973).

STAROBINSKY, A. A. and CHURILOV, S. M., "Amplification of Electromagnetic and Gravitational Waves Scattered by a Rotating 'Black Hole'," *Zh. Eksp. & Teor. Fiz.* **65**:3–11 (1973); English translation in *Sov. Phys.—JETP* **38**:1–5 (1974).

TEITELBOIM, C., "How Commutators of Constraints Reflect Spacetime Structure," *Ann. Phys.* **79**:542–557 (1973a).

TEITELBOIM, C., "The Hamiltonian Structure of Spacetime," Doctoral Dissertation, Princeton University, 1973b.

TEUKOLSKY, S. A., "Rotating Black Holes: Separable Wave Equations for Gravitational and Electromagnetic Perturbations," *Phys. Rev. Lett.* **29**:1114–1118 (1972).

TEUKOLSKY, S. A., "Perturbations of a Rotating Black Hole. I. Fundamental Equations for Gravitational, Electromagnetic, and Neutrino-Field Perturbations," *Astrophys. J.* **185**:635–647 (1973).

TEUKOLSKY, S. A., "Perturbations of a Rotating Black Hole," Doctoral Thesis, California Institute of Technology, 1974.

TEUKOLSKY, S. A. and PRESS, W. H., "Perturbations of a Rotating Black Hole. III Interaction of the Hole with Gravitational and Electromagnetic Radiation," *Astrophys. J.* **193**:443–461 (1974).

THORNE, K. S., "Nonspherical Gravitational Collapse," *Magic without Magic: John Archibald Wheeler* (J. R. KLAUDER, ed.), Freeman, San Francisco, California, pp. 231–251, 1972.

VAN DEN HEUVEL, E., "Discussion of Binary Stars," in *Astrophysics and Gravitation: Proceedings of the Sixteenth Conference in Physics of the University of Brussels, September 1973,* Editions de l'Universite de Bruxelles, 1040 Bruxelles, Belgium, 1974.

VISHVESHWARA, C. V., "Stability of the Schwarzschild Metric," Doctoral Dissertation, University of Maryland, 1968.

VISHVESHWARA, C. V., "Stability of the Schwarzschild Metric," *Phys. Rev.* **D1**:2870–2879 (1970).

WAGONER, R. V., "Big-Bang Nucleosynthesis Revisited," *Astrophys. J.* **179**:343–360 (1973).

WALD, R. M., "On Perturbations of a Kerr Black Hole," *J. Math. Phys.* **14**:1453–1461 (1973).

WALD, R. M., "Energy Limits on the Penrose Process," *Astrophys. J.* **191**:231–233 (1974).

WEINSTEIN, J., "Forced Perturbations of the Reissner–Nordstrøm Geometry," A. B. Senior Thesis, Princeton University, 1975.

WHEELER, J. A., "Geons," *Phys. Rev.* **97**:511–536 (1955).

WHEELER, J. A., "The Universe in the Light of General Relativity," *Monist* **47**:40–76 (1962).

WHEELER, J. A., "Geometrodynamics and the Issue of the Final State," in *Relativity, Groups, and Topology* (C. DEWITT and B. S. DEWITT, eds.), Gordon and Breach, New York, pp. 315–520, 1964.

WHEELER, J. A., "Superspace and the Nature of Quantum Geometrodynamics," in *Battelle Recontres: 1967 Lectures in Mathematics and Physics* (C. DEWITT and J. A. WHEELER, eds.), W. A. Benjamin, New York, pp. 242–307, 1968.

WHEELER, J. A., "Superspace," in *Analytic Methods in Mathematical Physics* (R. G. GILBERT and R. NEWTON, eds.), Gordon and Breach, New York, pp. 335–378, 1970.

WHEELER, J. A., "Mechanisms for jets," in *Study Week on Nuclei of Galaxies* (D. J. K. O'CONNELL, ed.), Vatican City, 1970, *Pontificiae Academiae Scientarum Scripta Varia, No. 35,* North Holland, Amsterdam, pp. 539–567, 1971a.

WHEELER, J. A., "Transcending the Law of Conservation of Leptons," in *Atti del Convegno Internazionale sul Tema: The Astrophysical Aspects of the Weak Interactions, Cortona 1970, Quaderno N. 157,* Accademia Nazionale dei Lincei, Roma, pp. 133–164, 1971b.

WHEELER, J. A., "From Relativity to Mutability," in *The Physicist's Conception of Nature*

(J. MEHRA, ed.), Reidel, Dordrecht, Holland, pp. 202–247, 1973; reprinted in *Revista Mexicana de Fis.* **23**: 1–57 (1974).

WHEELER, J. A., "The Black Hole," in *Astrophysics and Gravitation: Proceedings of the Sixteenth Conference on Physics at the University of Brussels, September 1973,* Editions de l'Universite de Bruxelles, 1040 Bruxelles, Belgium, 1974.

WIGNER, E. P., "Epistemological Perspective on Quantum Theory," *Contemporary Research in the Foundations and Philosophy of Quantum Mechanics* (C. A. HOOKER, ed.), Reidel, Dordrecht, Holland, pp. 369–385, 1973.

ZEL'DOVICH, YA. B., "Observations in a Universe Homogeneous in the Mean," *Astronom. Zh.* **41**: 19–24 (1964); English translation in *Sov. Astronom.—AJ* **8**: 13–16 (1964).

ZEL'DOVICH, YA. B., "Generation of Waves by a Rotating Body," *Zh. Eksp. & Teor. Fiz. Pis. Red.* **14**: 270–272 (1971); English translation in *JETP Lett.* **14**: 180–181 (1971).

ZEL'DOVICH, YA. B., "Creation of Particles and Antiparticles in an Electric and Gravitational Field," in *Magic without Magic: John Archibald Wheeler* (J. R. KLAUDER, ed.), W. H. Freeman, San Francisco, California, pp. 277–288, 1972a.

ZEL'DOVICH, YA. B., "Amplification of Cylindrical Electromagnetic Waves Reflected from a Rotating Body." *Zh. Eksp. & Teor. Fiz.* **62**: 2076–2081 (1972b); English translation in *Sov. Phys.—JETP* **35**: 1085–1087.

ZEL'DOVICH, YA. B. and NOVIKOV, I. D., *Relativistic Astrophysics, Vol. I: Stars and Relativity,* University of Chicago Press, Chicago, Illinois, 1971.

ZERILLI, F. J., "Gravitational Field of a Particle Falling in a Schwarzschild Geometry Analyzed in Tensor Harmonics," *Phys. Rev.* **D2**: 2141–2160 (1970a).

ZERILLI, F. J., "Effective Potential for Even-Parity Regge–Wheeler Gravitational Perturbation Equations," *Phys. Rev. Lett.* **24**: 737–738 (1970b); correction in REES, RUFFINI and WHEELER, 1974, appendix A-7.

ZERILLI, F. J., "Perturbation Analysis for Gravitational and Electromagnetic Radiation in a Reissner–Nordstrøm Geometry," *Phys. Rev.* **D9**: 860–868 (1974).